Chemistry of Nanostructures

Chemistry of Nanostructures

Edited by
Lindy Bowman

WILLFORD PRESS

www.willfordpress.com

Published by Willford Press,
118-35 Queens Blvd., Suite 400,
Forest Hills, NY 11375, USA

ISBN: 978-1-68285-566-9

Cataloging-in-Publication Data

Chemistry of nanostructures / edited by Lindy Bowman.
  p. cm.
Includes bibliographical references and index.
ISBN 978-1-68285-566-9
1. Nanostructured materials. 2. Nanostructures. 3. Nanochemistry.
3. Nanotechnology. I. Bowman, Lindy.
TA418.9.N35 C44 2019
620.5--dc23

For information on all Willford Press publications
visit our website at www.willfordpress.com

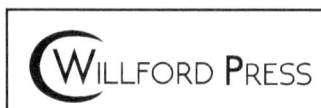

WILLFORD PRESS

Contents

Preface

Nanostructures are defined as structures that are intermediate between microscopic and molecular structures at nanoscale. These are classified depending on the number of dimensions, into nanostructured surfaces, nanotubes, spherical nanoparticles, etc. These have diverse applications across major domains of medicine, manufacturing, electronics, food science, etc. This book includes some of the vital pieces of studies on the chemistry of nanostructures that are being conducted across the globe. The diverse aspects of nanostructures are also glanced at and their applications as well as ramifications are looked at in detail. It also covers in detail some existent theories and innovative concepts related to the field. This book will prove to be immensely beneficial to students and researchers in the field of nanotechnology.

The researches compiled throughout the book are authentic and of high quality, combining several disciplines and from very diverse regions from around the world. Drawing on the contributions of many researchers from diverse countries, the book's objective is to provide the readers with the latest achievements in the area of research. This book will surely be a source of knowledge to all interested and researching the field.

In the end, I would like to express my deep sense of gratitude to all the authors for meeting the set deadlines in completing and submitting their research chapters. I would also like to thank the publisher for the support offered to us throughout the course of the book. Finally, I extend my sincere thanks to my family for being a constant source of inspiration and encouragement.

Editor

Voltammetric sensor based on Co_3O_4/SnO_2 nanopowders for determination of diltiazem in tablets and biological fluids

Abdol Mohammad Attaran[1] · Somaye Abdol-Manafi[1] · Mehran Javanbakht[2] · Morteza Enhessari[3]

Abstract In this work, an anodic stripping voltammetry for nanomolar determination of diltiazem with a chemically modified carbon paste electrode (CMCPE) containing Co_3O_4/SnO_2 nanopowders was studied. The accumulation potential and time were selected at, −0.2 V and 190 s, respectively. The electroanalytical performance of the CMCPE was evaluated regarding the carbon paste composition, the solution pH, the time and potential accumulation, and the potential interferences. The novel electrode showed linear response to diltiazem concentration range of 50–650 nM with a lowest detection limit value of 15 nM. The precisions for six consecutive determinations of 350 and 550 nM diltiazem were 3.2 and 2.5 %, respectively. It was demonstrated that the proposed method was free from most interference. Finally, the method was effectively applied to the determination of diltiazem in pharmaceutical tablets and biological samples.

Keywords Diltiazem · Anodic stripping voltammetry · Biological fluids · Cobalt oxide · Nanocatalyst

Introduction

Diltiazem, (+5)-5-[2-(dimethylamino) ethyl]-*cis*,2,3-dihydro-3-hydroxy-2-(*p*-methoxyphenyl-1,5-benzothiazepin-4(5H)-one acetate monohydrochloride (Fig. 1), is an important coronary vasodilator drug of the calcium channel-blocker, used in the therapy of heart disease and hypertension [1]. The drug has a mean biological half-life of 3–6 h in plasma [2]. The absolute bioavailability of the tablets ranges between 30 and 42 % largely due to presystemic hepatic metabolism [3, 4]. After absorption, diltiazem undergoes extensive hepatic metabolism through three major metabolic pathways, N-demethylation, O-deacetylation, and O-demethylation [2–5]. Diltiazem undergoes hydrolysis to desacetyl diltiazem in acidic solutions and is most stable at pH 5. The extrapolated room temperate shelf-life was 42 or 15.8 days at pH 5 or 2, respectively [6], as well as a stock solution of the drug was prepared daily during the measurements and protect from the light. Its therapeutic and toxic effects require very sensitive methods for determination of trace levels. Various analytical methods have been employed for this purpose, e.g. high-performance liquid chromatography in human cardiac tissue [7], in blood [8] and in human plasma [9–13], high-performance thin-layer chromatographic [14], gas chromatography [15], spectrophotometry [16, 17], liquid–solid extraction [18] and voltammetry [19]. Since employed techniques for determination of diltiazem in the presence of other drugs are mostly spectroscopy and HPLC, and these methods often are complex and consuming cost solvents, must easy-handled and simple-operating methods need be applied.

Recently, we used modified CPEs for trace analysis of some drugs and heavy metals in various environments [20–24]. Newly we showed that the application of nanopowders of Fe_2TiO_5 in CPEs catalyses the oxidation of salbutamol and enhances the peak current values in the differential pulse anodic stripping voltammetric (DPASV) determination of it at sub-nanomolar levels [25]. The developed method has also been satisfactorily applied for the

✉ Abdol Mohammad Attaran
amohammadattaran@gmail.com

[1] Department of Chemistry, Payame Noor University, Delijan, Iran

[2] Department of Chemistry, Amirkabir University of Technology, Tehran, Iran

[3] Department of Chemistry, Naragh Branch, Islamic Azad University, Naragh, Iran

determination of diltiazem in human serum and pharmaceutical samples.

In this study, we present an electroanalytical method based on anodic stripping voltammetry on a carbon paste electrode modified by Co_3O_4/SnO_2 nanopowders for determination of trace amount of diltiazem in pharmaceutical and biological samples. The factors that affect on the performance of the DPASV were optimized and the proposed method was applied in the determination of diltiazem. In accordance with our resulting data, the Co_3O_4/SnO_2 nanopowders offered several distinct advantages, including an extraordinary stability, reproducibility, low background response and a satisfactory detection limit for diltiazem.

Tin and cobalt-based oxides and complexes are universally known as inorganic or organometallic materials with wide electrocatalytic applications in modified electrodes, gas sensors, solar cells, and high-performance catalysts [26–29]. One of these usages is electro-catalytic role that applied in chemically modified electrodes. Some studies showed the electrocatalytic of Co or Sn-based materials such as cobalt phthalocyanine, [30] iron-doped cobalt oxide [31] or tin-modified palladium electrodes [32].

Experimental

Reagents and chemicals

All chemicals were of analytical grades and obtained from Merck (Germany). Diltiazem was purchased from Amin Pharmacy Corporation. Graphite powder (particle diameter 1–2 μm) and high-viscosity paraffin with high density was used as the pasting liquid for the carbon paste electrode purchased from Merck. Double distilled water was used throughout. Diltiazem stock solution (1.0 mM) was prepared by dissolving 0.0207 g of diltiazem separately in buffer phosphate solution in a 50 mL volumetric flask. Working solutions were prepared by appropriate dilution of the stock solution with buffer solution.

Apparatus

Electrochemical measurements were performed using a Metrohm 797 VA Potentiostat–Galvanostat. The experiments were carried out in a single-compartment three-electrode cell, at room temperature (25 °C). The counter electrode was a platinum wire, and an Ag/AgCl saturated KCl electrode was used as the reference electrode. Carbon paste electrodes modified with Co_3O_4/SnO_2 nanopowders (MCPEs) were used as the working electrode. The preconcentration step was conducted using a magnetic stirrer

at 200 rpm. The electrochemical experiments were performed in an electrolytic cell with 20 mL solutions. A digital pH meter (Metrohm 827) was used for the preparation of the buffer solutions in voltammetric experiments. A double water distiller with extremely pure distillation with conductivity of 1 μS/cm was used.

Procedures

Preparation of Co_3O_4/SnO_2 nanopowders

The Co_3O_4/SnO_2 nanopowders were prepared according to the similar procedure described previously [33]. For the synthesis of this compound, first 0.05 mol of cobalt acetate was mixed with 0.2 mol of melted stearic acid to produce cobalt stearate sol. In the next step 0.05 mol of tin tetrachloride was reacted to stoichiometric mol of *n*-butanol (in ice bath) to produce tin alkoxide solution. Then cobalt stearate sol was mixed with tin alkoxide solution for production of gel precursor. Afterwards the gel precursor was calcined to a temperature of 800 °C for 4 h. Finally during pulverization the Co_3O_4/SnO_2 nanocatalyst was obtained.

Fabrication of Co_3O_4/SnO_2 modified carbon paste electrode

Carbon paste electrodes (CPE) were prepared by mixing 68.9 % graphite powder with 8 % Co_3O_4/SnO_2 nanopowders and 23.1 % paraffin oil (ratio of C:Paraffin: Co_3O_4/SnO_2, 68.9:23.1:8, w/w) on mortar for at least 20 min to produce the modified carbon paste electrode (MCPE). The MCPE was finally obtained by packing the paste into an insulin syringe and arranged with a copper wire serving as an external electric contact. Appropriate packing was achieved by pressing the electrode surface against a weighing paper until a smooth surface was obtained.

General analytical procedure

Solutions (20 mL) containing appropriate amounts of diltiazem in 0.1 M phosphate buffer at pH 6.3 were transferred into the voltammetric cell. The differential pulse voltammograms were recorded by applying positive going potentials from 0.53 to 0.98 V. The voltammograms showed anodic peaks around 0.78 V corresponding to diltiazem of heights proportional to the concentrations in solution. The calibration curve was obtained by plotting anodic peak currents of diltiazem versus the corresponding concentrations. After each measurement, the MCPE was regenerated by pushing an excess of paste out of the tube, removing the excess, and mechanically polishing the electrode surface.

Real sample preparation for measurements

Twenty tablets were weighed accurately and finely powdered. A portion of the powder, equivalent to 0.232 g of diltiazem was transferred to a 100 mL volumetric flask and dissolved in approximately 90 mL of choloridric acid/water (1:10). The solution was then filtered and the first portion of the filtrate was discarded. An accurate measured volume of the filtrate was quantitatively diluted with buffer phosphate to yield a sample solution having a final concentration assumed to be 0.1 mM diltiazem. An aliquot was then transferred to a voltammetric cell containing 20 mL of buffer phosphate (pH 6.3) to yield a final concentration of 30 μM diltiazem.

For sample preparation of plasma, the plasma samples was diluted with 25 mM ammonium acetate (pH 5.0), then centrifuge 20 min at 8000 rpm to remove of proteins. Then 1 mL portion of plasma was added to 2 mL of acetonitril for remove excess of proteins. After mixing for 2 min and centrifugation for 10 min at 7000 rpm, the aqueous layer was discarded and the organic layer was evaporated to dryness under stream of nitrogen. The residues were reconstituted in phosphate buffer pH 6.3 so as the final concentration was in range (100–500 nM) and transferred to the voltammetric cell.

For preparation of urine sample, urine of a healthy person was filtered and was diluted with phosphate buffer (1:10). Other steps are similar to the plasma treatments.

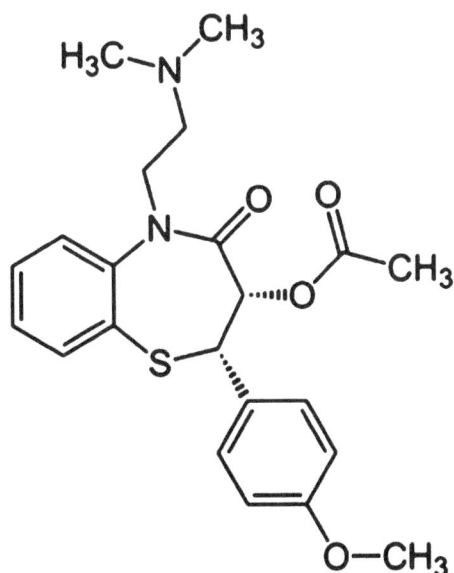

Fig. 1 Chemical structure of diltiazem

Results and discussion

Characterization of Co_3O_4/SnO_2 nanopowders

To analyse the crystal phases, XRD pattern was recorded. The obtained XRD pattern of Co_3O_4/SnO_2 nanopowders after heat treatment in 800 °C in air for 4 h is shown in Fig. 2. The pattern includes two main phases, cobalt oxide (JCPDS: 78-1969) and tin oxide (JCPDS: 77-0452) phase. The particle size of crystallites (L) has been estimated with Scherrer formula [34] [Eq. (1)]:

$$\left(L = \frac{K \times \lambda}{\Delta(2\theta) \times \cos(\theta)} \right) \tag{1}$$

where K is the form factor (equal to 0.9), $\lambda = 0.15418$ nM, 2θ is the peak position and $\Delta(2\theta)$ is the full width at half maximum of the diffraction peak in terms of radians.

In this way, we obtained the crystallite size of the nanopowders calcined at 800 °C about 42 nM.

Electrochemical behavior at CPE and MCPE

Cyclic voltammograms of diltiazem were recorded at the phosphate buffer (pH 6.3) at both bare CPE and CPE modified with 8 % (w/w) Co_3O_4/SnO_2 nanopowders. The voltammograms recorded in buffer solution at the bare CPE exhibited no any voltammetric peak over the entire pH range (Fig. 3, curve C) and in diltiazem solution (curve B). Whereas the voltammograms recorded in 100 nM diltiazem solution at the modified Co_3O_4/SnO_2 nanopowders electrodes exhibited a single irreversible anodic peak over the entire pH range. However, this peak was sharp and better developed in phosphate buffer of pH 6.3 when using CPE modified with 8 % (w/w) Co_3O_4/SnO_2 nanopowders (Fig. 3, curve A).

In addition, the electroactivity of the modified electrode is demonstrated in the differential pulse voltammograms (DPVs) observed for 100 nM diltiazem at the Co_3O_4/SnO_2 modified electrode and the bare CPE in 0.1 M phosphate buffer with pH 6.3 (Fig. 4). At the modified electrode, the oxidation peaks became well defined and appeared at 780 mV. This enhancement in current responses is a clear evidence of the electroactivity effect of the modified electrode towards the oxidation of diltiazem.

Composition and stability of the MCPE

The effect of the carbon paste composition on the voltammetric response of the electrode modified with Co_3O_4/SnO_2 nanopowders was evaluated by differential pulse voltammetry. Electrodes with different percent of modifier were prepared and examined for their voltammetric signals under identical conditions (Table 1). The

Fig. 2 XRD pattern of Co_3O_4/ SnO_2 nanopowders

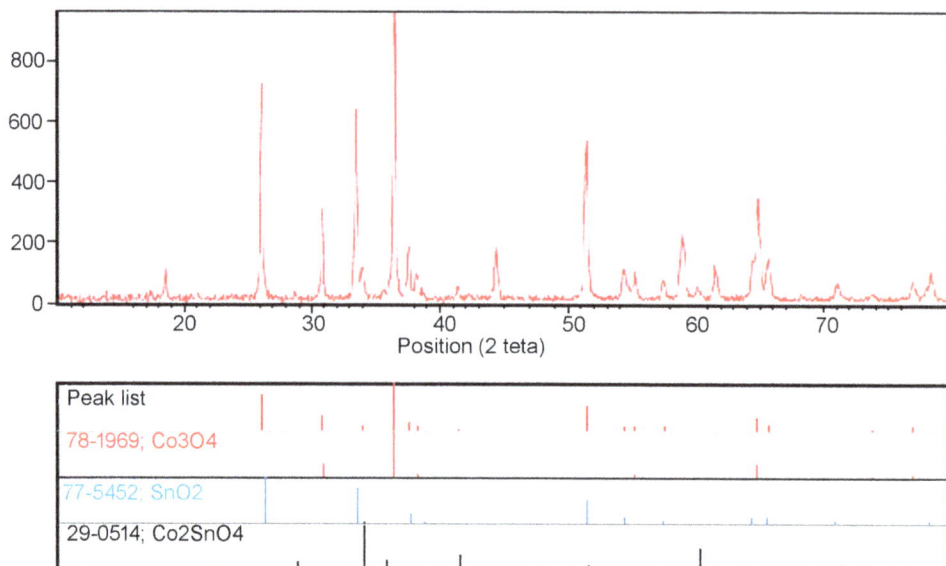

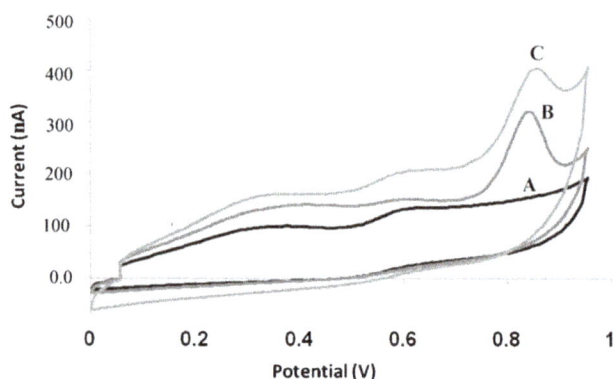

Fig. 3 Cyclic voltammograms of 0.1 M phosphate buffer solution (pH 6.3) recorded at the (*A*) modified CPE in the absence of diltiazem, (*B*) unmodified CPE with 1 µM diltiazem and (*C*) at the developed Co_3O_4/SnO_2 nanopowders modified CPE with 100 nM diltiazem

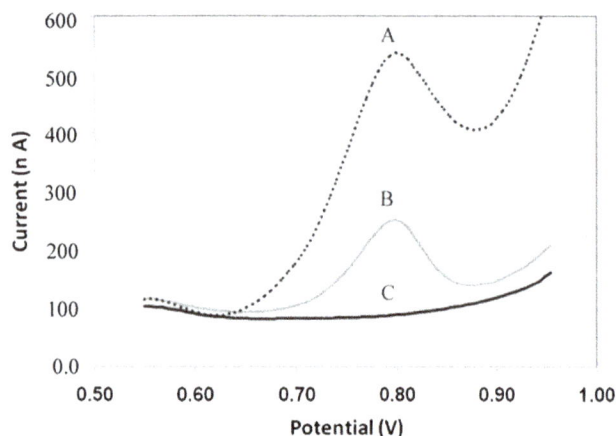

Table 1 Composition of the modified carbon paste electrodes and their DPV responses for 300 nM diltiazem in phosphate buffer 0.1 M

No.	Electrode composition			Peak current (nA)
	Modifier	Graphite powder	Paraffin oil	
CPE1	0	74.8	25.2	202
CPE2	2	73.3	24.7	250
CPE3	4	71.8	24.2	338
CPE1	6	70.3	23.7	406
CPE5	8	68.8	23.2	440
CPE6	10	67.3	22.7	436

maximum peak current was obtained for 8 % modified Co_3O_4/SnO_2 nanopowders in the paste. Higher concentrations (>10 %) showed a decrease in the peak current. This is presumably due to the reduction of conductive area at the electrode surface. According to these results, a carbon paste composition of 8 % modified Co_3O_4/SnO_2 nanopowders, 68.9 % graphite and 23.1 % paraffin oil was used in further studies.

Influence of accumulation time and accumulation potential

The influence of accumulation time is examined from 40 to 240 s. The DPV peak current of diltiazem increased with accumulation time increasing from 40 to 190 s, as shown in Fig. 5 for a 300 nM diltiazem solution. But when it exceeds 190 s, the peak current remains constant due to the surface saturation.

The influence of accumulation potential is examined from −0.1 to −0.5 V. However, with the potential increase

Fig. 4 Differential pulse voltammograms of buffer solution (pH 6.3) recorded (*A*) at the developed Co_3O_4/SnO_2 nanopowders modified CPE with 100 nM diltiazem in, (*B*) unmodified CPE with 1.0 µM diltiazem and (*C*) at the modified CPE in the absence of diltiazem

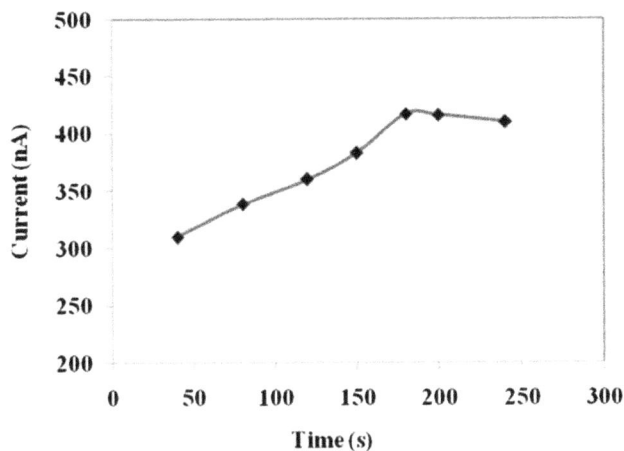

Fig. 5 The effect of accumulation time on the peak current of 300 nM diltiazem; with Co_3O_4/SnO_2 (8 %) modified CPE: accumulation potential, -200 mV (vs. Ag/AgCl); scan rate: 0.015 V s^{-1}; supporting electrolyte, 0.1 M phosphate buffer (pH 6.3)

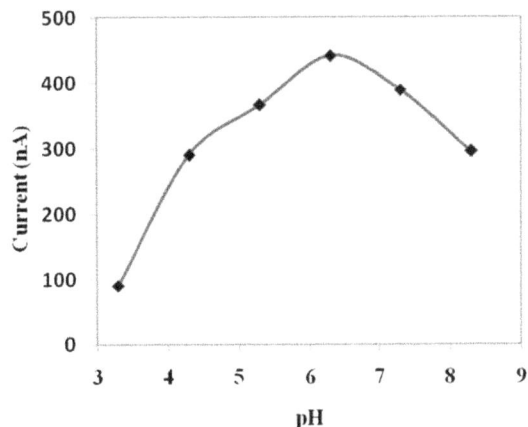

Fig. 6 Response profile for the Co_3O_4/SnO_2 nanopowders-MCPE for preconcentration solutions at different pH values, containing 300 nM diltiazem; other conditions as Fig. 4

to -700 mV more current to come, but since the peak current also increased in electrolyte, potential -200 mV as the optimal deposition potential due to the lower background and absence of any peaks in the electrolyte is selected.

The type of supporting electrolyte

The effect of different salt solutions (as supporting electrolyte) on DPV voltammogram of diltiazem was investigated. Sample solutions containing salts such as KCl, KNO_3, $NaNO_3$ and phosphate buffer in the concentration range of 0.01–0.2 mol L^{-1} were tested. The results revealed that the peak current and sensitivity of the method do not change significantly with type and salt concentration and the phosphate buffer was chosen as the best electrolyte.

The effect of pH on the peak current of diltiazem

The effects of the pH values of both the preconcentration and stripping solutions on the response of the Co_3O_4/SnO_2 nanopowders to diltiazem were investigated. The response of the MCPE is strongly affected by these conditions (the pH or the ionic strength) in the ion exchange reactions. The effect of pH on the electrode response was tested in solutions of 300 nM diltiazem prepared at different pHs. Figure 6 shows the effect of the pH of the preconcentration solution on the anodic stripping peak current of diltiazem, and shows that the pH has a strong influence on the preconcentration process. The anodic stripping peak current of diltiazem increased with increasing pH over the range of 3.3–6.3, and reached a maximum level at pH 6.3. During this process, diltiazem selectively enters paste due to simple electrostatic attractive forces, since it exists as

cation at pH 6.3. The anodic stripping peak current of diltiazem decreased with increasing pH at basic pH, which can be attributed to the deprotonation of diltiazem. The results indicated that the effect of the stripping solution pH on the response of Co_3O_4/SnO_2 nanopowders-MCPE to diltiazem is similar behavior to that seen for the preconcentration solution.

Linearity, limit of detection and limit of quantitation

Relationship between oxidation peak current magnitude and concentration of diltiazem was examined in phosphate buffer of pH 6.3 by the developed DP-ASV method utilizing the developed 8 % (w/w) Co_3O_4/SnO_2 nanopowdes modified CPE. A linear dynamic range of 50–650 nM diltiazem was obtained following its preconcentration onto the developed 8 % (w/w) MCPE electrode by adsorption/accumulation for 190 s at -0.2 V (Fig. 7). Its corresponding regression equation was: i_p (nA) $= 1.136$ C $+ 53.25$, $R^2 = 0.997$. Limits of detection (LOD) and quantitation (LOQ) of bulk diltiazem were estimated 15 nM and 50.1 nM diltiazem based on 3 S/N and 10 S/N, respectively. The results indicated the reliability of the developed DP-ASV method for the trace assay of bulk diltiazem. This method shows a LOD of lower or compatible with other works such as high-performance liquid chromatography (0.9 μM), [13] and voltammetric sensor (0.5 nM) [35].

Trueness and precision

Trueness and precision of the optimized DP-ASV method were evaluated by performing five replicate measurements for various concentrations of diltiazem (250, 350 and 550 nM) through intraday and inter-day assays following

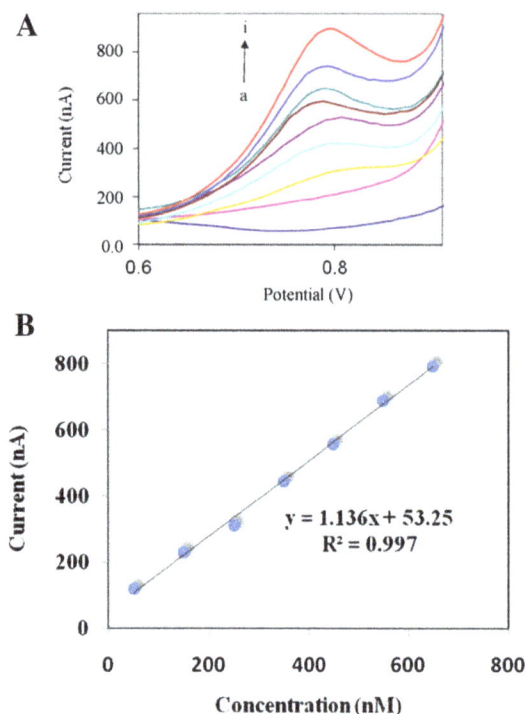

Fig. 7 Differential pulse voltammetric responses for successive additions of diltiazem from 50 (*a*) to 650 nM (*i*) concentration (**a**) and the calibration curve (**b**) at the Co_3O_4/SnO_2 (8 %) nanopowders modified CPE; other conditions: preconcentration time: 190 s; preconcentration potential: −200 mV; scan rate: 15 mV/s

preconcentration by adsorptive accumulation onto MCPE at −0.2 V for 190 s. To characterize the precision of the modified electrode, repetitive determinations of diltiazem were carried out at 350 and 550 nM concentrations at 0.1 M phosphate buffer. The results of five replicate measurements showed a relative standard deviation of 3.2 and 2.5 % (relative standard deviation), respectively, indicating that the results are reproducible. Also for determination of trueness as percentage relative error (Er %) was performed five replicate measurement for 250 nM diltiazem that was 0.16 %.

Interferences

The influence of various foreign species and co-formulated drugs on the determination of 100 nM diltiazem was investigated. The tolerance limit was taken as the maximum concentration of the foreign substances, which caused an approximately ±5 % relative error in the determination. The results showed that no interference occurred for the co-formulated drugs such as atenolol, phenobarbital and phenytoin and common substances and ions such as Na^+, Ca^{2+}, K^+, Ni^{2+}, Fe^{3+}, glucose, acid citric, urea, glycine, and ascorbic acid. The fact that Na^+, K^+, Ca^{2+}, acid citric, glucose, urea, glycine, atenolol and phenobarbital (a 1000-fold excess) and ascorbic acid (a 100-fold excess) and phenytoin (a tenfold excess) were not interference is of particular significance.

Analytical application

Assay of tablet

The validity of the proposed voltammetric method was investigated by assaying diltiazem in tablets (each is labeled to contain 60 mg diltiazem per tablet). The recoveries were calculated by the calibration graph. The statistical calculations for the assay results show a good precision of the proposed method (Table 2).

Assay of diltiazem in spiked human urine

The proposed method was also successfully applied to the determination of diltiazem in spiked urine samples from healthy volunteers using calibration curve. This determination carried out at three different levels of concentrations (250, 300 and 420 nM) and six determinations were carried at each concentration level (Table 3). The mean apparent recoveries for the three concentrations were 101.4, 99.64 and 97.33 % with relative standard deviations of 4.9, 4.5 and 4.1 %, respectively.

Table 3 Apparent recovery of diltiazem from human urine

Sample	Added (nM)	Found (nM)	Apparent recovery (%)	RSD	E_r (%)
1	0	0	–	–	–
2	250	253.6	101.4	4.9	1.4
3	300	292.0	97.3	4.1	2.7
4	420	420	99.6	4.5	0.4

Average of four determinations

Table 2 Analysis of diltiazem in tablet

Company	Labeled (mg/tablet)	Found (mg/tablet)	Apparent recovery (%)	RSD	E_r (%)
Arya pharmaceutical co.	60	60.8	101.4	4.9	1.4
Sobhan Darou co.	60	58.4	97.3	4.1	−2.7

Average of three determinations

Table 4 Apparent recovery of diltiazem from human plasma

Sample	Added (nM)	Found (nM)	Apparent recovery (%)	RSD	E_r (%)
1	0	0	–	–	–
2	350	340.0	97.1	5.3	2.8
3	410	417.0	101.7	3.4	1.7
4	480	465.5	97.0	5.5	3.0

Average of four determinations

Assay of diltiazem in spiked human plasma

The proposed method was applied to the determination of diltiazem in spiked human serum using the calibration curve. The direct determination of diltiazem in human serum was found to be possible after dilution of the sample with the supporting electrolyte. The percentage recovery of the drug in serum, based on the average of six replicate measurements, is listed in Table 4. The values obtained for recovery are acceptable for biological fluids.

Conclusion

The Anodic stripping voltammetry method (ASV) for the quantitative determination of diltiazem with carbon paste electrode modified with Co_3O_4/SnO_2 nano-powder was found to be simple and highly sensitive over the reported data for the determination of diltiazem. It can be used successfully to stability-indicating assay the drug in dosage form as well as in urine and plasma at trace levels.

Acknowledgments We gratefully acknowledge the Payame Noor University (Delijan, Iran) and the Amirkabir University of Technology (Tehran, Iran) for supporting this work.

References

1. Feld, G., Singh, B.N.: Diltiazem-pharmacological properties and therapeutic uses. Hosp. Formul. **20**, 814 (1985)
2. Hermann, Ph, Morselli, P.L.: Pharmacokinetics of diltiazem and other calcium entry blockers. Acta Pharmacol. Toxicol. **57**, 10–20 (1985)
3. Kolle, E.U., Ochs, H.R., Vollmer, K.O.: Pharmacokinetic model of diltiazem. Arzneim.-Forsch. **33**, 972–977 (1983)
4. Rovei, V., Gomeni, R., Mitchard, M., Larribaud, J., Blatrix, C., Thebault, J.J., Morselli, P.L.: Pharmacokinetics and metabolism of diltiazem in man. Acta Cardiol. **35**, 35–45 (1980)
5. Bianchetti, G., Regazzi, M., Rondanelli, R., Ascalone, V., Morselli, P.L.: Bioavailability of diltiazem as a function of the administered dose. Biopharm. Drug Dispos. **12**, 391–401 (1991)
6. Suleiman, M.S., Abdulhameed, M.E., Najib, N.M., Muti, H.Y.: Effect of ultraviolet radiation on the stability of diltiazem. Int. J. Pharm. **50**, 71–73 (1989)
7. Laeer, S., Scholz, H., Uebeler, P., Neumann, J., Zimmermann, N.J.: Quantitation of diltiazem in human cardiac tissue using high-performance liquid chromatography. Chromatogr. Sci. **35**, 93–96 (1997)
8. Meehan, E., Kelly, J.G.: High-performance liquid chromatographic assay for diltiazem in small-volume blood specimens and application to pharmacokinetic studies in rats. J. Chromatogr. **729**, 297–300 (1996)
9. Morris, R.G., Saccoia, N.C., Jones, T.E.: Modified liquid chromatographic assay for diltiazem and metabolites in human plasma. Liq. Chromatogr. J. Relat. Technol. **19**, 2385–2394 (1996)
10. Shi, X.J., Zong, M.K., Wang, H.T., Zhang, J.H.: Studies on the chemical components of essential oils of Elsholtzia patrini Garcke. Yaowu Fenxi Zazhi **15**(23), 20–22 (1995)
11. Ascalone, V., Locatelli, M., Malavasi, B.J.: Determination of diltiazem and its main metabolites in human plasma by automated solid-phase extraction and high performance liquid chromatography: a new method overcoming instability of the compounds and interference problems. Chromatogr. B Biomed. Appl. **657**, 133–140 (1994)
12. Li, N., Li, C.-L., Lu, N.-W., Dong, Y.-M.: A novel micellar per aqueous liquid chromatographic method for simultaneous determination of diltiazem hydrochloride, metoprolol tartrate and isosorbide mononitrate in human serum. J. Chromatogr. B Biomed. Appl. **967**, 90–97 (2014)
13. Christensen, H., Carlson, E., Asberg, A., Schram, L., Berg, K.J.: A simple and sensitive high-performance liquid chromatography assay of diltiazem and main metabolites in renal transplanted patients. Clin. Chim. Acta **283**, 63–75 (1999)
14. Devarajan, P.V., Dhavse, V.V.: High-performance thin-layer chromatographic determination of diltiazem hydrochloride as bulk drug and in pharmaceutical preparations. J. Chromatogr. B **706**, 362–366 (1998)
15. Alebic-Kolbah, T., Plavsic, F.: Determination of serum diltiazem concentrations in a pharmacokinetic study using gas chromatography with electron capture detection. J. Pharm. Biomed. Anal. **8**, 915–918 (1990)
16. Sreedhar, K., Sastry, C.S.P., Narayane Reddy, M., Sankar, D.G.: Extractive spectrophotometric determinations of diltiazem hydrochloride. Indian Drugs **32**, 90–92 (1995)
17. Rahman, N., Azmi, S.N.H.: Spectrophotometric determination of diltiazem hydrochloride with sodium metavanadate. Microchem. J. **65**, 39–43 (2000)
18. Hubert, Ph, Chiap, P.J.: Automatic determination of diltiazem and desacetyldiltiazem in human plasma using liquid–solid extraction on disposable cartridges coupled to HPLC—part I: optimization of the HPLC system and method validation. J. Pharm. Biomed. Anal. **9**, 877–882 (1991)
19. Ghandour, M.A., Aboul Kasim, E., Ali, A.M.M., El-Haty, M.T., Ahmed, M.M.: Adsorptive stripping voltammetric determination of antihypertensive agent: diltiazem. J. Pharm. Biomed. Anal. **25**, 443–451 (2001)
20. Javanbakht, M., Divsar, F., Badiei, A., Fatollahi, F., Khaniani, Y., Ganjali, M.R., Norouzi, P., Chaloosi, M., Mohammadi Ziarani, G.: Determination of picomolar silver concentrations by differential pulse anodic stripping voltammetry at a carbon paste electrode modified with phenylthiourea-functionalized high ordered nanoporous silica gel. Electrochim. Acta **54**, 5381–5386 (2009)
21. Javanbakht, M., Khoshsafar, H., Ganjali, M.R., Norouzi, P., Adib, M.: Adsorptive stripping voltammetric determination of nanomolar concentration of cerium(III) at a carbon paste electrode modified by N'-[(2-Hydroxyphenyl)Methylidene]-2-Furohydrazide. Electroanalysis **21**, 1605–1610 (2009)
22. Javanbakht, M., Khoshsafar, H., Ganjali, M.R., Norouzi, P., Badei, A., Hasheminasab, A.: Stripping voltammetry of cerium(III) with a chemically modified carbon paste electrode con-

taining functionalized nanoporous silica gel. Electroanalysis **20**, 203–206 (2008)

23. Javanbakht, M., Khosh safar, H., Ganjali, M.R., Badei, A., Norouzi, P., Hasheminasab, A.: Determination of nanomolar mercury(II) concentration by anodic-stripping voltammetry at a carbon paste electrode modified with functionalized nanoporous silica gel. Cur. Anal. Chem. **5**, 35–41 (2009)

24. Javanbakht, M., Fathollahi, F., Divsar, F., Ganjali, M.R., Norouzi, P.: A selective and sensitive voltammetric sensor based on molecularly imprinted polymer for the determination of dipyridamole in pharmaceuticals and biological fluids. Sens Actuators B: Chem **182**, 362–367 (2013)

25. Attaran, A.M., Javanbakht, M., Fathollahi, F., Enhesari, M.: Determination of salbutamol in pharmaceutical and serum samples by adsorptive stripping voltammetry on a carbon paste electrode modified by iron titanate nanopowders. Electroanalysis **24**, 2013–2020 (2012)

26. Taylor, D., Fleig, P.F., Page, R.A.: Characterization of nickel titanate synthesized by sol–gel processing. Thin Solid Films **408**, 104–110 (2002)

27. Martinez, A.A.: Cerium–terbium mixed oxides as potential materials for anodes in solid oxide fuel cells. J. Power Sources **151**, 43–51 (2005)

28. Chaubal, N.S., Sawant, M.R.: Synergistic role of aluminium in stabilization of mixed metal oxide catalyst for the nitration of aromatic compounds. Catal. Commun. **7**, 443–449 (2006)

29. Xiong, W., Kale, G.M.: Novel high-selectivity NO2 sensor incorporating mixed-oxide electrode. Sens. Actuators, B **114**, 101–108 (2006)

30. Li, B., Wang, M., Zhou, X., Wang, X., Liu, B., Li, B.: Pyrolyzed binuclear-cobalt-phthalocyanine as electrocatalyst for oxygen reduction reaction in microbial fuel cells. Bioresour. Technol. **193**, 545–548 (2015)

31. Zhang, J., Wang, X., Qin, D., Xue, Z., Lu, X.: Fabrication of iron-doped cobalt oxide nanocomposite films by electrodeposition and application as electrocatalyst for oxygen reduction reaction. Appl. Surf. Sci. **320**, 73–82 (2014)

32. Antonin, V.S., Assumpção, M.H.M.T., Silva, C.M., Parreira, L.S., Lanza, M.R.V., Santos, M.C.: Synthesis and characterization of nanostructured electrocatalysts based on nickel and tin for hydrogen peroxide electrogeneration. Electrochim. Acta **109**, 245–251 (2013)

33. Enhessari, M., Parviz, A., Ozaee, K., Karamali, E.: Magnetic properties and heat capacity of CoTiO$_3$ nanopowders prepared by stearic acid gel method. J. Exp. Nanosci. **5**, 61–68 (2010)

34. Enhessari, M., Parviz, A., Karamali, E., Ozaee, K.: Synthesis, characterisation and optical properties of MnTiO3 nanopowders. J. Exp. Nanosci. **7**, 327–335 (2012)

35. Gevaerd, A., Caetano, F., Oliveira, P., Zarbin, A., Bergamini, M., Marcolino-Junior, L.: Thiol-capped gold nanoparticles: influence of capping amount on electrochemical behavior and potential application as voltammetric sensor for diltiazem. Sens. Actuators B: Chem. **220**, 673–678 (2015)

2

Synthesis and characterization of silica nanostructures for cotton leaf worm control

Haytham A. Ayoub[1,2] · Mohamed Khairy[1] · Farouk A. Rashwan[1] ·
Hanan F. Abdel-Hafez[2]

Abstract Herein, silica nanostructures with various physicochemical characteristics were synthesized via surfactant-assisted methods. Potent entomotoxic effects of silica nanostructures were explored against cotton leaf worm (*Spodoptera littoralis*) for the first time by utilizing surface contact and feeding bioassay protocols. The mortality of the treated larvaes by surface contact was faster than feeding bioassay method. The results showed that the surface characteristics and particle size of silica nanostructures could effectively control their entomotoxic effects compared to commercial silica or even organic pesticides. It was also observed that the dead bodies of the insects became extremely dehydrated due to the damage of insect cuticular water barrier as a result of abrasion. Furthermore, the physical mode of action of silica nanostructures makes insects is unlikely to become physiologically resistant; hence, silica nanostructures can be efficiently used as a valuable tool in *S. littoralis* management programs.

Keywords Silica nanoparticles · Mesoporous materials · Nanocides · Plant protection · *Spodoptera littoralis*

Introduction

Nanomaterial research was prompted recently to develop novel nano-technological products, processes, and applications that promise special merits for saving human health

✉ Mohamed Khairy
mohamed.khairy@science.sohag.edu.eg

[1] Chemistry Department, Faculty of Science, Sohag University, Sohag 82524, Egypt

[2] Plant Protection Research Institute, ARC, Dokki, Giza, Egypt

and environment. The sustainable potentials of nanotechnology are oriented to save of raw materials, energy and water as well as reduce greenhouse gases and hazardous wastes to improve the future. As a result, the application of nanomaterials in the area of plant sciences (i.e., nutrients and/or pest control) has been extensively investigated to overcome the expected increases of global population without negative impacts to environment and/or public health [1–3]. For example, Ag, CuO, MgO, and ZnO nanoparticles were presented as effective antimicrobial agents [4–8]. Amorphous silica nanoparticle (SiO_2 NPs) showed biological entomotoxic effects, although it is inert in nature and considered as a biocompatible material according to US Food and Drug Administration [9]. To the best of our knowledge, few papers have been reported to utilize such inorganic NPs alone and/or combined with organic ingredient for plant protection. Therefore, it is meaningful to investigate the pesticidal behaviors of the inorganic NPs or their formulations to reduce harmful organic pesticide usage, enhance durability and efficiency as well as overcome the physiological resistance of the pests [10–13].

Mesoporous metal oxides in nanometer scale have received much attention in the field of research because of their unique physical and chemical properties in terms of large surface areas, hydrothermal stability in organic and inorganic solution phases, and reduced densities [14, 15]. Therefore, several of mesoporous metal oxide nanoparticles have been recently synthesized with controlled compositions, structures, and morphologies due to their composition-, structure-, and morphology-dependent properties and applications [16–18]. Among various metal oxides, mesoporous silica nanostructures were extensively studied for their scientific and technological interests including separations, catalysis, sensing, and drug delivery

[19–22]. Different mesoporous nanostructures, monoliths, fibers, spheres, tubes, thin films, and ellipse, were synthesized by utilizing organic/inorganic synthetic chemistry [20–25]. However, considerable efforts have been devoted to control the intrinsic features and enable specific functionality which is selected to align with targeted applications. Despite this significant progress, most of the synthesis procedures are too delicate which render them less viable for large-scale industrial applications.

Spodoptera littoralis is considered as one of the most serious and destructive pest not only for cotton plant, but also for other vegetable, ornamental, and field crops in Egypt. Previously, we have investigated the effect of commercial SiO_2 on the second larval instar of *S. littoralis*. However, the key factors that control SiO_2 nanostructure entomotoxic efficiency were not recognized yet [12, 13]. In the present study, SiO_2 nanostructures with various shapes and sizes have been synthesized by utilizing surfactant-assisted methods [26]. Triton X100 (TX-100), cetyltrimethylammonium bromide (CTAB), and Polyvinylpyrrolidone (PVP) have been used as soft template. The entomotoxic effects of SiO_2 nanostructures against *S. littoralis* are investigated by applying surface contact and feeding bioassay methods. Importantly, the synthesis conditions of SiO_2 nanostructures are played significant role not only for control their physical properties, but also revealed variable entomotoxic effects against *S. littoralis*. The synthesized SiO_2 nanostructures showed promising inorganic-based pesticide in *S. littoralis* management programs compared to commercial SiO_2 materials.

Experimental section

Chemicals

All chemicals were used as received without further purification. Tetra methylorthosilicate (TMOS; 98%), ethanol; 96%, ammonium hydroxide (NH_4OH; 28%), triton X100 (TX-100), cetyltrimethylammonium bromide (CTAB), and polyvinylpyrrolidone (PVP) were obtained from Sigma–Aldrich Company LTD. Silica precipitated (SiO_2) was purchased from Fisher chemicals company, UK. All solutions were prepared using bidistilled water with resistivity more than 18.2 $M\Omega$ cm.

Insects

Laboratory strain of cotton leaf worm (*S. littoralis*) was cultured on leaves of the castor oil plant (*Ricinus communis L*) under constant laboratory conditions of 25 ± 2 °C and $65 \pm 5\%$ relative humidity (RH) in Plant Protection

Research Institute, ARC, Dokki, Giza, Egypt. The second instar larval stage of the insect was used in the entomotoxic effect evaluations.

Synthesis of silica nanostructures

SiO_2 nanostructures were synthesized by utilizing modified Stöber method. Three different surfactants of TX-100, CTAB, and PVP have been used [26].

SiO_2-TX

A mixture solution (50 mL) of TX-100, ethanol, NH_4OH, and H_2O with the final ratio (0.2:4.6:1:4) was stirred for 30 min at 50 °C. A 3.0 mL of TMOS was injected dropwise to the mixture with constant rate of 10 μL min^{-1}. The final solution was diluted with 100 mL H_2O and left under continuous stirring for 42 h. The formed silica sample was filtrated, washed several times by mixture solution of H_2O and ethanol, and left to dry at 60–80 °C. Finally, the remaining surfactant was removed by thermal treatment at 550 °C for 6 h.

SiO_2-CTAB

A 2 g of CTAB mixed with 5 mL of TMOS. The mixture was transferred quantitatively to 250 mL round bottom flask containing a mixture solution of 3.2 mL NH_4OH, 20 mL H_2O, and 143 mL ethanol. The flask contents were heated to 100 °C and left under constant stirring for 24 h. The formed silica sample was filtrated, washed several times with a mixture of water and ethanol, and then left to dry at 80 °C overnight.

SiO_2-PVP

A mixture solution (50 mL) of PVP, ethanol, NH_4OH, and H_2O with the final molar ratio (0.2:4.6:1:4) was formed and stirred for 30 min at 50 °C. The mixture solution was diluted by 100 mL H_2O. A 3.0 mL of TMOS was introduced with constant rate of 10 μL min^{-1} and then left it under constant stirring for 42 h. The formed silica sample was filtrated, washed several times with mixture of H_2O and ethanol, and left to dry at 60–80 °C. The remaining surfactant was removed by thermal treatment at 550 °C for 6 h.

Characterization of SiO_2 nanostructures

The morphology of the SiO_2 samples was investigated using scanning electron microscopy (SEM, JEOL model 5400 LV). The SiO_2 powders were ground and fixed onto a specimen stub using double-sided carbon tape. To obtain high-resolution micrographs, a 10 nm Pt film was coated on the SiO$_2$ powder using anion sputtering (Hitachi E-1030) at room temperature. The SEM was operated at 20 keV to obtain high-resolution SEM images. Transmission electron microscopy (TEM) of SiO$_2$ samples was performed using a JEOL-JEM microscope model 2100. TEM was conducted at an acceleration voltage of 200 kV to obtain a lattice resolution of 0.1 nm. TEM images were recorded using a CCD camera. The SiO$_2$ sample was dispersed in ethanol solution using an ultrasonic bath, and then dropped on a copper grid. Prior to inserting the samples into the TEM column, the grid was vacuum dried for 20 min.

The textural surface properties and pore size distribution were determined by N$_2$ adsorption/desorption isotherms at 73 K with a NOVA 3200 apparatus, USA. The specific surface area (S_{BET}) was calculated using the Brunauer–Emmett–Teller (BET) method with multipoint adsorption data from the linear segment of the N$_2$ adsorption isotherm. The pore size distribution was determined from the analysis of desorption branch of isotherm using Barrett–Joyner–Halenda (BJH) method. FTIR spectra of silica samples were recorded between 4000 and 400 cm^{-1} on a Bruker ALPHA FT-IR spectrometer. A little amount of each sample was scratched off and ground with KBr to press pellets for recording their FTIR spectrum.

Bioassays

Surface contact method

The silica samples were uniformly distributed on the bottom surface of plastic containers at doses of 0.25, 0.5, and 1.0 mg cm^{-2}. Silica-free container was used in the control set. The containers were covered with muslin cloth to allow aeration. Each treatment was repeated three times. About

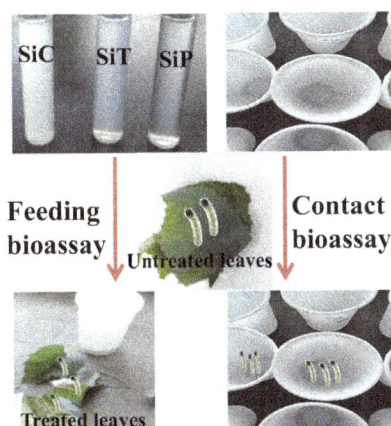

Scheme 1 Representation of feeding and contact method for cotton leaf worm control

ten second instars larvae of *S. littoralis* were introduced to the container after distributing of the material (Scheme 1). All bioassays were performed at 25 ± 2 °C and relative humidity (RH) of 65 ± 5%. Mortality of the insect was checked at particular days and the corrected percentages were statistically computed using Probit analysis program [27].

Feeding method

Feeding entomotoxic effect was performed via leaf dip bioassay method. Leaf discs of castor oil plant were prepared and dipped in a solution of 250, 500, and 1000 ppm of silica samples. The untreated leaf discs of castor oil plant were used in control set. The containers were covered with muslin cloth to allow aeration. Each treatment was repeated three times. About ten second instars larvae of *S. littoralis* were introduced in each container (Scheme 1). All bioassays were performed at 25 ± 2 °C and RH of 65 ± 5%. Insect mortality was checked at particular days. Their corrected percentages were statistically computed using Probit analysis program [27].

Biochemical analysis

The castor oil leaves were treated with LC50 of SiO$_2$-TX, SiO$_2$-CTAB, and SiO$_2$-PVP nanostructures via feeding bioassay method for 3 days. Then, the survived *S. littoralis* were collected and allowed to grow on untreated/normal leaves until to reach the sixth instar larvae. The insects were homogenized for biochemical analysis in a chilled glass Teflon tissue homogenizer (ST-2 Mechanic-Preczyina, Poland). The supernatant was kept in a deep freezer at −20 °C for further biochemical analysis. Total carbohydrates, total proteins, total lipids, lactate dehydrogenase (LED), phenol oxidase activity, and chitinase activity were estimated. The absorbance measurement of colored substances or metabolic compounds was performed using double beam UV/Vis spectrophotometer (Spectronic 1201, Milton Roy Co., USA).

Results and discussion

Physicochemical Characteristics of SiO$_2$ samples

Surfactant-assisted method was employed to control the morphology of silica nanostructures. Three different synthesis compositions have been used to form micro-emulsion phases. Figure 1 shows SEM and TEM micrographs of synthesized SiO$_2$ nanostructures in the presence of different structure directing agents: TX-100, CTAB, and PVP. In the presence of TX-100, the SEM image reveals the

Fig. 1 SEM and TEM images of SiO$_2$-TX (**a**, **b**), SiO$_2$-CTAB (**c**, **d**), and SiO$_2$-PVP (**e**, **f**) samples prepared using surfactant-assisted method

formation of monolithic particles with plate-like morphology and average particle size about 5–10 μm (Fig. 1a). Aggregated spherical nanoparticles are formed in the presence of CTAB and PVP (Fig. 1c, e). Furthermore, the TEM micrographs exhibit surfactant-dependent size and morphology of SiO$_2$ nanostructures. Figure 1b shows monolith silica particles with a worm-like mesoporous network over a large area of monolithic structure. SiO$_2$-CTAB sample shows aggregated sphere-like particles with an average diameter of 175 nm. However, PVP surfactant reveals uniform and dispersed SiO$_2$ nanostructures with sphere-like morphology; the average diameter is about 33 nm. Such different sizes and shapes of silica nanostructures might offer several interesting physicochemical characteristics that provide more information for designing new inorganic pesticide formulations as well as leading to find out the key factors which control the interactions of SiO$_2$ nanostructures with insect body.

N$_2$ adsorption/desorption isotherms of silica nanostructures showed different surface areas and pore size distribution curves (Fig. 2 a). According to IUPAC classification; SiO$_2$-TX presents type IV isotherm and H2-type hysteresis loop with the highest surface area (S_{BET}) of 634.5 m^2 g^{-1}. A well-defined steepness of adsorption/desorption branch features cage-like mesopore with uniform cavities. However, SiO$_2$-CTAB and SiO$_2$-PVP revealed type IV isotherm with pronounced H3 hysteresis

loop which could characterize slit-like mesopore entrance. The desorption branch of silica isotherms showed a stepwise behavior ended at the limiting pressure of the hysteresis closure, implying non-uniformity of the mesopore openings associated with pore constrictions and/or ink-bottle pores with narrow necks. The specific surface areas of SiO$_2$-CTAB and SiO$_2$-PVP are $S_{BET} = 66$ and 158 m^2 g^{-1}, respectively. The lower surface area of SiO$_2$-CTAB is attributed to the presence of CTAB surfactant in mesoporous architectures network. Hence, the SiO$_2$-CTAB sample did not treat thermally like SiO$_2$-TX and SiO$_2$-PVP to remove CTAB template. The existence of CTAB increases the hydrophilicity of SiO$_2$ surface and thus changes the interfacial region between the solid SiO$_2$ particle and the insect body wall. The SiO$_2$ nanostructures exhibit uniform pore size distribution of 28.1, 8.5, and 9.76 nm for SiO$_2$-TX, SiO$_2$-CTAB, and SiO$_2$-PVP, respectively. The mesoporous network of SiO$_2$ nanostructures might be more suitable for the adsorption of small and large biomolecules in case of SiO$_2$ contact with the insect body wall [22, 23]. These features are very useful to describe the entomotoxic effects based on surface area and/or lipid- or protein-dependent mechanism.

The chemical state of the SiO$_2$ nanostructures and commercial SiO$_2$ was confirmed using Fourier transform infrared spectroscopy FTIR. The FTIR spectra of SiO$_2$ samples in the region 4000–400 cm^{-1} are explored

Fig. 2 a Nitrogen adsorption/desorption isotherm and their corresponding pore size distribution curves for SiO₂-TX, SiO₂-PVP, and SiO₂-CTAB. **b** FTIR spectra of commercial SiO₂, SiO₂-TX, SiO₂-PVP, and SiO₂-CTAB

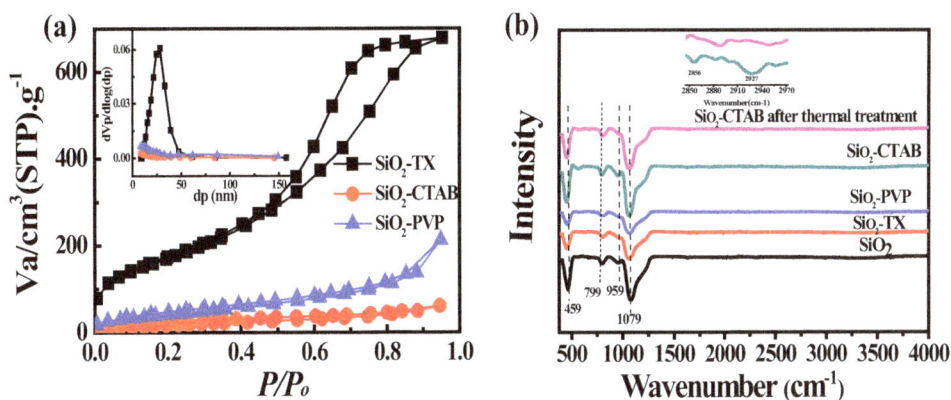

(Fig. 2b). The FTIR spectra of synthesized SiO₂ nanostructures are almost the same as commercial SiO₂. The strong and weak bands are centered at 1079, 799, and 459 cm⁻¹; these bands are attributed to the Si–O–Si bond corresponding to bending and stretching vibrations, respectively [28]. The absorption bands at 2856 cm⁻¹ and at 2927 cm⁻¹ for SiO₂-CTAB sample are assigned to –CH₂ group of CTAB. Furthermore, the absorption band at 959 cm⁻¹ of SiO₂-CTAB is attributed to the hydroxyl group (Si–OH); this band disappears after thermal treatment at 550 °C for 6 h [29]. These results suggested the existence of hydroxyl groups on the SiO₂ surfaces.

Cotton leaf worm control

Insecticide bioassay is often used to estimate the median lethal dose (LD50) or median lethal concentration (LC50) with associated 95% confidence intervals from a dose–response model. The LD50 or LC50 is the amount of insecticide required to kill 50% of a given population or strain under the specified conditions. Several standard methods such as topical application, residual or surface contact, immersion, and feeding bioassays have been commonly employed to determine the relationship between pesticides administered amount (i.e., dose or concentration) and its response magnitude for living organisms [30]. In this study, we have focused on surface contact and feeding bioassay methods to evaluate the entomotoxic effect of SiO₂ nanostructures against *S. littoralis*.

Surface contact bioassay

The contact entomotoxic effect of the synthesized SiO₂ nanostructures has been explored and compared to commercial silica materials (Scheme 1). The SiO₂ powders that were uniformly distributed on the bottom of plastic vessels contain different doses of 0.25, 0.5, and 1.0 mg cm⁻² at 25 ± 2 °C. The *S. littoralis* mortality was checked at certain days, as shown in Fig. 3. Interestingly, the commercial

SiO₂ material has no entomotoxic effects during the first 3 days and the corresponding contact mortality was evaluated to be zero. After that, the mortality was increased up to 100% in the 11th day of treatment (Fig. 3a). However, the synthesized SiO₂ nanostructures showed an efficient *S. littoralis* control with high mortality from the first exposure days. In the first day, the mortality percentage was estimated to be 29.6, 29.1, and 28.1% for SiO₂-TX, SiO₂-CTAB, and SiO₂-PVP, respectively. The mortality was increased up to 96.6, 93.3, and 90.0% in the third day of exposure. The median lethal doses (LD50) were calculated to be 0.61, 0.58, and 0.57 mg cm⁻² after 3 days compared to 0.495 mg cm⁻² for commercial SiO₂ after 7 days (Fig. 3). There are several mechanisms for inorganic materials toxicity which have been recently reported including generation of reactive oxygen species, oxidative stress, membrane disruption, protein unfolding, and/or inflammation. The results showed that the surface area of SiO₂ nanostructures has no significant key role in the absorption of lipids present in insect cuticle. Therefore, the dehydration of the insect body caused by SiO₂ dust was mainly due to acute damage of the cuticular water barrier as a result of abrasion, and the insects began to lose the water from their bodies and died due to desiccation [31, 32]. Significantly, the surface contact bioassay is independent of particle size and morphology but mainly controlled by the hydrophilicity of the SiO₂ surfaces. The presence of small amount of surfactant in SiO₂-CTAB does not have any effect on the mortality values, indicating that the hydrophilic surfaces of synthesized SiO₂ nanostructures might only enhance the entomotoxic rate (Fig. 3c).

Feeding bioassay

Feeding experiment was performed by dipping of castor oil leaf in SiO₂ suspension (Scheme 1). Three suspension solutions containing 250, 500, and 1000 ppm of SiO₂ nanostructures were prepared in the presence of 0.1% (V/V) Triton X-100 at 25 °C as recommend in pesticide

Fig. 3 Accumulative mortality
at indicated days of *S. littoralis*
exposed to different doses of
commercial SiO$_2$, SiO$_2$-TX,
SiO$_2$-PVP, and SiO$_2$-CTAB via
surface contact bioassay

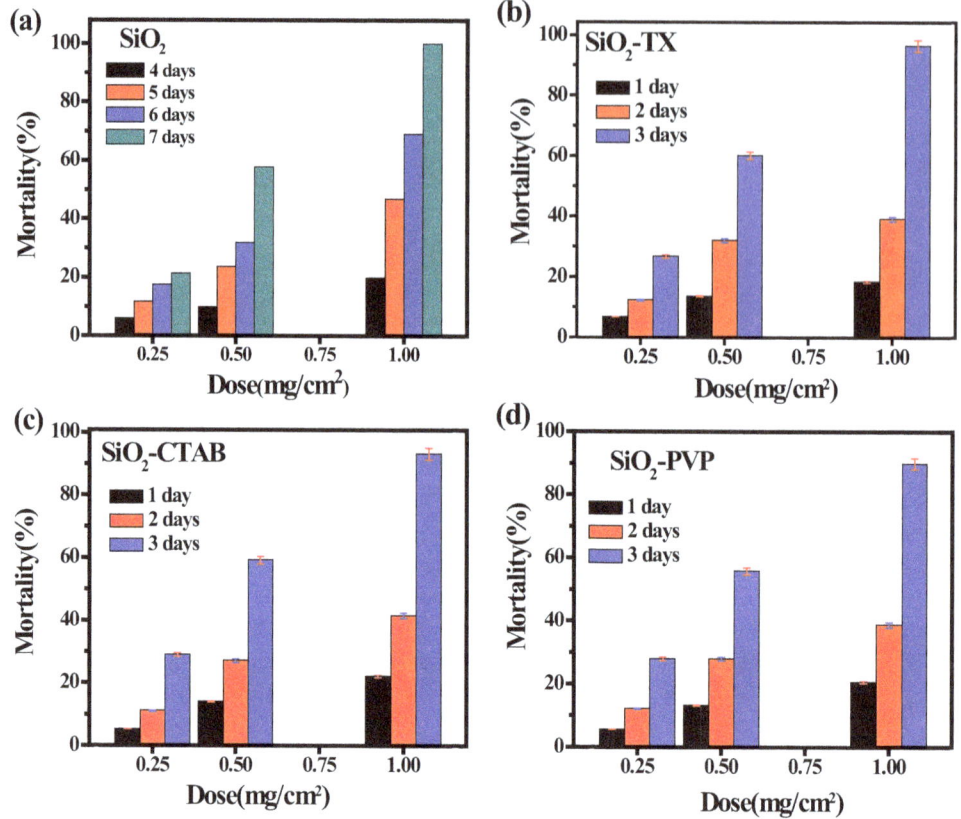

Fig. 3 Accumulative mortality at indicated days of *S. littoralis* exposed to different doses of commercial SiO$_2$, SiO$_2$-TX, SiO$_2$-PVP, and SiO$_2$-CTAB via surface contact bioassay

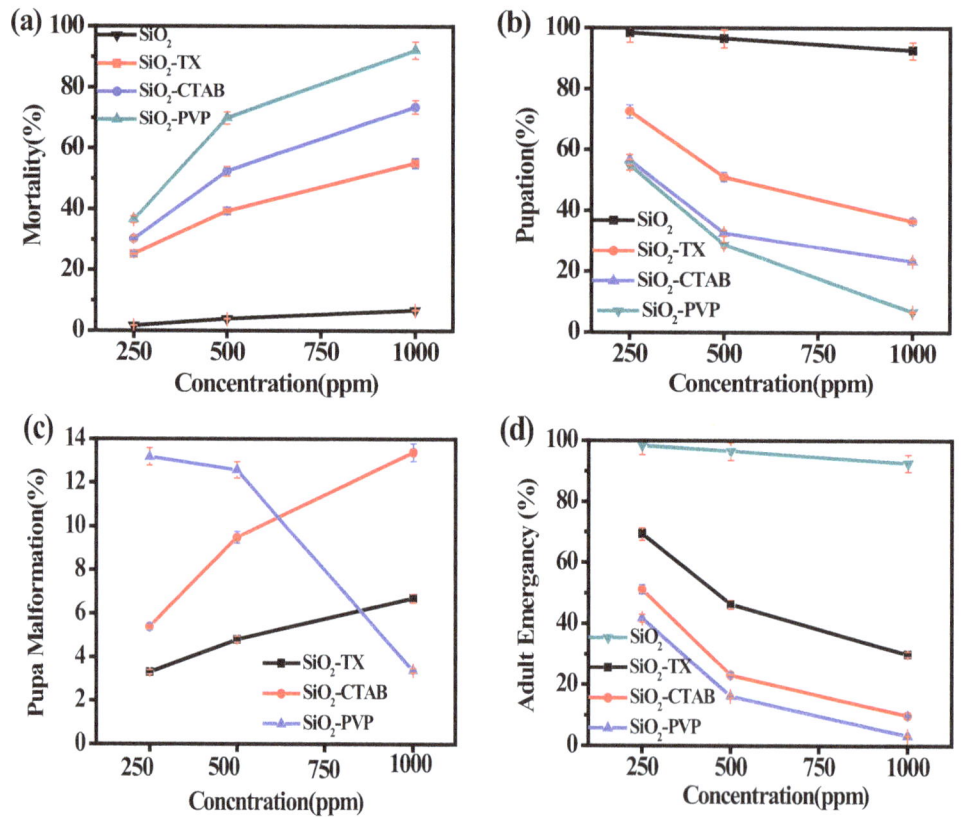

Fig. 4 Accumulative mortality after 11 days of the post treatments (**a**), pupation (**b**), pupa malformation (**c**), and adult emergence (**d**) of *S. littoralis* exposed to dried and treated leaf discs of castor oil plant with different concentrations of commercial SiO$_2$, SiO$_2$-TX, SiO$_2$-PVP, and SiO$_2$-CTAB via feeding bioassay

formulations [33]. Figure 4a shows feeding entomotoxic effect for commercial and synthesized SiO_2 nanostructures. After 11 days of the treatment, accumulative mortality and post emergence for commercial silica were closely to that of control set experiment even at higher concentrations, i.e., 1000 ppm. However, the synthesized SiO_2 nanostructures showed variable intensive entomotoxic effects. The accumulative mortalities using 1000 ppm were about 52.3, 73.1, and 92.2% for SiO_2-TX, SiO_2-CTAB, and SiO_2-PVP, respectively. The accumulative mortality of SiO_2-PVP was almost double of SiO_2-TX. This result revealed that the entomotoxic effect of the synthesized SiO_2 NPs is mainly dominated by particle size compared to surface characteristic effect as in the previous contact experiment. The corresponding LC50 values were calculated after 11 days to be 795, 464.2, and 327.7 ppm for SiO_2-TX, SiO_2-CTAB, and SiO_2-PVP, respectively (Fig. 4a). The LC50 of methomyl against S. littoralis is about 434.49 ppm after 1 day of exposure under the same experimental conditions. Methomyl is a broad-spectrum carbamate insecticide. It has been used in a wide range of agriculture products, although it was considered as highly toxic organic pesticide for birds and mammals [34]. Although SiO_2-TX has the highest surface area, it showed the lowest entomotoxic effect with LC50 = 795 ppm. This observation indicated that surface area is not important factor, especially there is no strong effect for the absorption of lipids present in insect cuticle. Intensive morphological changes of the dead sixth instar of treated larvae have been observed compared with that in control set (Fig. 5a). The biological aspects including pupation, pupa malformation, and adult emergency were investigated (Fig. 4b–d). It was found that the survived treated larvaes with commercial silica transferred to adult insects without any pupa malformation closely to the control set (Fig. 5b). This result indicated that the feeding of commercial silica has no entomotoxic effect against S.

littoralis in contrast with synthesized silica nanostructures. The pupa malformations were estimated to be 6.7, 13.4, and 3.4% for treated larvaes with SiO_2-TX, SiO_2-CTAB, and SiO_2-PVP, respectively. Such malformations in pupa stage affect on the final percentage of the adult emergence. Hence, the entire malformed pupas do not succeed to transform into adult insects. These results suggested that silica nanostructures are more effective on adults than larvaes. Therefore, the mortality is attributed to the impairment of the digestive tract and surface enlargement of the integument as a consequence of dehydration or blockage of spiracles and tracheas. Such intensive damage sorption and abrasion might be due to the generation of reactive oxygen radicals in aqueous suspensions of the synthesized silica nanoparticles. Both homolytic (Si·, SiO·) and heterolytic (Si^+, SiO^-) cleavages might take place in the silicon-oxygen bond through particle cracking in digestion tract. These radicals stabilized as a surface bound reactive oxygen species and then decay subsequently [35]. Therefore, biochemical analyses have been performed to investigate the exposed particle surfaces which could interact with the insect.

Biochemical impacts

The biochemical changes in living organisms that exposed to insecticides provide some clues of their mode of actions. Total carbohydrate content of the sixth instars of S. littoralis treated with LC50 of SiO_2 nanostructures was estimated (Fig. 6a). A significant decrease can be observed: −52.7, −32.43, and −27.71% for SiO_2-TX, SiO_2-CTAB, and SiO_2-PVP, respectively. The reduction in total soluble carbohydrate content of the treated larvaes could be attributed to metamorphic changes in larvaes. Hence, the carbohydrate content supplies the body with glucose and provides an energy source for synthesis of larvae and adult tissues, especially the cuticle [36, 37]. Therefore, the deficiency in carbohydrates content has been leading to pupa malformation (Fig. 5b). Furthermore, a marked diminution in total proteins content of the sixth instars of S. littoralis: −47.6, −44.11, and −23.36% treated with SiO_2-TX, SiO_2-CTAB, and SiO_2-PVP, respectively (Fig. 6b). Such reduction of total protein content might be attributed to protein leakage during intoxication that caused a significant lack in body weight, conversion of protein to amino acids, and degradation of protein to release energy or the direct effect of the tested materials on the amino acids transport of the cell [38].

The change in the total lipid content of the sixth instars of S. littoralis treated with LC50 of silica nanostructures was presented in Fig. 6c. The total lipid content was slightly decreased: −28.95, −5.26, and −2.63% for SiO_2-

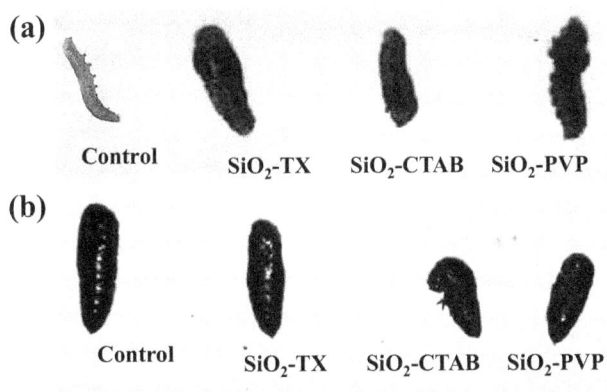

Fig. 5 Photographic images of **a** sixth instars and **b** pupa malformation for S. littoralis exposed to 1000 ppm of SiO_2-TX, SiO_2-PVP, and SiO_2-CTAB by feeding bioassay

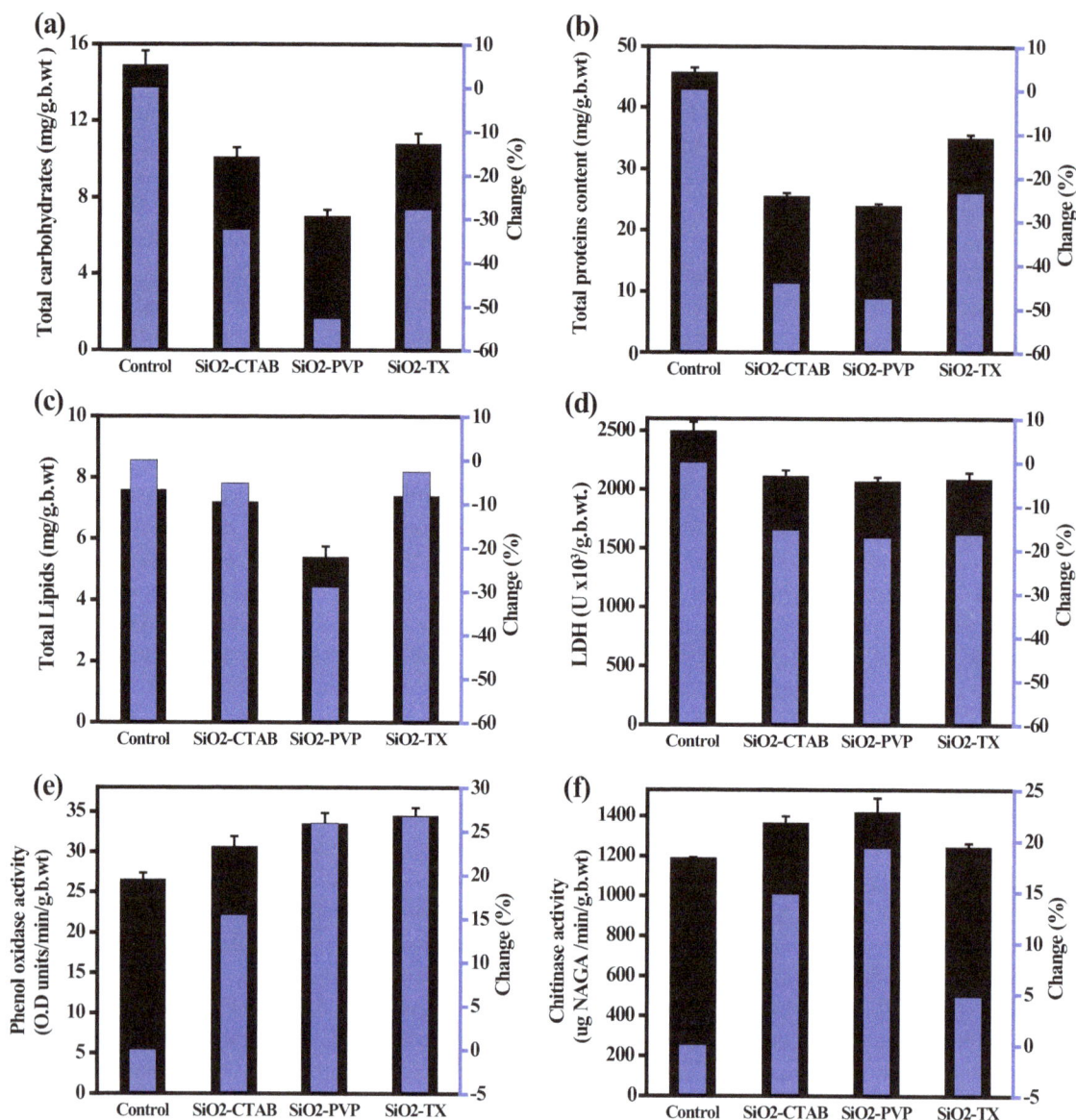

Fig. 6 Total carbohydrates (**a**), total protein content (**b**), total lipids (**c**), LDH (**d**), phenol oxidase activity (**e**), and chitinase activity (**f**) of the sixth instars of *S. littoralis* treated with LC50 of iO$_2$-TX, SiO$_2$-PVP, and SiO$_2$-CTAB via feeding bioassay

TX, SiO$_2$-CTAB, and SiO$_2$-PVP, respectively. This result clearly confirmed our suggestion; there is no strong effect for the absorption of lipids present in insect cuticle. Although SiO$_2$-TX sample has the highest surface area, it has the lowest effect on total lipids content (Fig. 2a). Therefore, the dehydration of the insect body caused by surface reactive silica particles was mainly due to the acute damage of the cuticular water barrier as a result of abrasion [39]. Furthermore, lactate dehydrogenase (LDH) activity of *S. littoralis* in the sixth instars was also decreased: −17.18, −15.41, and −16.42% for SiO$_2$-TX, SiO$_2$-CTAB, and SiO$_2$-PVP, respectively (Fig. 6d). This observation supports a physical mode of action of SiO$_2$ nanostructures with a weak chemical stress upon *S. littoralis* in contrast with

the traditional organic pesticides which caused a strong chemical stress [40].

Phenol oxidase (PO) is an important component of *S. littoralis* immune systems. Figure 6e illustrates the phenol oxidase activity of *S. littoralis* treated with silica nanostructures. It was found that the PO activity was increased in the following: +25.94, +15.42, and +26.7% for SiO$_2$-TX, SiO$_2$-CTAB, and SiO$_2$-PVP, respectively. In general, PO activity is correlated with the resistance of the insect to pathogens [41]. The lower activity of phenol oxidase using SiO$_2$-CTAB is attributed to the presence of surfactant-coated silica particles. Furthermore, the chitinase activity was also increased (Fig. 6f): +19.32, +14.78, and +4.79% for SiO$_2$-TX,

SiO$_2$-CTAB, and SiO$_2$-PVP, respectively. The hyperchitinase activity might be attributed to chitin synthesis inhibitor or may be a secondary effect for the reduced activity of β-ecdysone metabolizing enzymes followed by β-ecdysone accumulation [42, 43]. In general, the mentioned biochemical analyses provide a clear evidence for the hypothesis of physical mode action of SiO$_2$ NPs due to the impairment of the digestive tract and generation of reactive SiO$_2$ surfaces in the aqueous formulations. So far, the application of silica NPs in inorganic pesticide program will offer many advantages over the traditional organic pesticide control as:

1. The insects are unlikely to become genetically selected or physiologically resistance [42, 43].
2. The inorganic formulations based on nanomaterials will reduce the environment risk. Over than 2.8 million tons of organic pesticides are used annually over the world. It has been estimated that only 0.1% of applied pesticides reach the target pests, leaving the bulk of the pesticides in water supplier and soil.
3. Saving the water resources particularly in case of no testing for toxic residue levels in developing countries. Energy and raw materials that might be used for cleaning and decontamination of the water will be utilized.
4. Development of sustainable strategy for cotton leaf worm control that enhances the productivity of such essential crops in Mediterranean area.

Conclusion

SiO$_2$ nanostructures have been synthesized via surfactant-assisted method and utilized for cotton leaf worm control. The entomotoxic effects of SiO$_2$ NPs were explored by surface contact and feeding bioassay methods. Interestingly, the commercial silica did not show any entomotoxic effect on treated larvaes by feeding bioassay and slower entomotoxic effect in surface contact experiment. However, the synthesized silica nanostructures exhibited intensive entomotoxic effect in both bioassay methods. It was observed that the dead bodies of the insects became extremely dehydrated due to the damage of insect cuticular water barrier as a result of abrasion. This hypothesis for the physical mode of action makes the use of nanopesticide stronger and insects are unlikely to become genetically selected or having no physiological resistant to such a type of materials. The usage of inorganic nanopesticides might open a new avenue for plant protection that definitely enhances environmental and economic benefits compared to toxic and persistent organic pesticides.

References

1. Wang, X., Liu, Y., Zhang, H., Shen, X., Cai, F., Zhang, M., Gao, Q., Chen, W., Wang, B., Tao, S.: The impact of carbon nanotubes on bioaccumulation and translocation of phenanthrene, 3-CH$_3$-phenanthrene and 9-NO$_2$-phenanthrene in maize (Zea mays) seedlings. Environ. Sci. Nano. **3**, 818–829 (2016)
2. Gogos, A., Knauer, K., Bucheli, T.: Nanomaterials in plant protection and fertilization: current state, foreseen applications, and research priorities. J. Agric. Food Chem. **60**, 9781–9792 (2012)
3. Raliya, R., Franke, C., Chavalmane, S., Nair, R., Reed, N., Biswas, P.: Quantitative understanding of nanoparticle uptake in watermelon plants. Front. Plant Sci. **7**, 1288 (2016)
4. Reed, R.B., Zaikova, T., Barber, A., Simonich, M., Lankone, R., Marco, M., Hristovski, K., Herckes, P., Passantino, L., Fairbrother, D.H., Tanguay, R., Ranville, J.F., Hutchison, J.E., Westerhoff, P.K.: Potential environmental impacts and antimicrobial efficacy of silver- and nanosilver- containing textiles. Environ. Sci. Technol. **7**, 4018–4026 (2016)
5. Byeon, J.H.: Rapid green assembly of antimicrobial nanobunches. Sci. Rep. **6**, 27006 (2016)
6. Purwajanti, S., Zhou, L., Nor, Y.A., Zhang, J., Zhang, H., Huang, X., Yu, C.: Synthesis of magnesium oxide hierarchical microspheres: a dual-functional material for water remediation. ACS Appl. Mater. Inter. **38**, 21278–22128 (2015)
7. Khan, M.F., Ansari, A.H., Hameedullah, M., Ahmad, E., Husain, F.M., Zia, Q., Baig, U., Zaheer, M.R., Alam, M.M., Khan, A.M., Al Othman, Z.A., Ahmad, I., Aliev, G.M.A.G.: Sol–gel synthesis of thorn-like ZnO nanoparticles endorsing mechanical stirring effect and their antimicrobial activities: potential role as nanoantibiotics. Sci. Rep. **6**, 27689 (2016)
8. Jingzhe, X., Zhihui, L., Ping, L., Yaping, D., Yi, C., Qingsheng, W.: A residue-free green synergistic antifungal nanotechnology for pesticide thiram by ZnO nanoparticles. Sci. Rep. **4**, 5408 (2014)
9. US Food and Drug Administration's Website. GRN No. 321: Synthetic amorphoussilica, http://www.accessdata.fda.gov/scripts/fcn/gras_notices/GRN000321.pdf. 21 May 2014
10. Gerstl, Z., Nasser, A., Mingelgrin, U.: Controlled release of pesticides into soils from clay–polymer formulations. J. Agric. Food Chem. **46**, 3797–3809 (1998)
11. Xiang, Y., Wang, N., Song, J., Cai, D., Wu, Z.: Micro-nanopores fabricated by high-energy electron beam irradiation: suitable structure for controlling pesticide loss. J. Agric. Food Chem. **22**, 5215–5219 (2013)
12. Osman, H.H., Abdel-Hafez, H.F., Khidr, A.A.: Comparison between the efficacy of two nano-particles and effective microorganisms on some biological and biochemical aspects of *Spodoptera littoralis*. IJAIR. **3**, 1620–1626 (2015)
13. Spodoptera litura Fabricious: European and mediterranean plant protection organization (EPPO). EPPO Bull. **9**, 142–146 (2008)
14. Walcarius, A., Sibottier, E., Etienne, M., Ghanbaja, J.: Electrochemically assisted self-assembly of mesoporous silica thin films. J. Nat. Mater. **6**, 602–608 (2007)
15. El-Safty, S.A., Khairy, M., Shenashen, M.A., Elshehy, E., Warkocki, W., Sakai, M.: Optical mesoscopic membrane sensor layouts for water- free and blood-free toxicants. Nano Res. **8**, 3150–3163 (2015)
16. Mahmoud, B.G., Khairy, M., Rashwan, F.A., Foster, C.W., Banks, C.E.: Self-assembly of porous copper oxide hierarchical nanostructures for selective determinations of glucose and ascorbic acid. RSC Adv. **6**, 14474–14482 (2016)
17. Khairy, M., El-Safty, S.A.: Mesoporous NiO nanoarchitectures

for electrochemical energy storage: influence of size, porosity, and morphology. RSC Adv. **3**, 23801–23809 (2013)

18. Largeot, C., Portet, C., Chmiola, J., Taberna, P.-L., Gogotsi, Y., Simon, P.: Relation between the ion size and pore size for an electric double-layer capacitor. J. Am. Soc. **9**, 2730–2731 (2008)

19. El-Safty, S.A., Shenashen, M.A., Ismael, M., Khairym, M.: Mesocylindrical Aluminosilica monolith biocaptors for size-selective macromolecule cargos. Adv. Funct. Mater. **22**, 3013–3021 (2012)

20. Chen, Y., Chen, H., Ma, M., Chen, F., Guo, L., Zhang, L., Shi, J.: Double mesoporous silica shelled spherical/ellipsoidal nanostructures: synthesis and hydrophilic/hydrophobic anticancer drug delivery. J. Mater. Chem. **21**, 5290–5298 (2011)

21. Khairy, M., El-Safty, S.A., Shenashen, M.A., Elshehy, E.A.: Hierarchical inorganic–organic multi-shell nanospheres for intervention and treatment of lead-contaminated blood. Nanoscale. **5**, 7920–7927 (2013)

22. Niu, Z., Kabisatpathy, S., He, J., Lee, L.A., Rong, J., Yang, L., Sikha, G., Popov, B.N., Emrick, T.S., Russell, T.P., Wan, Q.: Synthesis and characterization of bionanoparticle -Silica composites and mesoporous silica with large pores. Nano Res. **2**, 474–483 (2009)

23. El-Safty, S.A., Shenashen, M.A., Khairy, M.: Bioadsorption of proteins on large mesocage-shaped mesoporous alumina monoliths. Colloid Surf. B **103**, 288–297 (2013)

24. Xu, P., Chen, C., Li, X.: Mesoporous-silica nanofluidic channels for quick enrichment/extraction of trace pesticide molecules. Sci. Rep. **5**, 17171 (2015)

25. El-Safty, S.A., Shenashen, M.A., Khairy, M.: Optical detection/collection of toxic Cd (II) ions using cubic Ia3d aluminosilica mesocage sensors. Talanta **98**, 69–78 (2012)

26. StÖber, W., Fink, A., Bohn, E.: Controlled growth of monodisperse silica spheres in the micron size range. J. Colloid Interface Sci. **26**, 62–69 (1968)

27. Finney, D.J.: Probit analysis, 3rd edn, p. 333. Cambridge Univ. Press, London (1971)

28. Zhang, X., Xia, B., Ye, H., Zhang, Y., Xiao, B., Yan, L., Lv, H., Jiang, B.: One-step sol–gel preparation of PDMS–silica ORMOSILs as environment-resistant and crack-free thick antireflective coatings. J. Mater. Chem. **22**, 13132–13140 (2012)

29. Zhang, M., Wu, Y., Feng, X., He, X., Chen, L., Zhang, Y.: Fabrication of mesoporous silica-coated CNTs and application in size-selective protein separation. J. Mater. Chem. **20**, 5835–5842 (2010)

30. Zhu, K.Y.: Insecticide bioassay. Encyclopedia of entomology, pp. 1974–1976. Springer, Berlin (2008)

31. Debnath, N., Das, S., Seth, D., Chandra, R., Bhattacharya, SCh., Goswami, A.: Entomotoxic effect of silica nanoparticles against *Sitophilus oryzae* (L.). J. Pest. Sci. **84**, 99–105 (2011)

32. Debnath, N., Mitra, S., Das, S., Goswami, A.: Synthesis of surface functionalized silica nanoparticles and their use as entomotoxic nanocides. Powder Technol. **221**, 252–256 (2012)

33. Memarizadeh, N., Ghadamyari, M., Adeli, M., Talebi, K.: Linear–dendritic copolymers/indoxacarb supramolecular systems: biodegradable and efficient nano-pesticides. Environ. Sci. Process. Impacts **16**, 2380–2389 (2014)

34. Abdien, S.A., Ahmed, M.A.I., AbduAllah, G.A.M., Ezz El-Din, H.A.: Potential evaluation of certain conventional pesticides on fourth instar larvae of cotton leaf worm, *Spodoptera littoralis* (Boisd.) (Lepidoptera: Noctuidae) under laboratory conditions. Adv. Environ. Biol. **10**, 282–287 (2016)

35. Fubini, B., Hubbard, A.: Reactive oxygen species (ROS) and reactive nitrogen species (RNS) generation by silica in inflammation and fibrosis. Free Radical Biol. Med. **34**, 1507–1516 (2003)

36. Chippenadale, G.M.: In: Rockstein, M. (ed.) The function of carbohydrates in insect life processes Biochemistry of insects, pp. 2–54. Academic press, New York (1978)

37. El-Sheikh, T.A.A., Rafea, H.S., El-Aasar, A.M., Ali, S.H.: Biological and biochemical effects of *Bacillus thuringiensis*, *Serratia marcescens* and Teflubenzuron on cotton leafworm, Egypt. J. Agric. Res. **91**, 1327–1345 (2013)

38. Rawi, S.M., El-Gindy, H., Haggag, A.M., Aboul El-Hassan, A., Abdel-Kader, A.: New possible molluscicides from Calendula mircantha officinalis and Ammi majus plants. I. Physiological effect on *B. alexandrina* and *B. trucatus*. J. Egypt Ger. Soc. Zool. **16**, 49–75 (1995)

39. Knight, J.A., Anderson, S., Rawle, J.M.: Chemical basis of the sulfo-phospho-vanillin reaction for estimating total serum lipids. Clin. Chem. **18**, 199–202 (1972)

40. Abdel-Aziz, H.S.: Effect of some insecticides on certain enzymes of *Spodoptera littoralis*, Egypt. J. Agric. Res. **92**, 501–512 (2014)

41. Nigam, Y., Maudlin, I., Welburn, S., Ratcliffe, N.A.: Detection of phenoloxidase activity in the hemolymph of tsetse flies, refractory and susceptible to infection with *Trypanosoma brucei* rhodesiense. J. Invertebr. Pathol. **69**, 279–281 (1997)

42. Waterhouse, D.F., Hockman, R.H., Mckellar, J.W.: An investigation of chitinase activity in cockroach and termite extracts. J. Insect Physiol. **6**, 96–112 (1961)

43. Yu, S.J., Terriere, L.C.: Ecdysin metabolism by soluble enzymes from three species of Diptera and its inhibition by the insect growth regulatar TH-6040. Pestic. Biochem. Physiol. **7**, 48–55 (1977)

3

A computational study of nitramide adsorption on the electrical properties of pristine and C-replaced boron nitride nanosheet

Mahdi Rezaei-Sameti[1] · Neda Javadi Jukar[1]

Abstract The aims of this work is to scrutinize the structural, physical and electrical properties of nitroamine (NH_2NO_2) adsorption on the outer and inner surface of pristine and C-replacing boron nitride nanosheet (BN nanosheet), using density functional theory methods at cam-B3LYP/6-31G (d) level of theory. Inspections of determined results represent that the adsorption of nitramide on the outer surface of pristine and C-replaced BN nanosheet is exothermic and on the inner surface it is endothermic. The deformation energy of system displays that the geometry and structure of BN nanosheet and nitramide in the BN nanosheet/NH_2NO_2 complex change significantly from the original state, whereas the quantum parameters and gap energy of the BN nanosheet/NH_2NO_2 system alter slightly from the original state. The nuclear magnetic resonance and molecular electrostatic potential consequences exhibit that in the BN nanosheet/NH_2NO_2 complex, the highest density of electrons is concentrated surrounding the NH_2NO_2 molecule.

Keywords Boron nitride nanosheet · C replaced · NH_2NO_2 adsorption · Density functional theory · Molecular electrostatic potential

Introduction

After the discovery of the carbon nanotube, extensive researches have been done to find other nanomaterials and nanoconfigurations. One of the most interesting configurations of carbon is graphene. It is a crystalline allotrope of carbon with two-dimensional properties. Due to its fascinating structural and electronic properties, it is suitable for providing electronic nanosheets or nanosensors [1–7]. The band gap of graphene sheet is almost zero and this property is useful to making electronic devices. In recent years, functionalizing and doping methods have been extensively used to increase the efficiency of graphene sheet in electronic and nanodetector applications [8–11]. In recent years, many researches have focused on finding other nanosheet compounds such as silicon carbide, boron phosphide, aluminum nitride, zinc oxide, gallium nitride, aluminum phosphide, beryllium oxide and boron nitride [12–18]. Boron nitride nanosheet was discovered in 2005 and, due to its structural and fundamental properties, it is useful in making nanochips, nanosensors, optoelectronic devices, nanoscale device technology and nanoadsorbent [19–30]. On the other hand, BN nanosheet with a band gap in the range of 4.2–6 eV is notably used to make a sensitive sensor of toxic and hazardous compounds in industries [12, 27, 28, 31–35].

Nitroamine (NH_2NO_2) is the simplest nitramine compound that is greatly used in military explosive, propellant and fuel applications. In recent years, extensive theoretical and experimental investigations have been done on the interaction of NH_2NO_2 compound with nanomaterials and nanotubes [36, 37]. In the current work, the structural, physical, and electrical properties of BN nanosheet in the presence of C-replaced and NH_2NO_2 molecule were studied using the DFT theory. To find the appropriate and suitable adsorption sites, many different configurations and orientations of

✉ Mahdi Rezaei-Sameti
mrsameti@gmail.com; mrsameti@malayeru.ac.ir

[1] Department of Applied Chemistry, Faculty of Science, Malayer University, Malayer 65174, Iran

NH$_2$NO$_2$ molecule on the B and N sites of nanosheet were examined, and all configuration models were optimized using B3LYP/3-21G level of theory. From all optimized models, the ten (10) stable and suitable models were selected for this work and then the selected models were optimized again using cam-B3LYP/6-31G (d) level of theory. The geometrical, chemical reactivity, quantum parameters, adsorption, deformation energy, and molecular electrostatic potential for all selected models are determined and analyzed. The obtained results may be useful for performing an adsorbent or nanosensor for detecting nitramine molecule.

Computational section

For denoting the adsorption of NH$_2$NO$_2$ molecule on the surface of pristine and C-replaced BN nanosheet, we define the A and B models, respectively. The a, b, c, d and e indexes are used for adsorption of NH$_2$NO$_2$ from H, N(NH$_2$), O, N(NO$_2$) head and parallel orientation on the boron position of the BN nanosheet, respectively (see Fig. 1). The f, g, h, i and j indexes are used for adsorption of NH$_2$NO$_2$ molecule from H, N(NH$_2$), O, N(NO$_2$) head and parallel orientation on the nitrogen position of the BN nanosheet, respectively.

The A-a to B-j models are optimized by using the DFT method at cam-B3LYP level of theory using the 6-31G (d) base set [38] when performing the GAMESS suite of programs [39]. The pristine and C-replaced BN nanosheet and the A-a to B-j models before and after the optimizing process are given in Figs. 1 and 2. The ends of the nanosheet at all systems are saturated by hydrogen atoms for preventing the dangling bonds at the edges of the nanosheet. The adsorption energy (E_{ads}) [40–42] of NH$_2$NO$_2$ molecule on the surface of pristine, C-replaced BN nanosheet is calculated by:

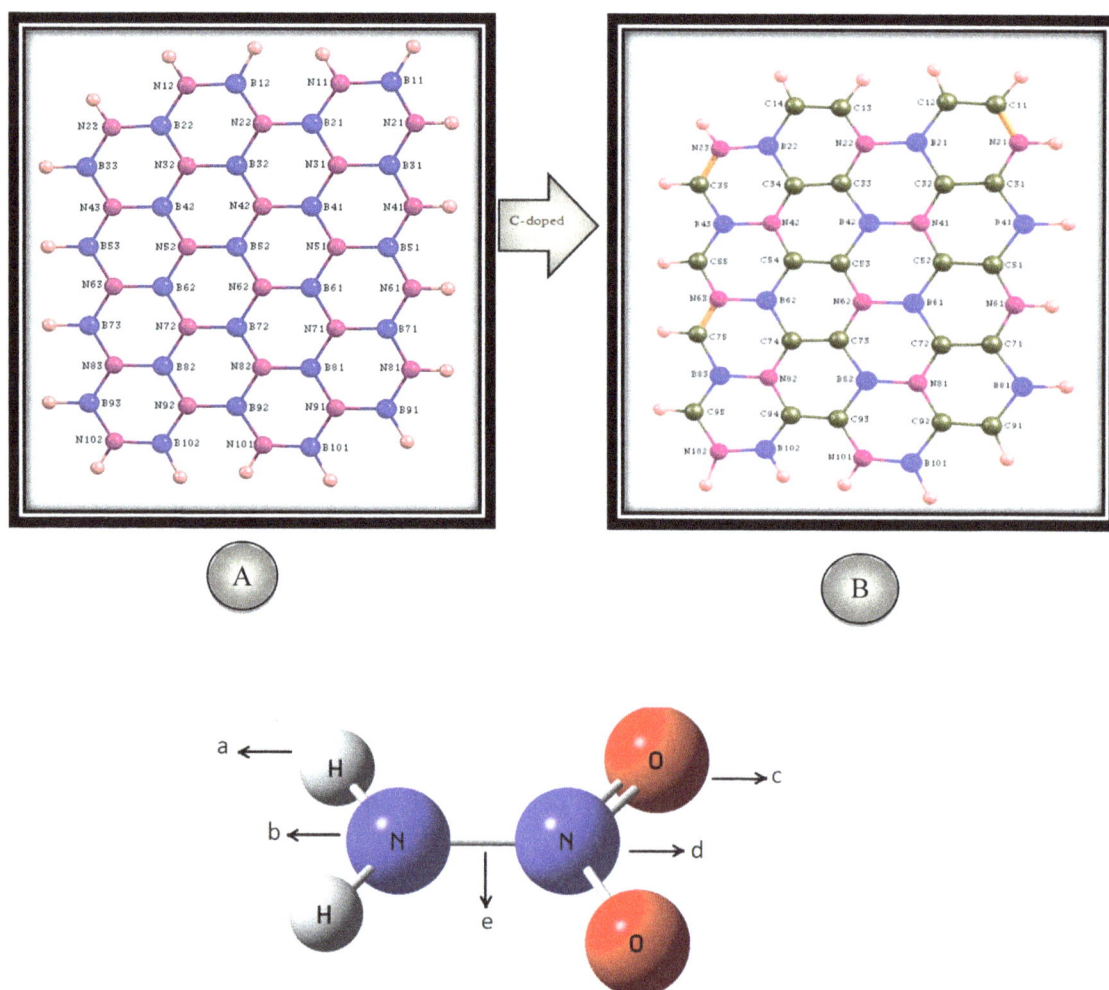

Fig. 1 2D views of the pristine and C-doped BN nanosheets

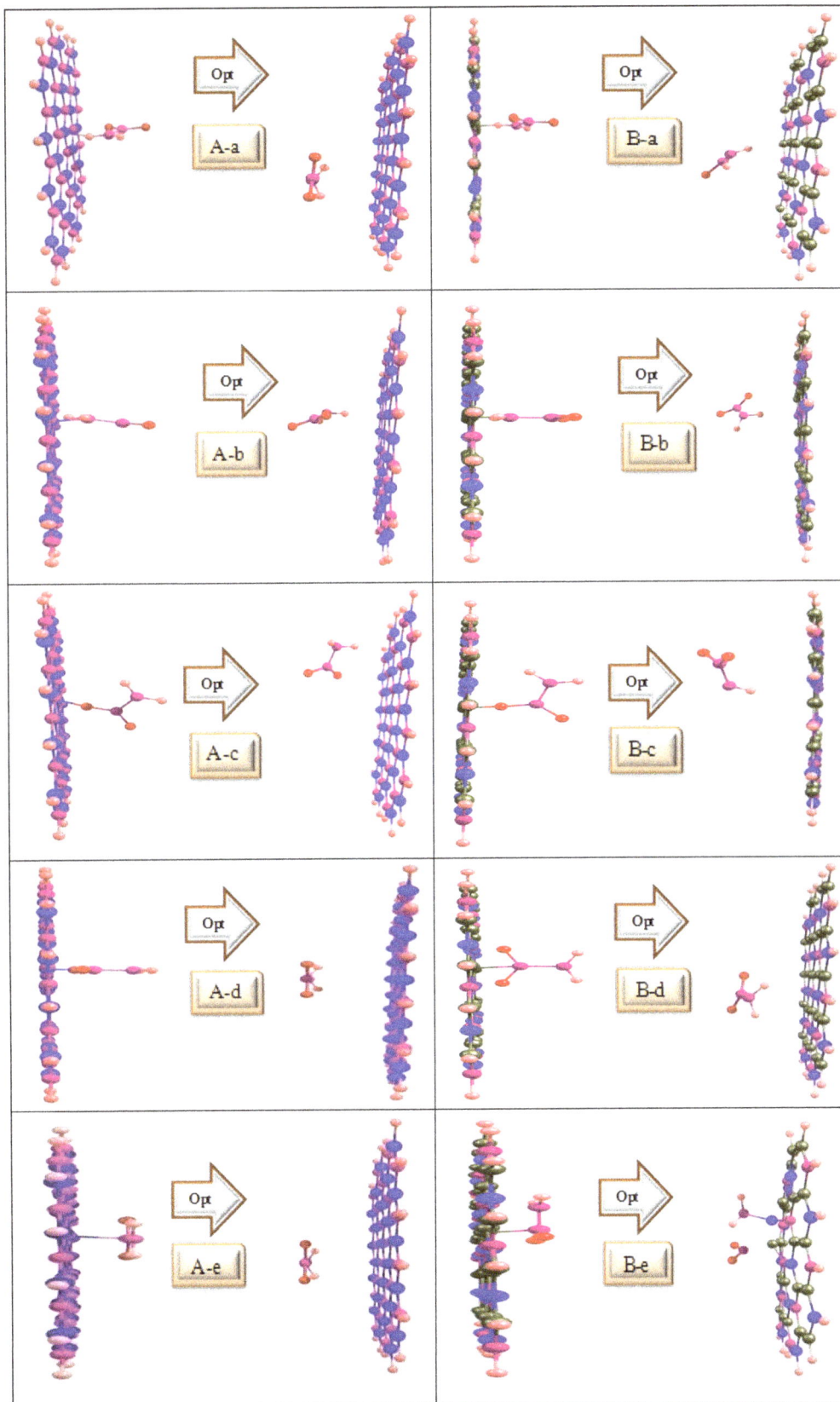

Fig. 2 2D views of NH$_2$NO$_2$ adsorption on the surface of pristine and C-doped BN nanosheets (A-a to B-j models); in all models the *left configuration* is before optimization

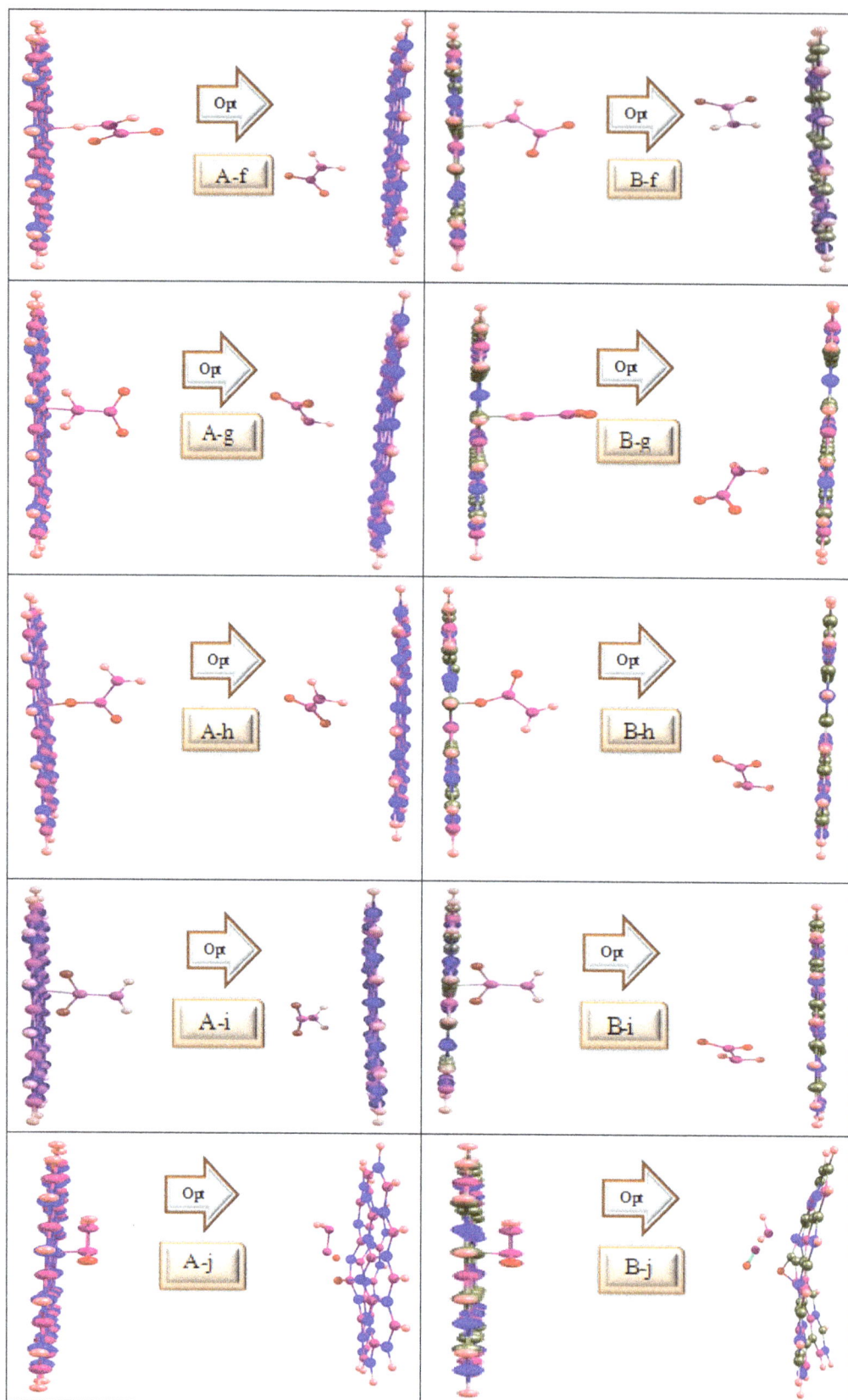

Fig. 2 continued

$$E_{\text{ads}} = E_{\text{BN−nanosheet/NH}_2\text{NO}_2} - (E_{\text{BN−nano sheet}} + E_{\text{NH}_2\text{NO}_2}) + \text{BSSE}. \tag{1}$$

$E_{\text{BN−nano sheet/NH}_2\text{NO}_2}$, $E_{\text{BN−nano sheet}}$ and $E_{\text{NH}_2\text{NO}_2}$ are the total potential energy of the BN nanosheet/NH$_2$NO$_2$ system, BN nanosheet and NH$_2$NO$_2$, respectively, and the BSSE is the base set superposition error. The deformation energy (E_{def}) of the BN nanosheet, NH$_2$NO$_2$ molecule, BN nanosheet/NH$_2$NO$_2$ complex and binding energy (E_{bin}) for the A-a to B-j systems are calculated by Eqs. (2–5) [43].

$$E_{\text{def BN− nanosheet}} = E_{\text{BN− nanosheet pure}} - E_{\text{BN− nanosheet in system}}, \tag{2}$$

$$E_{\text{def−NH}_2\text{NO}_2} = E_{\text{NH}_2\text{NO}_2 \text{ pure}} - E_{\text{NH}_2\text{NO}_2 \text{ in system}}, \tag{3}$$

where $E_{\text{BN-nanosheet in system}}$ and $E_{\text{NH}_2\text{NO}_2}$ are the energy of the BN nanosheet and NH$_2$NO$_2$ molecule in the BN nanosheet/NH$_2$NO$_2$ system, when NH$_2$NO$_2$ and BN nanosheet are absent, respectively:

$$E_{\text{bin}} = E_{\text{BN−nanosheet/NH}_2\text{NO}_2} - (E_{\text{BN−nanosheet in system}} + E_{\text{NH}_2\text{NO}_2 \text{ in system}}), \tag{4}$$

$$E_{\text{total def}} = E_{\text{ads}} - E_{\text{bin}}. \tag{5}$$

The E_{totaldef}, $E_{\text{defBN-nanosheet}}$ and $E_{\text{def NH}_2\text{NO}_2}$ are the deformation energy of the BN nanosheet/NH$_2$NO$_2$ system, BN nanosheet and NH$_2$NO$_2$ molecule in their optimized structure.

The highest occupied molecular orbital (HOMO) and the lowest unoccupied molecular orbital (LUMO) energies are used for determining the gap energy (E_g), density of state (DOS) plots, Fermi level energy (E_{FL}) and work functions ($\Delta\varphi$) (see Eqs. 6–8):

$$E_{\text{gap}} = E_{\text{LUMO}} - E_{\text{HOMO}}, \tag{6}$$

$$E_{\text{FL}} = (E_{\text{HOMO}} + E_{\text{LUMO}})/2, \tag{7}$$

$$\Delta\phi = E_{\text{HOMO}} - E_{\text{FL}}. \tag{8}$$

The chemical potential (μ) and electronegativity of nanosheet (χ), global hardness (η) and softness (S) of nanosheet, index of electrophilicity (ω) and charge transfer parameters (ΔN) of the A-a to B-j adsorption systems are calculated using Eqs. (9–14):

$$\mu = -(I + A)/2, \tag{9}$$

$$\chi = -\mu, \tag{10}$$

$$\eta = (I - A)/2, \tag{11}$$

$$\omega = \mu^2/2\eta, \tag{12}$$

$$S = 1/2\eta, \tag{13}$$

$$\Delta N = \left(-\frac{\mu}{\eta}\right), \tag{14}$$

where $A = -E_{\text{LUMO}}$ and $I = -E_{\text{HOMO}}$ are the electron affinity and ionization potential, respectively [40–42].

Table 1 The adsorption energy, deformation energy, binding energy, (Kcal mol^{-1}), distance between the BN nanosheet and NH$_2$NO$_2$ (Å) and dipole moment of the BN nanosheet/NH$_2$NO$_2$ (debye) for the A-a to B-j models

Model	E_{ads}	$E_{\text{def BN-nanosheet}}$	$E_{\text{def NH}_2\text{NO}_2}$	E_{bin}	$E_{\text{def (total)}}$	Distance	Dipole moment
A-a	−6.85	0.29	−1.75	−5.39	−1.46	3.20	2.14
A-b	−6.24	0.38	−1.74	−4.88	−1.36	3.28	2.71
A-c	−6.85	0.35	−1.75	−5.46	−1.39	3.29	5.96
A-d	−6.90	0.24	−1.76	−5.38	−1.52	3.47	2.12
A-e	−5.42	0.98	−1.55	−4.85	−0.57	3.62	2.17
B-a	−8.56	−0.70	−1.55	−6.31	−2.25	3.53	9.29
B-b	−8.94	−0.68	−1.54	−6.71	−2.23	3.40	7.78
B-c	−8.38	−0.60	−1.53	−6.23	−2.15	3.69	12.38
B-d	−8.54	−0.72	−1.50	−6.31	−2.23	3.45	10.12
B-e	16.47	73.60	116.10	−173.22	189.69	1.51	13.84
A-f	−5.16	1.06	−1.57	−4.66	−0.50	3.39	2.66
A-g	−5.18	1.10	−1.57	−4.71	−0.47	3.38	6.33
A-h	−5.84	1.09	−1.56	−5.38	−0.46	3.34	5.97
A-i	−5.84	0.77	−1.51	−5.10	−0.74	3.65	3.07
A-j	135.97	163.25	186.95	−214.22	350.19	1.77	12.55
B-f	−9.12	−0.67	−1.55	−6.89	−2.23	3.39	7.96
B-g	−10.48	−0.74	−1.42	−8.31	−2.17	3.46	6.30
B-h	−8.72	−0.77	−1.51	−6.44	−2.28	3.59	9.17
B-i	−8.79	−0.75	−1.50	−6.53	−2.26	3.60	8.83
B-j	26.76	107.44	171.19	−251.87	278.63	1.63	13.50

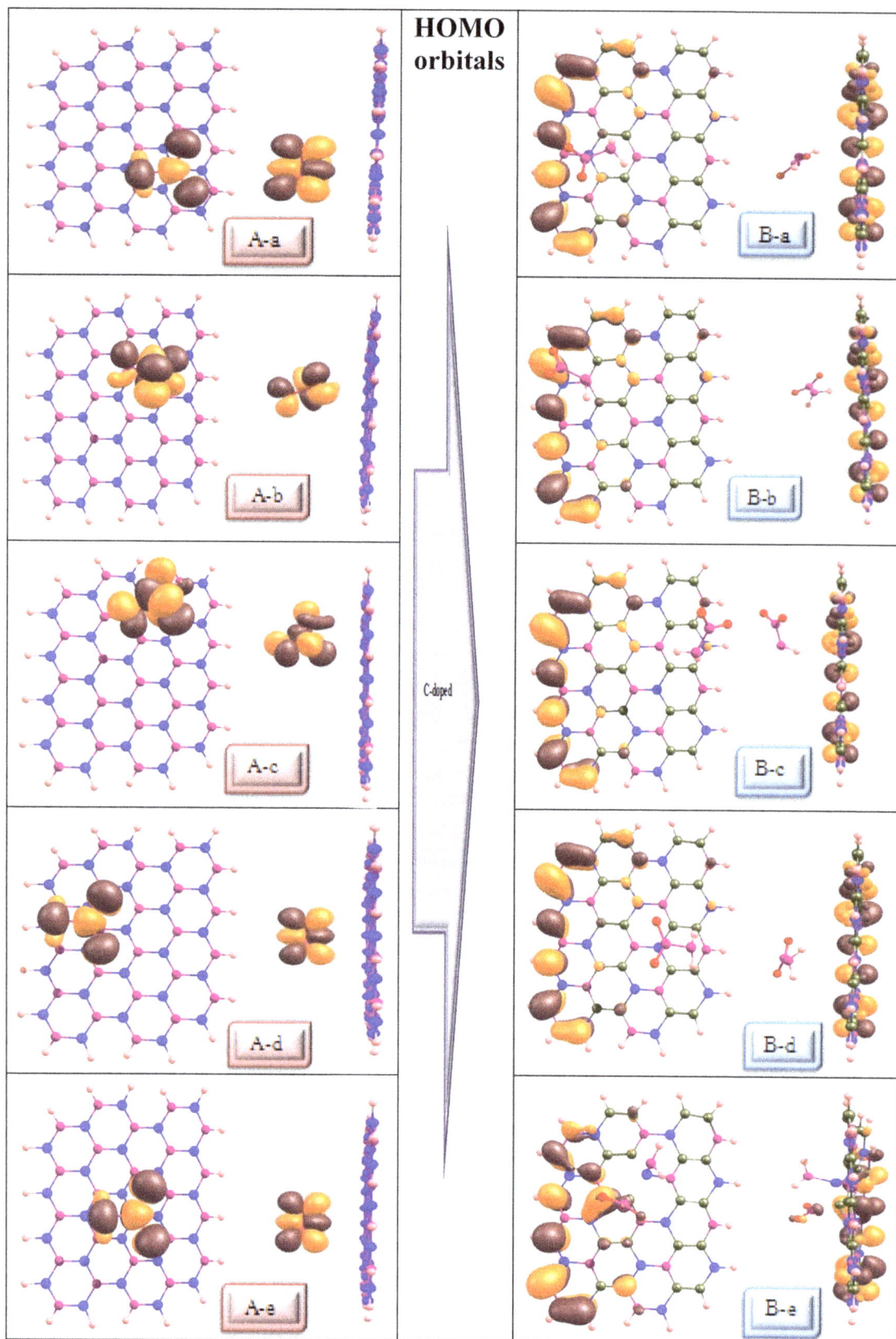

Fig. 3 Plots of HOMO and LUMO orbital structures for NH_2NO_2 adsorption on the surface of pristine and C-doped BN nanosheets (A-a to B-j models, see Fig. 2)

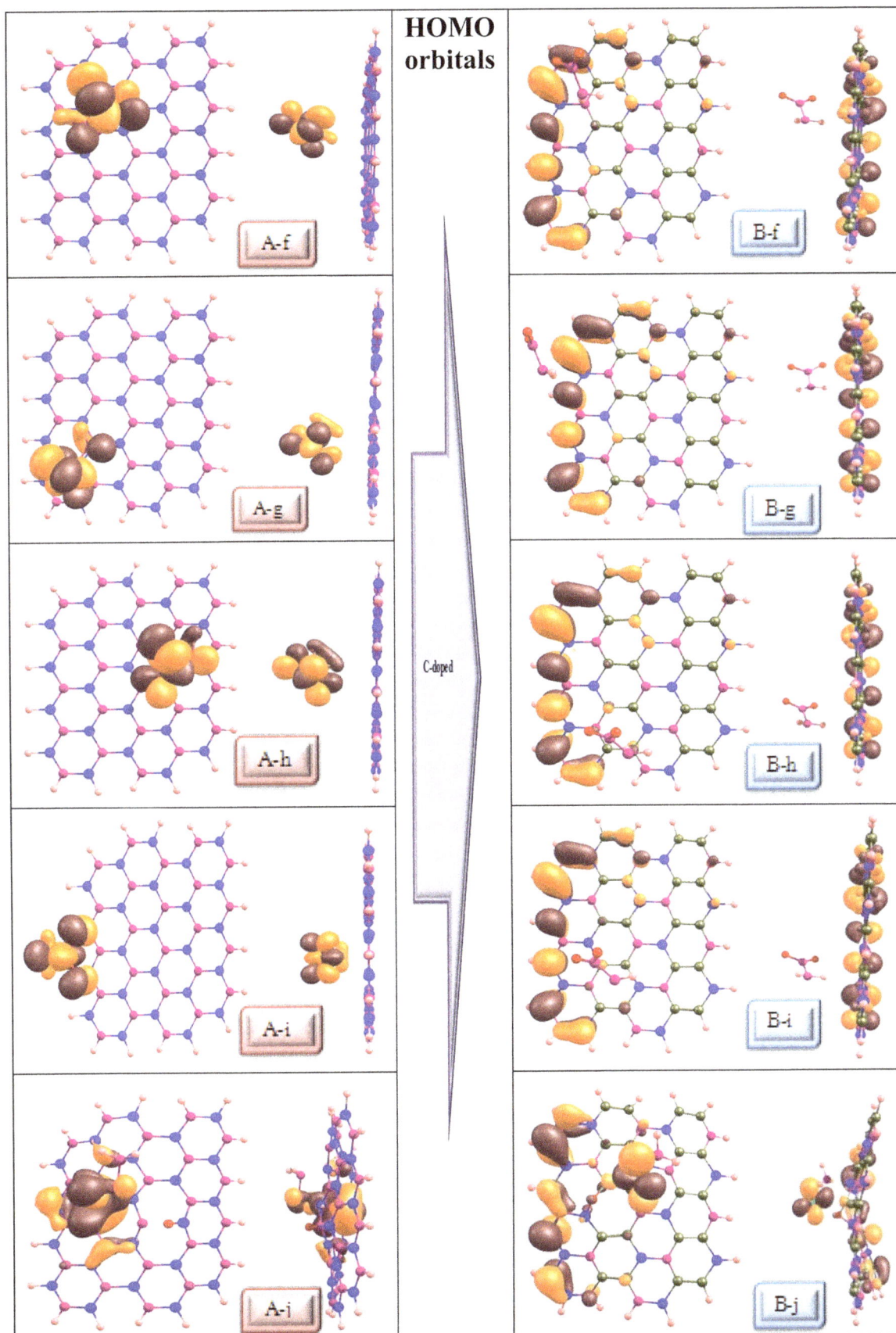

Fig. 3 continued

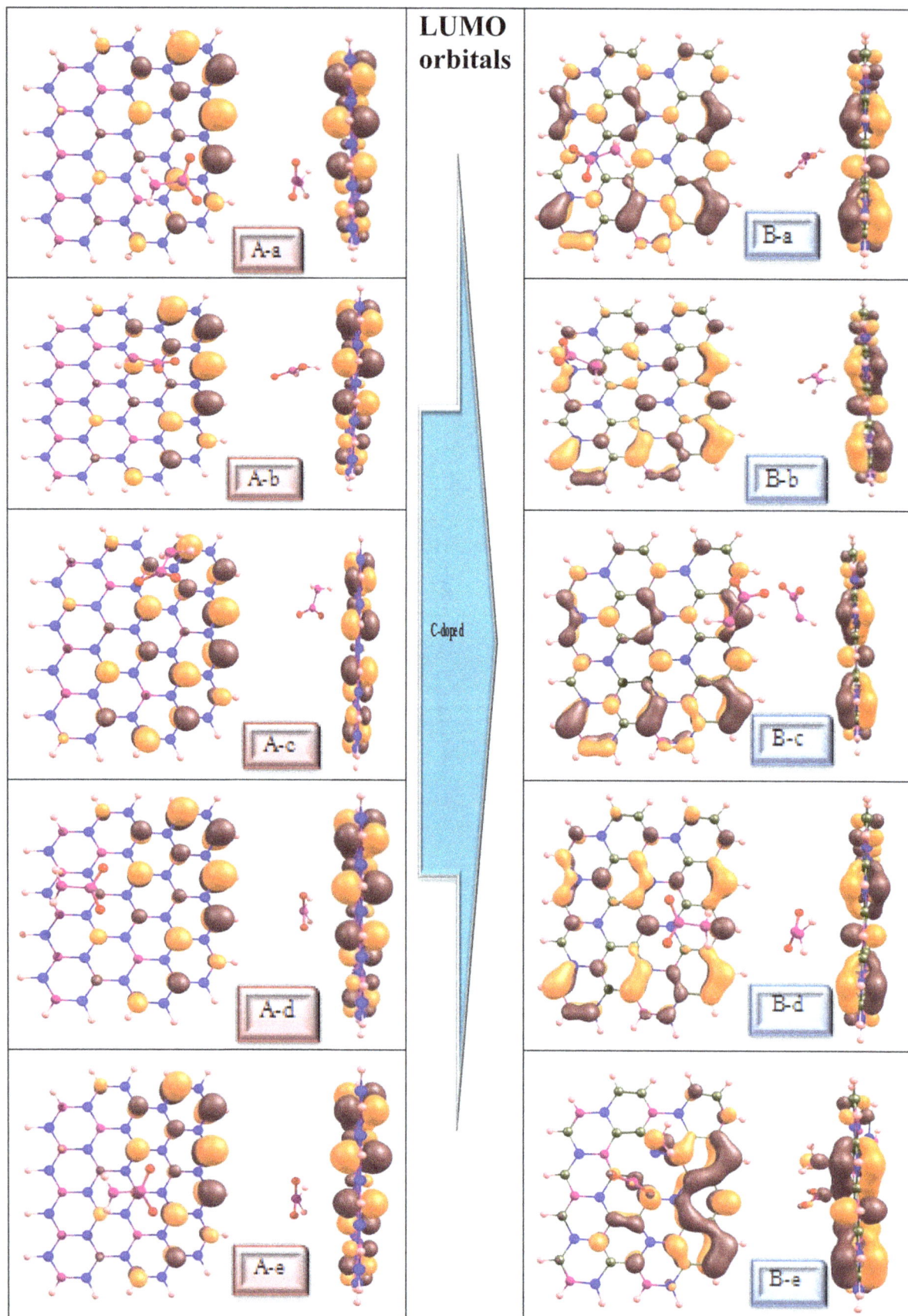

LUMO orbitals

C-doped

Fig. 3 continued

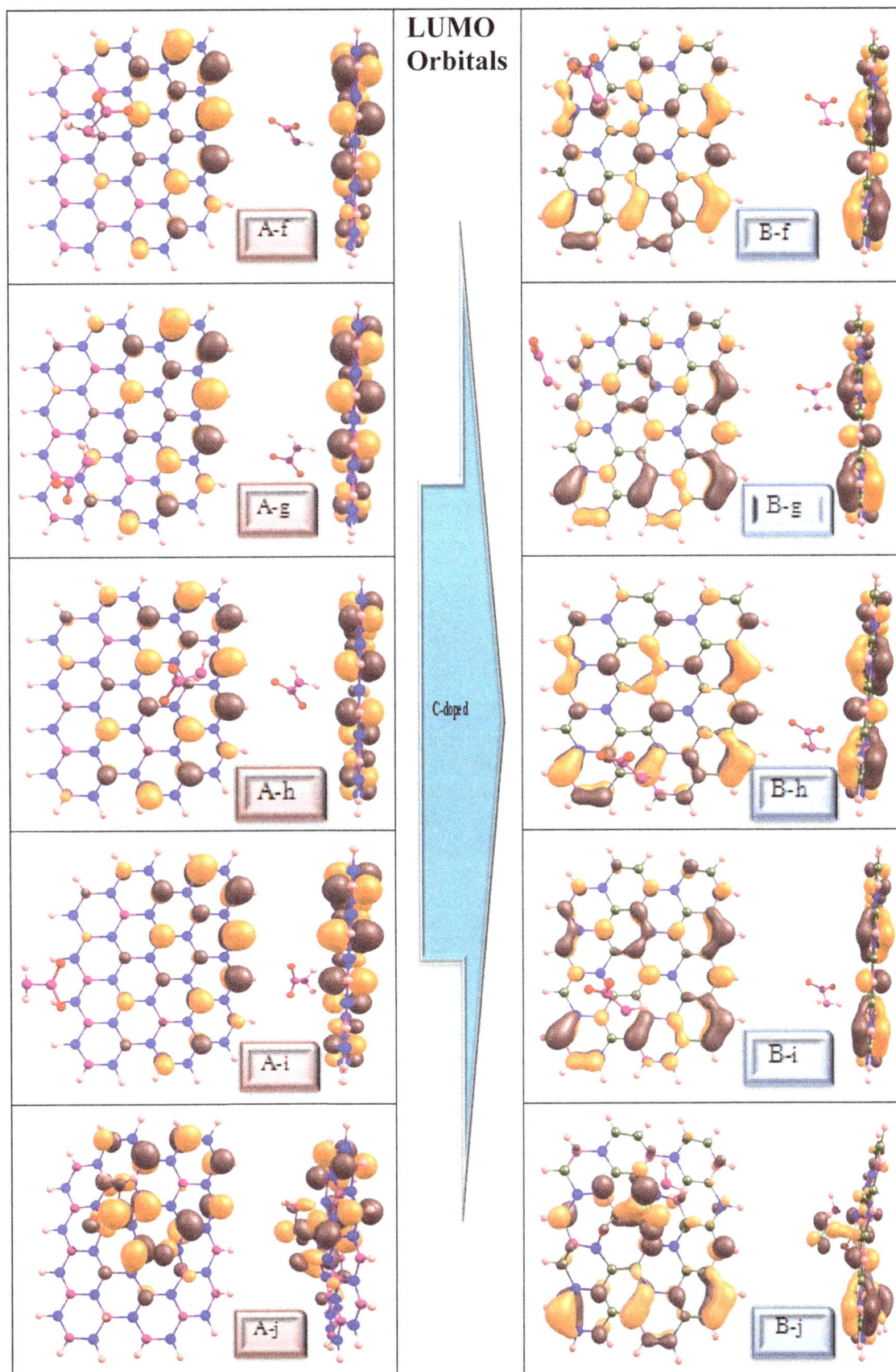

LUMO Orbitals

C-doped

A-f   A-g   A-h   A-i   A-j

B-f   B-g   B-h   B-i   B-j

Fig. 3 continued

Table 2 NBO and Mulliken charge of nitramide on the surface of the BN nanosheet for the A-a to B-j models, and the $\rho_1 = 0$ for pure NH_2NO_2

Model	$\Delta\rho_{NBO}$	$\Delta\rho_{Mulliken}$
A-a	−0.012	−0.006
A-b	−0.001	−0.023
A-c	−0.001	−0.021
A-d	−0.016	−0.011
A-e	−0.820	−0.009
B-a	−0.001	−0.014
B-b	−0.002	−0.005
B-c	−0.004	−0.023
B-d	−0.002	−0.020
B-e	−0.578	−0.517
A-f	−0.001	−0.021
A-g	−0.001	−0.020
A-h	−0.002	−0.023
A-i	−0.012	−0.007
A-j	−0.832	−0.856
B-f	−0.003	−0.005
B-g	−0.018	−0.055
B-h	−0.006	−0.025
B-i	−0.001	−0.019
B-j	−0.644	−0.532

Results and discussion

Structural and geometrical properties

Herein, the A model of the BN nanosheet contains 24B, 24N and 18H atoms (B24N24H18, see Fig. 1) and in the B model 24C atoms are superseded in the layers 1, 3, 5, 7 and 9 in spite of B and N sites of nanosheet (B12N12C24H18, see Fig. 1). From the optimized results, the B–N bond length of the A model is 1.44 Å and the bond lengths of B–N, B–C, N–C and C–C in the B model are 1.45, 1.39, 1.54 and 1.38 Å, respectively (see Fig S1 in supplementary data).

As seen in Fig S2 in supplementary data, the B–N, B–C and N–C bond lengths and the <B–C–N bond angle of the A-a to B-j models after adsorbing NH_2NO_2 molecule change slightly from the unadorned models.

The optimized results display that in the B-e model, the NH_2NO_2 molecule dissociates to NH_2 and NO_2 parts, and in this form the NH_2 part is adsorbed on the nanosurface and the NO_2 part goes away from the nanosheet. In the A-j model, the NH_2NO_2 molecule dissociates into 2O atoms and NNH_2 part, and also in this model the 2O atoms are chemically adsorbed on the surface of the nanosheet and the NNH_2 part goes away from the nanosheet. Also in the B-j model, the NH_2NO_2 molecule dissociates into NH_2, NO and O parts; the O atom is chemically connected on the surface of the nanosheet and other parts go away from the nanosheet. On the other hand, in the other adsorption

models after optimizing the NH_2NO_2/BN nanosheet complex, the NH_2NO_2 molecule goes away from the nanosheet and so the type of adsorption process of these systems is physorption.

According to calculated results in Table 1, the average distance between the NH_2NO_2 molecule and the BN nanosheet in the B-e, A-j and B-j models is 1.51, 1.77 and 1.63 Å, respectively. By replacing C atoms, the geometrical structure of the BN nanosheet is dramatically deformed and so the average distance of the NH_2NO_2/nanosheet decreases.

Comparison results show that the adsorption energy of the B-e, A-j and B-j models is positive due to dissociation of the NH_2NO_2 molecule on the surface of the nanosheet, whereas the E_{ads} values of the other models are negative and these adsorbed models are exothermic. Also, the B-g model with $E_{ads} = -10.48$ kcal/mol is the most stable adsorption model (see Table 1). The obtained results demonstrate that the E_{ads} of the NH_2NO_2/nanosheet system depends on C-replaced positions and NH_2NO_2 molecule orientation. The C-replacing atoms increase the adsorption of the NH_2NO_2 molecule on the surface of the nanosheet. These results confirm that C-replaced BN nanosheet is a good compound for adsorbing NH_2NO_2 molecule. In all studied models (the A-a to B-j models), the basis set superposition error (BSSE) is in the range 0.0005–0.001 kcal/mol. In the A-a to B-j models, the binding energy (E_{bin}) values are negative and the NH_2NO_2 molecule is adsorbed on the surface of the BN nanosheet. The maximum binding energy is seen in the B-e, A-j and B-j models with −173.22, −214.22 and −257.87 kcal/mol, respectively.

To understand the deforming structures of the BN nanosheet and NH_2NO_2 molecule in the BN nanosheet/NH_2NO_2 complex, the deformation energy of the BN nanosheet, NH_2NO_2 molecule and BN nanosheet/NH_2NO_2 complex is calculated using Eqs. (2–4) and the results are listed in Table 1.

Comparisons of structural results indicate that significant curvature in the geometry of nanosheet and NH_2NO_2 molecule occurs in the B-e, A-j and B-j adsorption models. On the other hand, the deformation energy of the BN nanosheet (E_{def} BN nanosheet) for the B-a, B-b, B-c, B-d, B-f, B-g, B-h and B-i models is negative and for the A-a to A-j models positive. The negative value of deformation energy shows that the structures of NH_2NO_2 molecule and BN nanosheet are deformed spontaneously from the original state and all deformed structures are stable from the thermodynamic point of view. However, the deformation energies of the B-e, A-j and B-j models for NH_2NO_2 molecule, BN nanosheet and NH_2NO_2/BN nanosheet complex are positive and so the deformation of these models is endothermic and needs more energy. The order

Table 3 Calculated quantum parameters for adsorption of nitramide on the surface of the pristine and C-replaced BN nanosheet for the A-a to B-j models

Properties/eV	A-a	A-b	A-c	A-d	A-e	B-a	B-b	B-c	B-d	B-e
$E(\text{LUMO})$	−1.36	−1.22	−1.26	−1.47	−1.52	−2.78	−2.80	−2.86	−2.78	−3.17
$E(\text{HOMO})$	−6.38	−6.41	−6.49	−6.31	−6.34	−4.01	−4.01	−4.05	−4.01	−4.11
E (gap)	5.02	5.19	5.23	4.83	4.82	1.23	1.21	1.19	1.23	0.94
I	6.38	6.41	6.49	6.31	6.34	4.01	4.01	4.05	4.01	4.11
A	1.36	1.22	1.26	1.47	1.52	2.78	2.80	2.86	2.78	3.17
μ	−3.87	−3.81	−3.87	−3.89	−3.93	−3.39	−3.40	−3.45	−3.39	−3.64
X	3.87	3.81	3.87	3.89	3.93	3.39	3.40	3.45	3.39	3.64
η	2.51	2.59	2.61	2.41	2.41	0.61	0.60	0.59	0.61	0.47
$S/(\text{eV})^{-1}$	0.20	0.19	0.19	0.20	0.20	0.81	0.82	0.84	0.81	1.06
E_{FL}	−3.87	−3.81	−3.87	−3.89	−3.93	−3.39	−3.40	−3.45	−3.39	−3.64
$\Delta\Phi$	−2.51	−2.59	−2.61	−2.41	−2.41	−0.61	−0.60	−0.59	−0.61	−0.47
ω	2.98	2.80	2.87	3.13	3.21	9.37	9.58	10.08	9.41	14.09
ΔN	−1.54	−1.47	−1.48	−1.61	−1.63	−5.52	−5.62	−5.84	−5.55	−7.74

Properties/eV	A-f	A-g	A-h	A-i	A-j	B-f	B-g	B-h	B-i	B-j
$E(\text{LUMO})$	−1.38	−1.39	−1.33	−1.45	−1.85	−2.80	−2.87	−2.77	−2.76	−2.87
$E(\text{HOMO})$	−6.38	−6.48	−6.53	−6.35	−5.91	−4.02	−4.03	−4.01	−4.00	−4.20
E (gap)	5.00	5.09	5.20	4.90	4.06	1.22	1.16	1.24	1.24	1.33
I	6.38	6.48	6.53	6.35	5.91	4.02	4.03	4.01	4.00	4.20
A	1.38	1.39	1.33	1.45	1.85	2.80	2.87	2.77	2.76	2.87
μ	−3.88	−3.93	−3.93	−3.90	−3.88	−3.41	−3.45	−3.39	−3.38	−3.53
X	3.88	3.93	3.93	3.90	3.88	3.41	3.45	3.39	3.38	3.53
η	2.50	2.54	2.60	2.45	2.03	0.61	0.58	0.62	0.62	0.66
$S/(\text{eV})^{-1}$	0.19	0.19	0.19	0.20	0.24	0.81	0.86	0.80	0.80	0.75
E_{FL}	−3.88	−3.93	−3.93	−3.90	−3.88	−3.41	−3.45	−3.39	−3.38	−3.53
$\Delta\Phi$	−2.50	−2.54	−2.60	−2.45	−2.03	−0.61	−0.58	−0.62	−0.62	−0.66
ω	3.01	3.04	2.97	3.10	3.70	9.53	10.26	9.26	9.21	9.39
ΔN	1.55	−1.54	−1.51	−1.59	−1.91	−5.59	−5.94	−5.46	−5.45	−5.31

of deformation energy in the B-e, A-j and B-j models is: E_{def} (A-j) > E_{def} (B-j) > E_{def} (B-e). Furthermore, the deformation energy of the NH_2NO_2 molecule and total deformation energy of the A-a, A-b, A-c, A-d, A-e, A-f, B-a, B-b, B-c, B-d and B-f models are negative. The total deformation energy of the B model is lower than that of the A models, thereby replacing of carbon atoms alter the structure of the BN nanosheet/NH_2NO_2 complex.

Dipole moment is one another properties of system that it used to detect the nature of reactivity and impurity atoms effect of system. The dipole moment of the C-replaced models (B-a to B-j models) are more than pristine models (A-a to A-j models). The replacing carbon atoms increase dipole moment and reactivity of nanosheet (see Table 1), and also the orientations of NH_2NO_2 molecule on the surface of BN nanosheet change the dipole moment values. The dipole moment of A-g model (6.33 debye) and B-e model (13.84 debye) are significantly more than other those models. It is notable that with increasing the dipole moment at the A-a to B-j models the adsorption and interaction energy of system increase.

Analysis HOMO and LUMO orbitals

For understanding the structural and electrical properties of BN nanosheet/NH_2NO_2 complex, the HOMO and the LUMO orbitals of the A-a to B-j models are determined and results are given in Fig. 3.

The HOMO orbital density of the A-a, A-b, A-c, A-d, A-e, A-f, A-g, A-h and A-i models are uniformly localized surrounding NH_2NO_2 molecule, whereas at the A-j model the HOMO orbital density is localized around NH_2NO_2 molecule and near adsorption position of nanosheet. While with replacing C atoms in the B-a, B-b, B-c, B-d, B-e, B-f, B-g, B-h and B-i models the HOMO orbital density are distributed on the end layers surface of nanosheet, and also in the B-j model the HOMO orbital density is localized surrounding NH_2NO_2 and end layers of nanosheet.

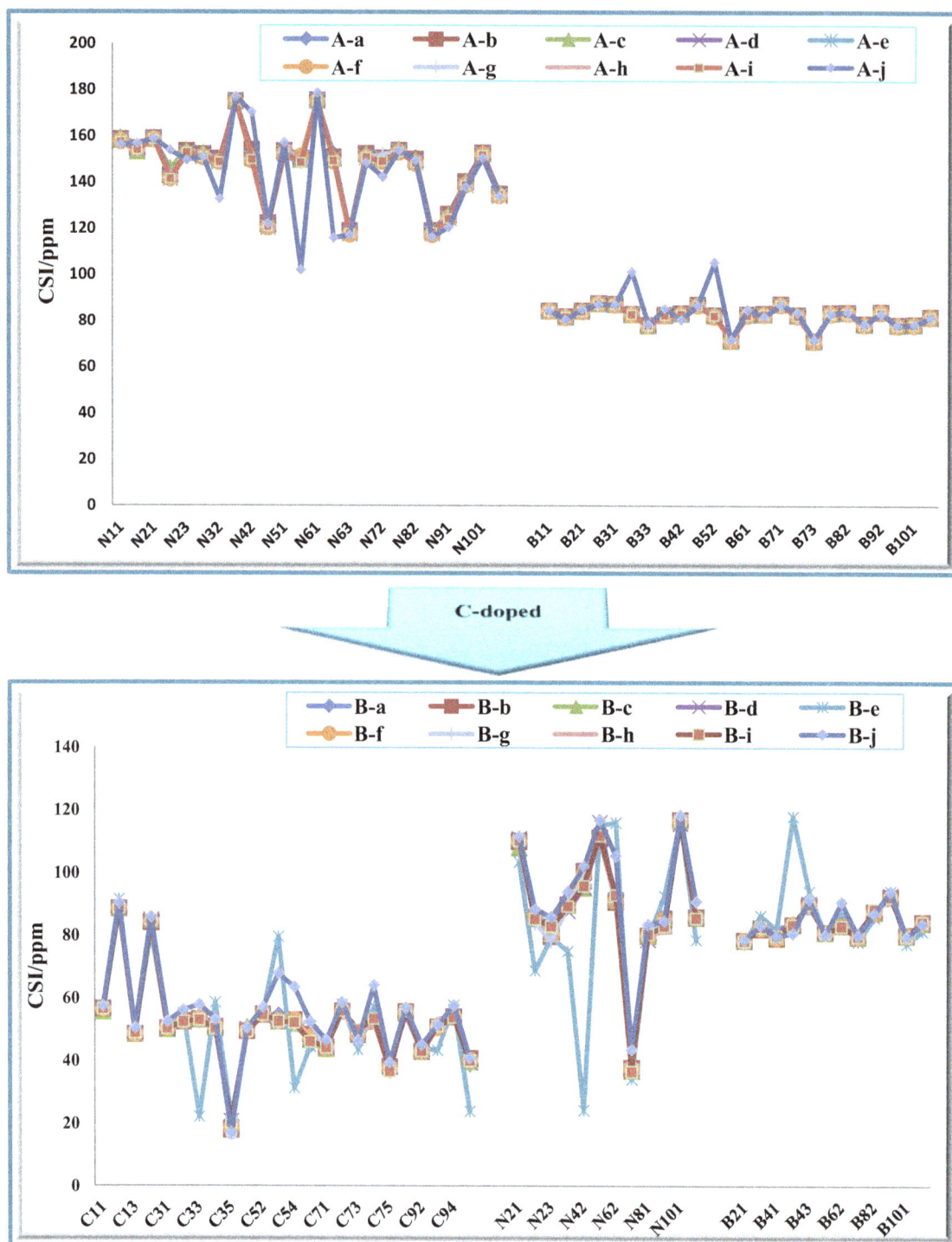

Fig. 4 The *CSI plots* for NH$_2$NO$_2$ adsorption on the surface of pristine and C-doped BN nanosheet (A-a to B-j models, see Fig. 2)

It is remarkable that the LUMO orbital densities at all adsorption models except A-j and B-j models are distributed uniformly around nanosheet. These results confirm that the surface of nanosheet is a good surface for electrophilic reaction. The negative values of $\Delta\rho(\Delta\rho = \rho_2 - \rho_1)$ of NBO and Mulliken charge for NH$_2$NO$_2$ molecule confirm that the

NH$_2$NO$_2$ molecule is an electron acceptor and decrease the charge densities around nanosheet from original state (see Table 2).

The calculated gap energy (E_g) for the A-a to B-j models are tabulated in Table 3. The E_g energy of the A-a to A-j models is in range 4.06–5.23 eV; these results are in agreement with other research [32]. The E_g energy of the B-a to B-j models is in range 0.94–1.33 eV and is lower than the A-a to A-j models. Due to replacing carbon atoms, the conductivity and reactivity of the B-a to B-j models are more than the A-a to A-j models, and also the B-a to B-j models are a good semiconductor in electronic applications.

The density of state (DOS) plots for the A-a to B-j adsorption models display that at the A-a, A-b, A-c, A-d, A-e, A-f, A-g and A-h models in the HOMO and LUMO regions there are 12 and 14 peaks, respectively (see Fig S3). In the B-a, B-b, B-c, B-d, B-f, B-g, B-h and B-i in the HOMO and LUMO region, there are 16 and 17 peaks, respectively. In the A-j, B-e and B-j models due to disso-ciation of NH_2NO_2 molecule in the HOMO and LUMO regions, we observe 18 and 19 peaks, respectively. Com-parison results exhibit that in the LUMO region, more maxima peaks are observed than in the HOMO region.

By replacing C atoms, two small peaks are seen in the gap region and then due to these unfolded peaks the E_g values of the BN nanosheet decrease significantly com-pared to pristine BN nanosheet models. So, the conduc-tivity of the C-replaced BN nanosheet is more than that of the pristine model of BN nanosheet.

According to the calculated quantum parameters of the BN nanosheet/NH_2NO_2 in Table 3, the global hardness (η) values of the A-a to A-j models and the B-a to B-j models are in the range 2.03–2.61 and 0.47–0.66 eV, respectively. The global hardness of C-replaced BN nanosheet is lower than that of pristine BN nanosheet and so the reactivity of C-replaced BN nanosheet is more than that of pristine BN nanosheet. This fact confirms that the C-replaced BN nanosheet is more active than the pristine models for adsorbing and detecting NH_2NO_2 molecules.

The electrochemical potential (μ) results reveal that on replacing C atoms and adsorbing NH_2NO_2 molecules, the electrochemical potential of the system changes slightly from the original state.

The NBO and Milliken charge around the NH_2NO_2 molecule ($\Delta\rho_{NBO}$ and $\Delta\rho_{mulk}$) and the maximum amount of electronic charge (ΔN) for the A-a to B-j models are neg-ative (see Tables 2, 3). The negative values of $\Delta\rho_{NBO}$, $\Delta\rho_{mulk}$ and ΔN at all adsorption models demonstrate that in the BN nanosheet/NH_2NO_2 system, the BN nanosheet has a donor electron effect and NH_2NO_2 molecule has an acceptor electron effect. Due to this effect, the charge density is localized around the NH_2NO_2 molecule.

The work function ($\Delta\Phi$) is the minimum energy to remove an electron from a solid surface to outside of the compound and it is used to describe the emitted electron current densities.

The work function values of the A-a to A-j models and the B-a to B-j models are in the range of −2.03 to

−2.61 eV and −0.47 to −0.66 eV, respectively (see Table 3). A comparison of results shows that the work function of C-replaced BN nanosheet/NH_2NO_2 complex is lower than that of pristine models and so the emitted electron current densities of the B-a to B-j models are lower than those of the A-a to A-j models.

Nuclear magnetic resonance parameters

In this section, the evaluated nuclear magnetic resonance (NMR) parameters are used to understand the charge density of the material. The nuclear magnetic resonance parameters are generated by the response of electrons to an external uniform magnetic field and the spin–spin interac-tion. By using the gauge included atomic orbital (GIAO) approach [40–42], the chemical shielding isotropic (CSI) and chemical shielding anisotropic (CSA) parameters for ^{11}B, ^{15}N and ^{13}C sites are calculated using Eqs. (15, 16):

$$CSI = \frac{1}{3}\left(\sigma_{xx} + \sigma_{yy} + \sigma_{zz}\right) \tag{15}$$

$$CSA = \sigma_{zz} - \frac{1}{2}\left(\sigma_{xx} + \sigma_{yy}\right) \quad \sigma_{zz} > \sigma_{yy} > \sigma_{xx}. \tag{16}$$

The calculated CSI and CSA parameters of the A-a to B-j models are given in Tables S3–S6 in supplementary data. The NMR spectrum and the CSI plots for the A-a to B-j models are shown in Fig. 4 and Fig S4 in supplementary data. It is evident that the CSI values of N atoms are more than those of C and B atoms due to more electronegativity of the N atom with respect to B and C atoms. In the A-a to B-j models, the CSI values of N42, N51, N61, B32, B52, C12, C14, C53, C72, C74, C91, C94, N61, N101 and B43 sites are more than those of other sites of the nanosheet, whereas the CSI values of N32, N43, N52, N62, N83, N52, B73, C13, C35, C54, C75, C95, N42 and N63 sites are lower than those of other sites of the nanosheet. Therefore, the electrostatic behavior of the BN nanosheet surface varies significantly from the unabsorbed system. Compar-ison NMR spectrum of the A-a to B-j models (see Fig S4) shows that the shielding parameters of the A-j, B-e and B-j models are different from those of other adsorption models due to dissociation of the NH_2NO_2 molecule.

The natural bond orbital (NBO)

The natural bonding orbital (NBO) is one of the important parameters used to understand the effects of NH_2NO_2 adsorption on the structural and orbital properties of the nanosheet. Using the natural bonding orbital results, the interaction energy or second-order perturbation interaction energy $E^{(2)}$ [44] between the bond orbitals with antibond orbitals or Rydberg orbitals can be calculated with delo-calization $i \rightarrow j$:

$$E^{(2)} = q_i \, \frac{F_{ij}^2}{\varepsilon_j - \varepsilon_i} \, , \tag{17}$$

where F_{ij} and q_i are off-diagonal and donor orbital occupancy, respectively, and ε_i and ε_j are orbital energies. According to the NBO calculated results of Tables S7–S10 in supplementary data, the $E^{(2)}$ values of the A-a to A-j models are greater than those of B-a to B-j models. This result confirms that the stabilization energy between the donor and acceptor orbitals in the C-replaced BN nanosheet is significantly lower than that of the pristine models. Thereby by replacing C atoms, the activities and polarizability of the BN nanosheet increases significantly from the original state.

A comparison of the NBO results shows that the highest stabilization energy ($E^{(2)}$) in the A-a, A-b, A-c, A-d, A-e models is shown at transition between the donor orbital to acceptor orbital $\sigma B62 - N52 \rightarrow \sigma^* B52 - N52$ and for the A-f, A-g, A-h, A-i and A-j models at the transition between the donor orbital and acceptor orbital $\sigma B62 - N52 \rightarrow \sigma^* B52 - N52$. On the other hand, the highest stabilization energy ($E^{(2)}$) in the B-a to B-j models is shown at the transition between the donor orbital and acceptor orbital $\sigma C53 - B42 \rightarrow \sigma^* C54 - C53$.

Molecular electrostatic potential

The molecular electrostatic potential (MEP) calculated results are used to determine the size and shape, and the positive, negative and neutral regions of the molecule [45, 46]. These results are useful for determining the charge distribution around the compound surface and nucleophilic or electrophilic region of the compound. This means that for all A-a to B-j models, the MEP and contour potential plots are calculated around the adsorption position and all calculated results are shown in Fig S5 in supplementary data. Herein, the green and red colors represent the negative and positive charges or the electrophilic and nucleophilic regions, respectively. The calculated results point out in all A-a to B-j models a negative potential, green color, is shown around the NH_2NO_2 molecule, whereas a positive potential, red color, is localized around the BN nanosheet. These results are completely in agreement with the NBO, quantum and NMR results. Therefore in the A-a to B-e models, due to donor electron effect of the BN nanosheet, a low charge transfer occurs between the nanosheet and the NH_2NO_2 molecule. Comparison of MEP and contour potential plots of the A and B models demonstrate that with C-replacing atoms the charge transfer between the BN nanosheet and NH_2NO_2 decreases slightly compared with the pristine models.

Conclusions

In this work, we investigated the structural, physical and electrical properties of adsorption NH_2NO_2 molecule on the surface of the pristine and C-replaced BN nanosheet. In the A-a to B-j models, the adsorption energy of the NH_2NO_2/BN nanosheet system strongly depends on the adsorption orientation of the NH_2NO_2 molecule and C-replaced atoms. The adsorption energy of NH_2NO_2 molecule on the outer surface of the pristine and C-replaced BN nanosheet is negative, exothermic and favorable.

The E_g, NBO and global hardness results of C-replaced BN nanosheet are lower than those of the pristine models, and so the conductivity, polarizability and activity of C-replaced BN nanosheets are greater than those of pristine BN nanosheets. These results demonstrate that the C-replaced BN nanosheet models are good compounds to act as sensitive sensors and absorbers of the NH_2NO_2 molecule.

Acknowledgements The authors thank the Computational Information Center of Malayer University for providing the necessary facilities to carry out the research.

References

1. Novoselov, K.S., Geim, A.K., Morozov, S.V., Jiang, D., Zhang, Y., Dubonos, S.V., Grigorieva, I.V., Firsov, A.A.: Electric field effect in atomically thin carbon films. Science **306**, 666–669 (2004)
2. Geim, A.K., Novoselov, K.S.: The rise of graphene. Nat. Mater. **6**, 183–191 (2007)
3. Zhang, Y.B., Tan, Y.W., Stormer, H.L., Kim, P.: Experimental observation of the quantum Hall effect and Berry's phase in graphene. Nature **438**, 201–204 (2005)
4. Morozov, S.V., Novoselov, K.S., Katsnelson, M.I., Schedin, F., Elias, D.C., Jaszczak, J.A., Geim, A.K.: Giant intrinsic carrier mobilities in graphene and its bilayer. Phys. Rev. Lett. **100**, 016602–016604 (2008)
5. Novoselov, K.S., Jiang, Z., Zhang, Y., Morozov, S.V., Stormer, H., Zeitler, U., Maan, J.C., Boebinger, G.S., Kim, P., Geim, A.K.: Room-temperature quantum Hall effect in graphene. Science **315**, 1379 (2007)
6. Zhou, J., Wang, Q., Sun, Q., Jena, P.: Stability and electronic structure of bilayer graphone. Appl. Phys. Lett. **98**, 063108–5 (2011)
7. Pujari, B.S., Kanhere, D.G.: Density functional investigations of defect-induced mid-gap states in graphane. J. Phys. Chem. C **113**, 21063–21067 (2009)
8. Lu, G., Ocula, E.L., Chen, J.: Gas detection using low-temperature reduced graphene oxide sheets. Appl. Phys. Lett. **94**, 083111–083113 (2009)
9. Pashangpour, M., Bagheri, Z., Ghaffari, V.: A comparison of electronic transport properties of graphene with hexagonal boron nitride substrate and graphene, a first principle study. Eur. Phys. J. B **86**(269), 1–6 (2013)
10. Robinson, J.T., Perkins, F.K., Snow, E.S., Wei, Z.Q., Sheehan, P.E.: NO_2 and humidity sensing characteristics of few-layer graphene. Nano Lett. **8**, 3137–3140 (2008)

11. Yoon, H.J., Jun, D.H., Yang, J.H., Zhou, Z., Yang, S.S., Cheng, M.M.: Carbon dioxide gas sensor using a graphene sheet. Sens. Actuators B. Chem. **157**, 310–331 (2011)

12. Sakhavand, N., Shahsavari, R.: Synergistic behavior of tubes, junctions, and sheets imparts mechano-mutable functionality in 3D porous boron nitride nanostructures. J. Phys. Chem. C **118**(39), 22730–22738 (2014)

13. Zhou, J., Wang, Q., Sun, Q., Jena, P.: Stability and electronic structure of bilayer graphone. Appl. Phys. Lett. **98**, 063108/1–3 (2011)

14. Ahin, H., Ataca, C., Ciraci, S.: Magnetization of graphane by dehydrogenation. Appl. Phys. Lett. **95**, 222510/1–4 (2009)

15. Boukhvalov, D.W., Katsnelson, M.I., Lichtenstein, A.I.: Hydrogen on graphene: Electronic structure, total energy, structural distortions and magnetism from first-principles calculations. Phys. Rev. B **77**, 035427/1–3 (2008)

16. Nair, R.R., Ren, W., Jalil, R., Riaz, I., Kravets, V.G., Britnell, L., Blake, P., Schedin, F., Mayorov, A.S., Yuan, S., Katsnelson, M.I., Cheng, H.M., Strupinski, W., Bulusheva, L.G., Okotrub, A.V., Grigorieva, I.V., Grigorenko, A.N., Novoselov, K.S., Geim, A.K.: Fluorographene: a two-dimensional counterpart of teflon. Small **6**, 2877–2884 (2010)

17. Ahin, H., Topsakal, M., Ciraci, S.: Structures of fluorinated graphene and their signatures. Phys. Rev. B. **83**, 115432–115438 (2011)

18. Withers, F., Dubois, M., Savchenko, A.K.: Electron properties of fluorinated single-layer graphene transistors. Phys. Rev. B. **82**, 073403/1–6 (2010)

19. Li, J., Jin, P., Dai, W., Wang, C., Li, R., Wu, T., Tang, C.: Excellent performance for water purification achieved by activated porous boron nitride nanosheets. Math. Chem. Phys. **196**, 186–193 (2017)

20. Li, J., Jin, P., Tang, C.: Cr(III) adsorption by fluorinated activated boron nitride: a combined experimental and theoretical investigation. RSC. Adv. **4**, 14815–14821 (2014)

21. Park, C.H., Louie, S.G.: Energy gaps and Stark effect in boron nitride nanoribbons. Nano Lett. **8**, 2200–2203 (2008)

22. Zhang, Z., Guo, W.: Energy-gap modulation of BN ribbons by transverse electric fields: first-principles calculations. Phys. Rev. B. **77**, 075403–075405 (2008)

23. Sun, L., Li, Y., Li, Z., Li, Q., Zhou, Z., Chen, Z., Yang, Z., Hou, J.G.: Electronic structures of SiC nanoribbons. J. Chem. Phys. **129**, 174114/1–15 (2008)

24. Botello-Méndez, A.R., Lpez-Ur, F., Terrones, M., Terrones, H.: Magnetic behavior in zinc oxide zigzag nanoribbons. Nano Lett. **8**, 1562–1565 (2008)

25. Li, H., Dai, J., Li, J., Zhang, S., Zhou, J., Zhang, L., Chu, W., Chen, D., Zhao, H., Yang, J., Wu, Z.: Electronic structures and magnetic properties of GaN sheets and nanoribbons. J. Phys. Chem. C **114**, 11390–11395 (2010)

26. Zhang, X., Liu, Z., Hark, S.: Synthesis and optical characterization of single-crystalline AlN nanosheets. Solid. Stat. Commun. **143**, 317–320 (2007)

27. Lopez, A., Bezanilla, H.J., Terrones, H., Sumpter, B.H.: Electronic structure calculations on edge functionalised armchair boron nitride nanoribbons. J. Phys. Chem. C **116**(29), 15675–15681 (2012)

28. Anota, E.C., Juarez, A.R., Castro, M., Cocoletzi, H.H.: A density functional theory analysis for adsorption of the amine group on graphene and BN nanosheets. J. Mol. Model. **19**, 321–328 (2013)

29. Barsan, N., Koziej, D., Weimar, U.: Metal oxide-based gas sensor research: how to? Sci. Direct. **121**(1), 18–35 (2007)

30. Lin, Y.M., Valdes, G.A., Han, S.J., Farmer, D.B., Meric, I., Sun, Y., Wu, Y., Dimitrakopoulos, C., Grill, A., Avouris, P., Jenkins, K.A.: 100-GHz transistors from Wafer-scale epitaxial graphene. Science **332**, 1294–1297 (2010)

31. Zhao, J.Y., Zhao, F.Q., Ju, X.H., Gao, H.X., Zhou, S.Q.: Density functional theory studies on the adsorption of NH_2NO_2 on Al13 cluster. J. Clust. Sci. **23**, 395–410 (2012)

32. Bhattacharya, A., Bhattacharya, S., Das, G.P.: Band gap functionalization of BN sheet. Phys. Rev. B. **85**, 035415 (2012)

33. Zhang, Z., Guo, W., Dai, Y.: Stability and electronic properties of small boron nitride nanotubes. J. Appl. Phys. **105**, 084312/1–8 (2009)

34. Zheng, F., Zhou, G., Liu, Z., Wu, J., Duan, W., Gu, B.L., Zhang, S.B.: Half metallicity along the edges of zigzag boron nitride nanoribbons. Phys. Rev. B. **78**, 205415/1–4 (2008)

35. Neek-Amal, M., Beheshtian, J., Sadeghi, A., Michel, K.H., Peeters, F.M.: Boron nitride monolayer: a strain tunable nanosensor. J. Phys. Chem. C **117**(25), 13261–13267 (2013)

36. Li, Y., Zhou, Z., Zhao, J.: Functionalization of BN nanotubes with dichlorocarbenes. Nanotechnology **19**, 015202 (2008)

37. Wang, L., Yi, C., Zou, H., Xu, J., Xu, W.: On the isomerization and dissociation of nitramide encapsulated inside an armchair (5,5) single-walled carbon nanotube. Mater. Chem. Phys. **127**, 232–238 (2011)

38. Zhang, M.L., Ning, T., Zhang, S.Y., Li, Z.M., Yuan, Z.H., Cao, Q.X.: Response time and mechanism of Pd modified TiO_2 gas sensor. Mater. Sci. Semi. Proc. **17**, 149–154 (2014)

39. Schmidt, M.W., Baldridge, K.K., Boatz, J.A., Elbert, S.T., Gordon, M.S., Jensen, J.H., Koseki, S., Matsunaga, N., Nguyen, K.A., Su, S.J., Windus, T.L., Dupuis, M., Montgomery, J.A.: General atomic and molecular electronic structure system. J. Comp. Chem. **14**, 1347–1363 (1993)

40. Rezaei-Sameti, M., Yaghoobi, S.: Theoretical study of adsorption of CO gas on pristine and AsGa-doped (4,4) armchair models of BPNTs. Comput. Cond. Mater. **3**, 21–29 (2015)

41. Rezaei Sameti, M.: The effect of doping three Al and N atoms on the chemical shielding tensor parameters of the boron phosphide nanotubes: a DFT study. Phys. B **407**, 22–26 (2012)

42. Rezaei-Sameti, M., Samadi Jamil, E.: The adsorption of CO molecule on pristine, As, B, BAs doped (4,4) armchair AlNNTs: a computational study. J. Nanostr. Chem. **3**, 1–9 (2016)

43. James, C., Amalraj, A., Reghunathan, R., Hubert Joe, I., Jaya Kumar, V.S.: Structural conformation and vibrational spectroscopic studies of 2, 6-bis (p-*N*,*N*-dimethyl benzylidene) cyclohexanone using density functional theory. J. Raman Spect. **37**, 1381–1392 (2006)

44. Glendening, E., Reed, A., Carpenter, J., Weinhold, F.: NBO Version 31. GaussianInc., Pittsburg (2003)

45. Bulat, F.A., Toro-Labbé, A., Brinck, T., Murray, J.S., Politzer, P.: Quantitative analysis of molecular surfaces: areas, volumes, electrostatic potentials and average local ionization energies. J. Mol. Model. **16**(11), 1679–1691 (2010)

46. Bulat, F.A., Burgess, J.S., Matis, B.R., Baldwin, J.W., Macaveiu, L., Murray, J.S., Politzer, P.: Hydrogenation and fluorination of graphene models: analysis via the average local ionization energy. J. Phys. Chem. A **116**(33), 8644–8652 (2012)

4

Production of functional graphene by kitchen mixer: mechanism and metric development for in situ measurement of sheet size

Zulhelmi Ismail[1] · Abu Hannifa Abdullah[2] · Anis Sakinah Zainal Abidin[2] ·
Kamal Yusoh[2]

Abstract It has been reported that the production of defect free graphene is possible by the application of a kitchen mixer. Yet, we note that the natural-surfactant role in the exfoliation mechanism by a kitchen mixer has rarely been discussed. To investigate the possibility of graphene exfoliation in a bio-surfactant medium, we have produced graphene from the co-mixing of graphite and gum Arabic. Through the modelling of bulky graphite as a single composite disc, we have shown that the exfoliation of graphite crystal may be possible through rotational motion of graphite surface. In this paper, we also have developed two simple metric systems that were designed from the application of UV spectroscopy for in situ measurement of graphene sheet size after exfoliation step.

Keywords Functional graphene · Natural surfactant ·
Kitchen mixer · Mechanism · In situ measurement

✉ Zulhelmi Ismail
zulhelmiismail.ump@gmail.com

Abu Hannifa Abdullah
abuhaniffa@yahoo.com

Anis Sakinah Zainal Abidin
anissakinah14010@gmail.com

Kamal Yusoh
kamal@ump.edu.my

[1] Faculty of Manufacturing Engineering, Universiti Malaysia Pahang, 26600 Pekan, Pahang, Malaysia

[2] Faculty of Chemical Engineering and Natural Resources, Universiti Malaysia Pahang, 26300 Kuantan, Pahang, Malaysia

Introduction

Ever since graphene first isolation from pyrolytic graphite in 2004 [1], the interest of research society on this material is continuously growing. The advancing development of graphene research field is majorly assisted by the interest of society towards unique properties possesses by graphene. Due to the strong mechanistic properties of graphene, it has high potential for application as filler in polymer composite [2, 3]. High transparency of graphene meanwhile is suitable for manufacturing of transparent conductor [4]. In the membrane technology, super permeability of graphene to water is useful as material for desalination of sea water [5]. Excellent electrical conductivity of graphene meanwhile links graphene to advanced electronic applications such as conductive ink [6] and 2D wire [7].

For introduction of graphene towards industrial application, selection of synthesis route is extremely important. Bottom-up approach such as chemical vapour disposition [8] or epitaxial growth [9] is useful for preparation of high quality grade of graphene. However, the resulting yield from both methods is low and not economical for large scaled production of graphene. To reduce the preparation cost of graphene, a kitchen mixer can be used as exfoliating tool for graphite to graphene [10–14]. This method principally belongs to liquid-phase exfoliation class, which also includes vortex fluidic exfoliation [15, 16] and high-pressure driven exfoliation [17–19]. Application of kitchen mixer allows a facile preparation of graphene and the quality of produced graphene is even comparable with the quality of graphene prepared from the lab homogeniser [10]. Moreover, the application of toxic solvent [12] as exfoliating medium in this method can be easily replaced with water-based surfactant [11] or edible protein [13] for bio-friendly and sustainability process.

In this work, we are interested to investigate the theoretical explanation behind possible exfoliation of graphene in gum Arabic solution by the mixing action of a kitchen mixer [20]. In addition, we want to know the exact component in gum Arabic that may be responsible for stabilisation of graphene against reaggregation in water. Finally, we used UV spectroscopy and TEM to design a metric that may be suitable for rapid characterisation of functional graphene sheet size after exfoliation step. We do believe that our study will assist in the understanding of graphene exfoliation process for mass synthesis of graphene in the future work.

Experimental

Materials

Industrial graphite flakes and gum Arabic were purchased from Sigma-Aldrich (Malaysia) and were used as received. Kitchen mixer of Philips brand with nominal power of 800 W was obtained from a local retail.

Preparation and characterisations of graphene

We performed the exfoliation of graphene by a co-mixing of graphite (1 mg/ml) and gum Arabic (1 mg/ml) in a mixing volume of 400 ml. After centrifugation (Heraeus Pico Micro) at 1500 rpm (RCF: 330) for 45 min, we washed the resulting supernatant through multiple filtration steps with ultrapure water (Millipore) before the freeze-drying stage. For observation of sheet size change after the increase of exfoliation time, we extracted 1 ml of aliquots from the original volume after every paused interval (1, 3, 5 and 7 h). Absorption coefficient of resulting graphene meanwhile was computed from the redispersion of dried graphene at varied mass (0.03, 0.06, 0.09, 0.12, 0.15 mg/ml) and was followed by the absorbance measurement of graphene solution at 660 nm for each mass by UV spectroscopy (Shidmazu). TEM imaging and size measurement of graphene sheets were conducted on TEM Libra (Zeiss), while Raman measurement was performed on graphite and graphene using Witec Alpha 300R (532 nm). The functionalization study of prepared graphene was possible through the applications of XPS (Ultra Kratos) and IR (Perkin Elmer) on the dried graphene. The stability of exfoliated graphene in water was studied through redispersion of 3 mg of graphene mass into 10 ml of ultrapure water by bath sonication (Branson). Absorbance monitoring of graphene suspension was conducted in every 24 h for 6 days using UV spectroscopy. For comparison purpose, stability of graphene supernatant after washing stage was also observed from absorbance value within similar time duration. Contact angle measurement meanwhile was determined from the change of droplet size on suspended graphite and graphene using goniometer.

Results and discussion

Initial characterisation of graphene

Validation of graphene

We notice that the presence of highest UV peak for the spectra of washed and unwashed gum Arabic-graphene (G_{GA}) in Fig. 1a is at 269 nm. Interestingly, this is a strong indicator for graphene presence and as evidence for successful exfoliation of multilayer by a kitchen mixer [21]. To rule out the effect of gum Arabic presence on the absorbance of graphene, we note that the highest UV peak for gum Arabic is below 200 nm and definitely is not in the range of wavelength for graphene. We also performed a calculation from the Lambert–Beers law ($A = \alpha Cl$) for determination of the absorption coefficient (α) value for G_{GA} at 660 nm of spectroscopy wavelength.

As shown in Fig. 1b, the obtained value for α is 1210 mg^{-1} m^{-1} ml and this value is definitely in the range with the previous reported α for surfactant-based graphene [22]. However, we note that the determined value of α in this research work is very far from the proposed theoretical α value of 4237 mg^{-1} m^{-1} ml for pristine monolayer graphene [23]. As suggested by theory, the deviation of α may possibly caused by the preparation method, the presence of functional group on graphene sheet and mean thickness of multilayer graphene [24].

Morphological study of graphene

For visual conformation of graphene presence in the resulting black opaque supernatant, we used transmission electron microscopy (TEM) for imaging of graphene in the drop casted sample. As shown by the example TEM image of graphene in Fig. 2a (t_{mix}: 7 h), the appearance of semi-transparent sheets is proving the possibility of graphene preparation by a kitchen mixer and gum Arabic/water solution. Moreover, the absence of wrinkles graphene sheets is also suggesting that this method does not introduce any defect on the basal structural of graphene despite long processing duration of mixing [10, 11, 21]. The mean length, <L> for resulting graphene from 1, 3, 5 and 7 h meanwhile, was measured from TEM by the edge–edge evaluation of 100 visible graphene sheets and included in Fig. 8. For comparison purpose, we have also studied the length distribution of graphene by atomic force microscopy (AFM). It was found that the size of small graphene in

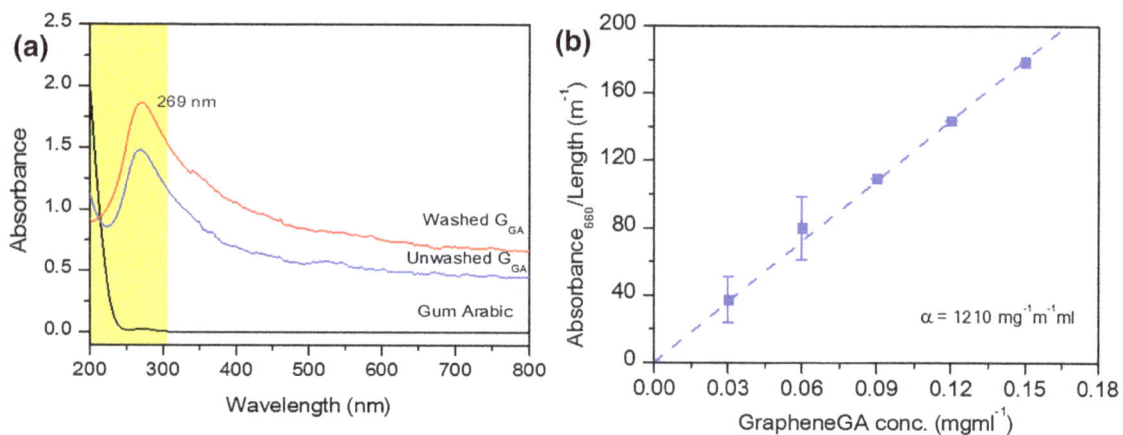

Fig. 1 a Spectrum of gum Arabic, unwashed graphene and washed graphene. b The absorption coefficient value, α, for graphene is extracted from the slope of plotted linear

Fig. 2 a High TEM magnification and b Low TEM magnification for exfoliated graphene after 7 h of mixing c AFM analysis on the similar sample was used to study statistical length and thickness distribution of graphene

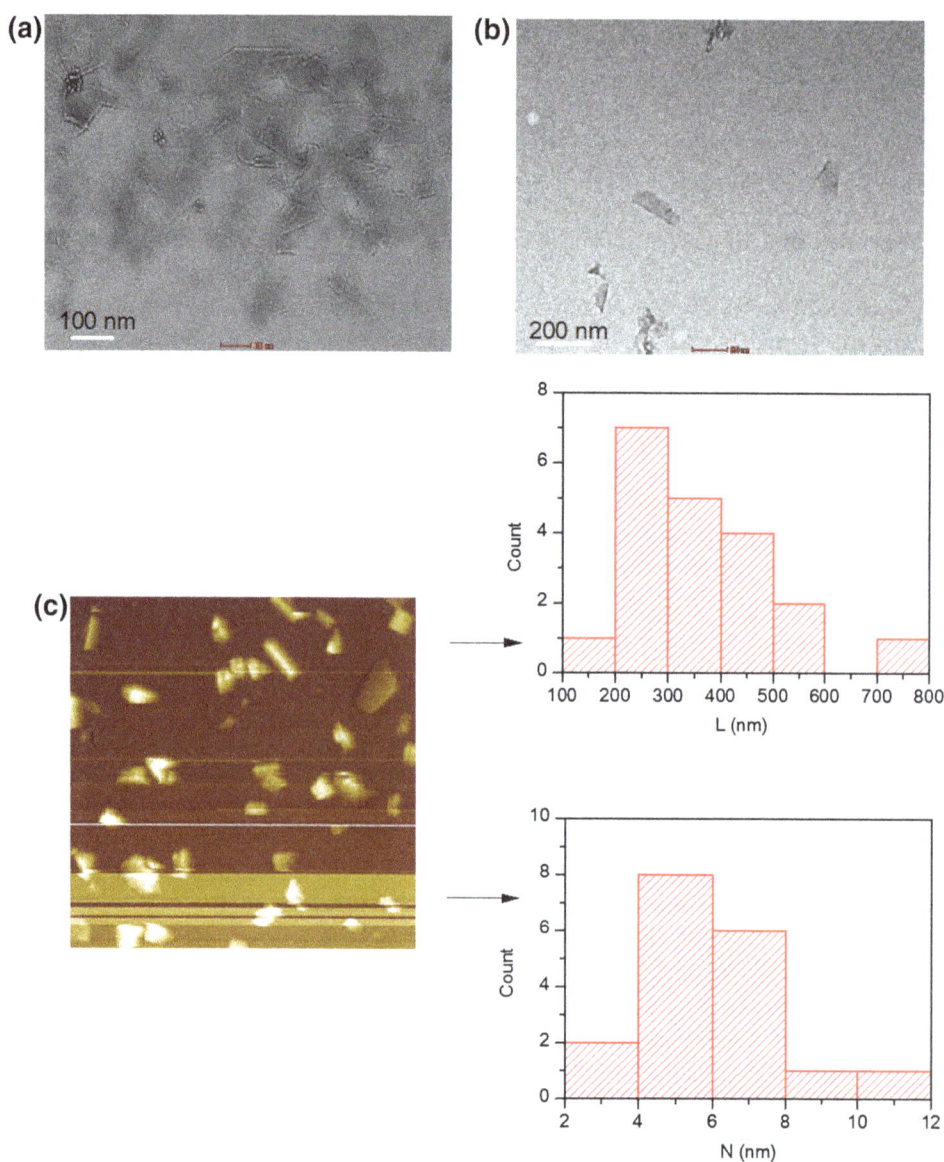

200–300 nm domain was more frequently measured by AFM than TEM in our work. As suggested by the previous report [11], the lack of small graphene count by TEM is normally caused by wide polydispersity of the sample as well as the high transparency of graphene at lower layers number. As only visible graphene sheets were considered for objective measurement of length by TEM, small size graphene that usually is almost invisible in the micrograph was not included in the counting process. As for the mean thickness, <N> of the produced graphene, the thickness of graphene was evaluated from 20 sheets with thickness value from five layers to twenty-nine layers of sheets (see Fig. 2b).

Defect evaluation of graphene

We conducted Raman spectroscopy measurement on the G_{GA} film for investigation of defect after mixing. A rectangle shape film with dimension of 2 cm × 2 cm was cut from the vacuum-filtered film. As shown from the example of Raman spectrum for graphene in Fig. 3, the presence of intense G band (1578 cm^{-1}), defect-associated D band (1347 cm^{-1}) and 2D band (2684 cm^{-1}) on the Raman spectrum of G_{GA} can be used to verify the presence of graphene on the vacuum-filtered membrane [10]. A significant Raman shift of G, D and 2D bands for the graphene from 1576, 1344 and 2701 cm^{-1} demonstrates the change of graphene structure after exfoliation from bulk graphite. The presence of D band in the graphene spectrum indicates a formation of defect in the graphene structure. We note that the mean intensity ratio of D and G bands (I_D/I_G) for the graphene is 0.42 and the value actually is not far off from the mean I_D/I_G of graphite (0.048). This observation shows that the defects presence in our exfoliated

graphene is not similar to the defects level in reduced graphene with I_D/I_G of 1 [25]. Furthermore, the low I_D/I_G for collected graphene can also be used as indicator that exfoliation during mixing action did not induce any significant defects on the graphene flakes as been suggested from previous TEM imaging study. We notice that for graphite the existing defect is not major and did not influence the resulting defects in exfoliated graphene. From the derived equation in the previous work by groups of Coleman [26], we also performed an estimation of mean length for resulting few-layers graphene using the obtained Raman data. Taken the value of FWHM for G-band as 21.66 cm^{-1}, we estimate that the mean length of produced graphene in this work was 529.1 nm with 20% of deviation.

Shear functionalization of graphene

We conducted X-ray photon spectroscopy (XPS) on the vacuum-filtered graphene film for the characterisation of graphene structure after exfoliation. Table 1 shows the composition difference between graphite and exfoliated graphene after the mixing. As shown in Fig. 4a, only oxygen (532.5 eV), silicon (103.5 eV) and carbon (284.5 eV) are present in the survey spectrum of initial graphite sample. After exfoliation, however, additional change, such as low nitrogen presence (400.5 eV) and increase of oxygen intensity, is observed for XPS analysed G_{GA} (see Fig. 4b). These chemical transformations of exfoliated graphene were possibly caused by the adsorption of gum Arabic on the graphene sheets [22]. Based on the calculated carbon to oxygen (C/O) ratio, we note that C/O value of G_{GA} is above the reported C/O for graphene oxide (2.1–2.6) but is actually far from the published C/O for reduced graphene oxide (7.1–10.3) [27]. High oxygen content in G_{GA} is indicating that large quantity of gum Arabic is required for stabilisation of graphene sheets against reaggregation to graphite in a liquid. To address this assumption, we investigated the chemical change of graphene from the presented carbon 1s (C1s) core of graphite and C1s core of graphene in Fig. 4c, d. In the original graphite, C=C sp^2 (284.5 eV) and C–C sp^3 (285.3 eV) are

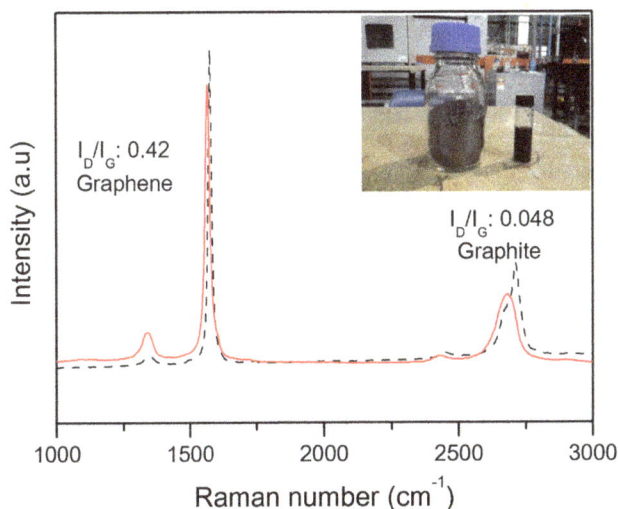

I_D/I_G: 0.42
Graphene

I_D/I_G: 0.048
Graphite

Fig. 3 Raman spectra for graphite and graphene

Table 1 Concentrations of atoms for G_{GA} and graphite from XPS study

Atomic composition in (%)					C/O ratio
Sample	C	O	Si 2p	N	
G_{GA}	69.97	23.63	4.23	2.17	2.96–3
Graphite	92.37	5.87	1.75	–	15.74–16

The C/O ratio was used to show the change of molecular structure for graphene after exfoliation by mixer

Fig. 4 **a** and **b** show the collected survey spectra from graphite and graphene while **c** and **d** are corresponding to the C1s core of graphite and graphene. **e** FTIR spectra of graphite, gum Arabic and graphene

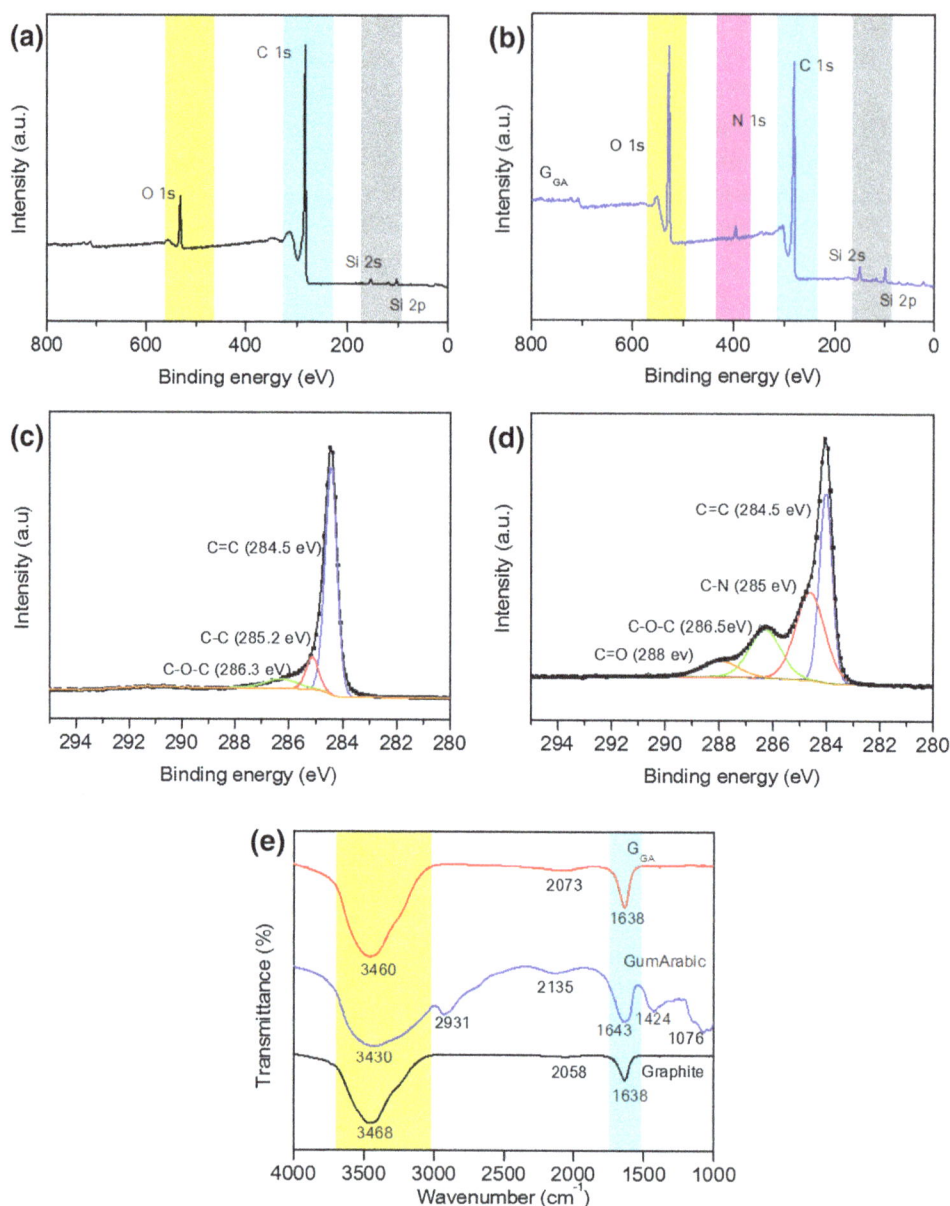

the only major groups that form the chemical structure of graphite. After exfoliation, the formations of, C–N, C–O–C and C=O groups besides the original C=C and C–C in the composition of exfoliated graphene are matched with the additional presence of three peaks at, 285, 286.5 and 288 eV [22]. This particular observation confirms that oxygen groups are majorly responsible for stability of graphene against reaggregation in gum Arabic solution.

Figure 4e shows the measured spectra of graphite, gum Arabic and G_{GA} from Infrared Spectroscopy (IR). The stretching vibrations at 1638 cm^{-1} correspond to C=O bonds in graphite and graphene. The broadening peak at 3460 cm^{-1} for G_{GA} and 3468 cm^{-1} for graphite meanwhile is originating from the presence of hydrogen bonded OH groups in both samples. The absorption band at

2931 cm^{-1} in IR spectrum of gum Arabic suggests the strong presence of sugar galactose, arabinose and rhamnose (Arabinogalactan) in the material [28]. Weak symmetric stretching at the band 1424 cm^{-1} is mainly due to the C=O of glucuronic acid while the absorption band at 1076 cm^{-1} highlights the plausible presence of alkene C–H bonds for existing polysaccharides in gum Arabic. We note that the disappearance of IR band at 1076 cm^{-1} in graphene spectrum may be due to the breaking of polysaccharide bonds during shear functionalization of gum Arabic with graphene. The loss of FTIR peak at 1424 and 2931 cm^{-1} for graphene meanwhile is attributed to the adsorption mechanism of gum Arabic on graphene that is involving carbohydrate and sugar groups. Stronger peaks for graphene at 3460 and 1638 cm^{-1} though indicate the increase of oxygen content in graphene structure after exfoliation by a kitchen mixer.

Yield and stability of exfoliated graphene in water

To study the yield concentration of produced graphene by gum Arabic-assisted exfoliation, we varied the mass of gum Arabic (1–5 g) and graphite (2.5 and 5.0 g) in 400 ml of volume. The kitchen mixer was operated at 15 min for each individual set of experiment. As shown in Fig. 5a, the mass of initial graphite is important to secure higher concentration yield of graphene. While the maximum yield of graphene for 2.5 g of graphite was only achieved at 3.0 g of gum Arabic, the increase of initial graphite to 5.0 g of graphite would improve the yield concentration even after incorporation 5.0 g of gum Arabic. For further investigation of this phenomenon, we replotted the data into a new graph with the concentration ratio of graphite to gum Arabic (C_{iG}/C_{GA}) taken as *x*-axis. It was found from Fig. 5b that the optimum production of graphene was only possible at the specific value of ratio ($C_{iG}/C_{GA} = 1$). As reported previously [11], this ratio is influential for efficient production of graphene by a kitchen mixer and must be considered prior of any exfoliation stage. In our case, the increase of graphite mass must be always accompanied by the similar mass increase of gum Arabic. The yield efficiency (%) for our method meanwhile was computed from the highest yield concentration data obtained in this study

(~ 17 μg/ml). We found that our yield value (0.136%) is comparable with the reported yield of graphene (0.1%) from shear exfoliation by homogenizer [10] and a kitchen mixer [11].

To evaluate the stability of produced graphene in water, we have to perform triple measurements of absorbance for each supernatant and dispersion of freeze-dried graphene at every 24 h for 6 days (see Fig. 5c). From the unchanged values of absorbance for supernatant throughout the measurement, it is suggested that the resulting graphene after exfoliation and centrifugation stage is very stable in water. However, we found that the redispersion of graphene into water would result in rapid sedimentation of sheets as been indicated from the continuous drop of absorbance within 6 days. We note, however, that our finding is consistent with the published result on the sedimentation study of polysaccharide-graphene after redispersion in water [29].

Mechanism in exfoliation of graphene by a kitchen mixer

Reduction of surface tension by gum Arabic

During the exfoliation of graphite to graphene, the presence of gum Arabic is required to promote the reduction of water surface tension. This is critical as it allows the surface tension of water to match the surface tension of graphene. By adjusting the surface tension value between gum

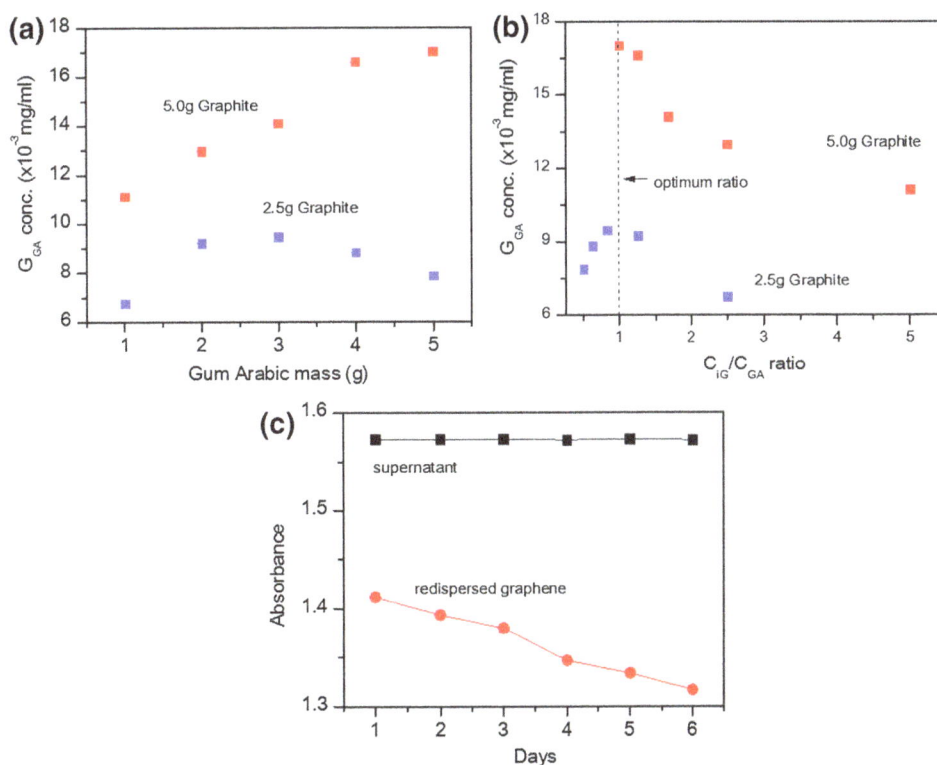

Fig. 5 **a** Effect of gum Arabic mass on the yield concentration of graphene **b** Determination of optimum ratio between graphite and gum Arabic mass **c** Stability of graphene in water after redispersion as compared to that of decanted supernatant

Arabic/water solution and graphite surface, the wetting capability of graphite is gradually increased from the adsorption of gum Arabic on graphene sheets [22]. As suggested by the data of graphene from XPS and IR, the adsorptions of chemical components from gum Arabic are mainly responsible for shear exfoliation and stabilisation of graphene in water. To confirm that the wetting of graphite is still the main mechanism for graphene exfoliation in a kitchen mixer, we have conducted contact angle measurements for both graphite and graphene. As shown in Fig. 6a, the contact angle of graphite (90°) is larger than that of graphene (75°) and is consistent with the previous reported work on sonication of graphene [22]. This is expected as it is known that graphite is originally hydrophobic in nature and the wetting mechanism is independent of selected preparation method. As suggested by previous work [30], the presence of surfactant in the chemical structure of

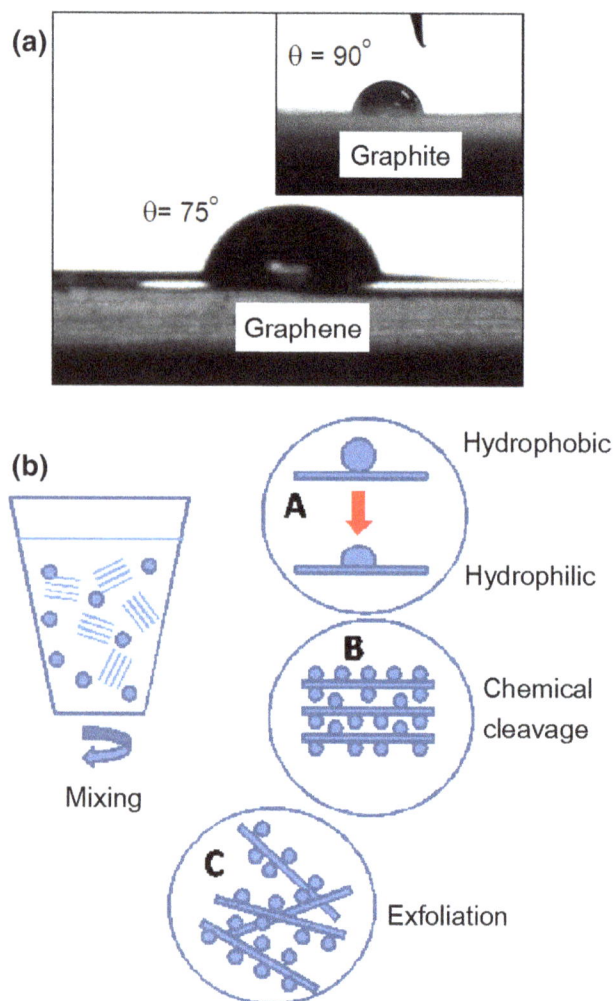

Fig. 6 a Contact angle measurement of water droplet on graphite and graphene film surface. The change of contact angle for graphene indicates the increase of wetting behaviour after adsorption of gum Arabic, **b** we describe exfoliation mechanism of graphite in gum Arabic in three stages

graphene through π to π mechanism can reduce the hydrophobicity of graphite and resulting in the exfoliation to graphene (see Fig. 6b).

From the 2nd scenario model proposed in the Paton's work [10], the reduction of water surface tension during exfoliation event can be investigated by the following expression:

$$\dot{\gamma}_{\min} = \frac{\left(\sqrt{E_{S,G}} - \sqrt{E_{S,L}}\right)^2}{\eta L}, \qquad (1)$$

where $\dot{\gamma}_{\min}$ is the minimum local shear rate required for exfoliation of graphene, $\sqrt{E_{S,G}}$ and $\sqrt{E_{S,L}}$ are the graphene and water surface energy, respectively, η is the dynamic viscosity of water whilst L is the corresponding length of exfoliated graphene at minimum shear rate. We note that the pre-determined value of maximum local shear rate in our exfoliation of graphene is $2.7 \times 10^4 \text{ s}^{-1}$, which was calculated by practically considering the 60% loss of motor power to surrounding [31]. This maximum shear rate surprisingly is relatively similar to the proposed value of $\dot{\gamma}_{\min}$ [10]. Therefore, we assume that it is safe to replace the value of L with the estimated L of 585.1 nm from our previous exfoliation stage. By taking the value of $E_{S,G}$ as 71 mJ/m^2 [10], η as 8.9×10^{-4} Pa s and $\dot{\gamma}_{\min}$ as $1 \times 10^4 \text{ s}^{-1}$, the actual surface energy of the water during exfoliation ($E_{S,L}$) was computed as $E_{S,L} = 70.99783$ mJ/m^2.

The surface tension of liquid, Γ, therefore can be expressed in terms of surface energy as [32]:

$$\Gamma = E_{S,L} - T \cdot S_{S,L}, \qquad (2)$$

where T is the temperature in Kelvin, whilst $S_{S,L}$ is the surface entropy for the liquid. As surface entropy of water is 0.07 mJ/m^{2}K, the required surface tension of water at room temperature is known as $\Gamma = 49.9978$ mJ/m^2. We note that this obtained value is actually below the surface tension of pure water at ambient (72.8 mJ/m^2) and is in range with the reported surface tension of NMP [10] and gum Arabic at 3 wt% of concentrations (55 mJ/m^2) [33]. Based on this observation, we suggest that the surface tension of water must be within this value prior the sliding of graphite layer. We found, however, that due to the dynamic change of water surface tension after surfactant addition, Eq. (2) is only applicable for determination of surface tension for solvent [34].

In principle, the adsorption of gum Arabic on graphene can be described by π–π stacking mechanism of surfactant on the interlayer graphite [30]. We suggest that the protein components of gum Arabic in arabinogalactan-protein (AGP) and glycoprotein (GP) are chemically attached on the graphite surface as the efficiency of solid–liquid interfaces is increased due to the amphiphilic nature of

AGP and GP components. Polar component of AGP and GP (polysaccharide) meanwhile attach themselves with water molecules in the medium [35]. Colloidal suspension of graphene provides the stability of gum Arabic-graphene after adsorption of gum Arabic on the surface. The assumption of protein as the responsible molecules for exfoliation of graphene is validated from the presented data in previous XPS and IR results. The addition of C=O and C–N in the sample after exfoliation stage indicates the adsorption of protein on the graphene surface.

Exfoliation model of graphite

Using Paton's model, which actually was developed for exfoliation mechanism of graphene in a solvent, we assume that the exfoliation of graphene is initiated after the adsorption of surfactant on the graphite sheets by wetting mechanism. This is of course plausible as we know that the exfoliation of graphene in surfactant medium is triggered after the adsorption of gum Arabic [22] or pyrenecarboxylic acid [30] on the graphite surface. With water as exfoliating medium, the adsorption of gum Arabic on graphite surface is only possible once the surface tension of water is reduced to match to that of graphite. It is predicted that the graphite hydrophobicity will gradually decrease towards hydrophilicity after the increase of gum Arabic presence on the graphite sheets. This increased hydrophilicity will further improve the wetting behaviour of graphite in water. The interfacial bonding energy between surfactant layer on the graphite and liquid becomes stronger than the interfacial bonding between graphite sheets by van der Waals forces. The exfoliation of graphite finally is initiated through sliding mechanism of top layer graphite from the driving forces by shear energy of a kitchen mixer.

In this work, however, we would like to discuss the role of rotational motion in the exfoliation mechanism of graphite by a kitchen mixer [12]. We note that the rotational motion aside translational may be applied on the graphite sheets during the pumping cycle of liquids in the mixing vessel. By considering the presence of τ and T, the mechanism schematic for exfoliation of graphite is graphically presented in Fig. 7.

To develop this simple model, we assume the geometry of graphite as a composition of multiple circular discs and each single disk is fixed at one end by the van der Waals (vdw) forces from the adjacent disc. Considering that the resulting shear flow of the fluid (moving parallel to each other but at different speed) is applied on the free end of graphite surface, the disk composite will be subjected to torsional forces due to the resultant rotational forces (T and T') on each end surfaces (scenario a). We note that the

cross-sectional plane of the discs will remain undistorted due to geometry of the composition [36]. However, as the top graphite crystal is only linked by weak vdw forces on one end, the resulting shearing stress (τ_{max}) on the graphite vertical will be used to slip the graphite further until it completes delamination stage (scenario b and c). Since $T = FL/2$ and F_{min} can be linked to the minimum local shear rate ($\dot{\gamma}_{min}$) of Newtonian fluid by [10]:

$$F_{min} = \eta\dot{\gamma}_{min}L^2, \tag{3}$$

the minimum required forces to initiate the exfoliation stage of graphene from the known values of γ_{min}, L and η can be practically computed. In our case, the resultant T_{min} is 6.586×10^{-19} Nm, which is extremely small value for rotational forces required on the graphite. The magnitude of shearing stress (τ) for exfoliation of individual graphite crystal then is calculated from the given T_{min} by the following equation [36]:

$$\tau = \frac{TL}{2J}, \tag{4}$$

where $J = 0.5\pi (L/2)^4$ is the polar moment of inertia for a solid disk and is established as 7.68×10^{-27} m. The applied shearing stress for slipping of graphite from its original plane is, therefore, known as 22.6 Pa. This value of τ is certainly not adequate to overcome the computed interlayer strength of graphite in dry state which is evaluated at 0.14 GPa [37]. This is where the critical role of wetting on graphite sheets is highlighted as the interlayer strength of graphite will be reduced greatly due to the low friction level of graphite in wet state [38]. Assuming that our graphite system remains elastic, we can determine the angle of twist produced during the slippage (scenario a) through the application of presented equation [36]:

$$\theta = \frac{Tt}{JG}, \tag{5}$$

where G is known as the reported modulus of rigidity for graphene (280 GPa) [39] and t is the thickness of single graphite crystal (0.335 nm) [2]. The resultant θ is 1.026×10^{-10} radians and this value is indicating that the plane of graphite crystal will rotate about 5.88° from its original axis in nano-domain to initiate a delamination stage. It is apparent that our proposed model could be used to describe the role of rotational motion during graphite delamination by a kitchen mixer. In spite of this possibility, we admit that the application of disc geometry to represent a single layer of graphite crystal in our proposed model might not be highly accurate. Moreover, we choose to neglect the possible rupture effect of shear flow on graphite crystal, which ultimately may be responsible for fragmentation mechanism of the sheets during exfoliation.

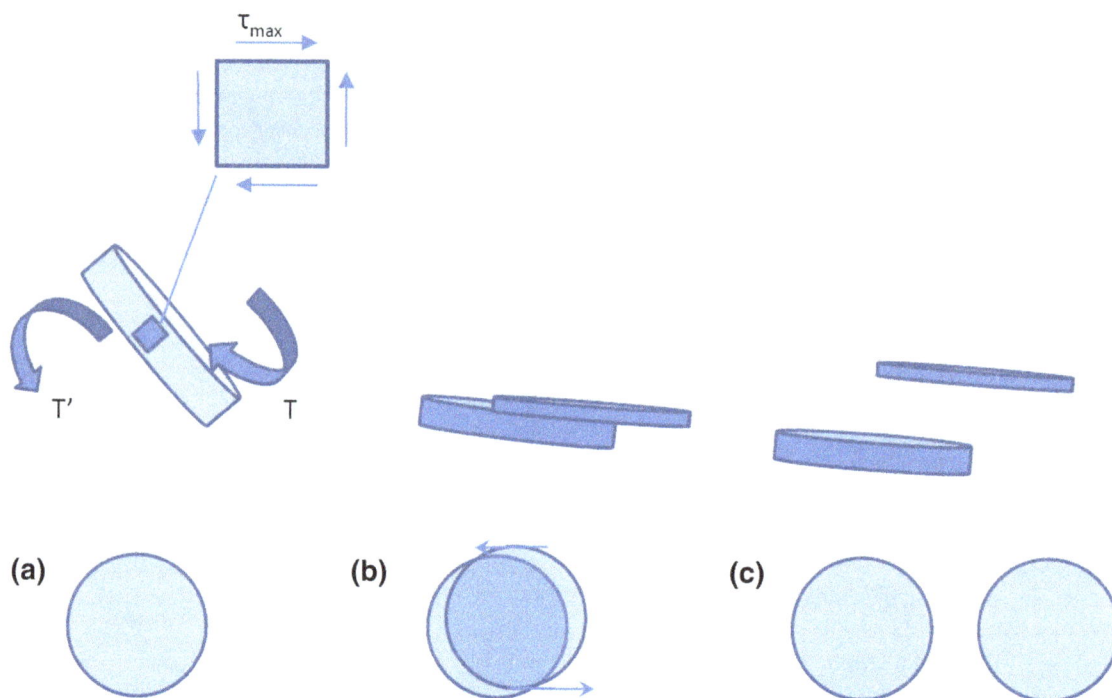

Fig. 7 Exfoliation schematic shows the role of rotation (T) and shear (τ) for exfoliation of graphite crystal

In situ measurement of graphene size by UV spectroscopy

Previous works on the sonication [40] and shear mixing of graphene [12, 13] have shown the effect of mixing on the size reduction of sheet. Even multiple empirical functions were developed in one work to demonstrate the possibility of UV and Raman for prediction of sheet size and thickness [26]. In their extensive work, we note, however, that the application of UV spectroscopy was used only to estimate the layer number but not the lateral size of graphene. Thus, we would like to propose a simple metric system for evaluation of graphite sheet size from measurement by UV spectroscopy in this paper.

To establish our metric, we note that the sheets size of graphene in our work is strongly dependent on the exfoliation time as shown in Fig. 8a–c. The fragmentation effect of shearing forces on graphene sheets is shown from the increase of graphene number in 300–400 nm range after exfoliation time of 7 h. In our work, we choose not to use centrifugation speed (ω) for size of selection of our graphene as $<L>$ and (N) of graphene are affected by ω value [41]. Instead we used variation of exfoliation time and constant ω to initiate the size change of graphene as only $<L>$ is strongly dependent on time and not $<N>$ of graphene sheets [12].

From the resulting UV spectra, the gradual increase of intensity for $\pi–\pi^*$ peak (268 nm) is indicating the strong of graphene presence after time variation in the supernatant.

We notice that there is no shift of $\pi–\pi^*$ peaks, which could be linked to the constant $<N>$ of graphene sheets in the supernatant [26]. Normalisation of UV spectrum at $\pi–\pi^*$ peak, however, reveals the difference in absorbance value from the shoulder (min) and peak point (max) location is parallel with the size increase of graphene sheets. We estimate that the difference between λ_{peak} and $\lambda_{shoulder}$ is (-30 nm) for easier determination of maximum and minimum absorbance of graphene. With this knowledge, the $<L>$ data were plotted as a function of the absorbance ratio between minimum and maximum of $\pi–\pi^*$ peak (Abs_{min}/Abs_{max}). Fitting of the data resulting in the metric of $<L>$ is:

$$<L> = 1030.1 \left(\frac{Abs_{min}}{Abs_{max}} \right)^{1.36}. \tag{6}$$

While this developed metric may be useful for estimation of sheet size after exfoliation stage, we note however that the value of Abs_{min}/Abs_{max} must not exceed 0.76. We evaluate the practicality of the proposed metric from the comparison of $<L>$ after 1 h exfoliation time, resulting in $<L>$ of 761 nm from the metric and $<L>$ of 943 nm from the imaging by TEM for $Abs_{min}/Abs_{max} = 0.8$. This limitation is possibly caused by the exponential change of graphene size with exfoliation duration that is lower than 180 min [12].

Based on the produced TEM data, the minimum size of our graphene is measured at 200 nm and we know that the value of Abs_{min}/Abs_{max} is beyond the scope of our metric.

Fig. 8 **a**, **b** and **c** show the change of graphene size after the increase of exfoliation time, (**d**) The increase of $\pi-\pi^*$ peak is indicating the improvement of graphene concentration in collected supernatant, (**e**) After normalisation at local maximum, we note a shift of absorbance value at UV shoulder, (**f**) Empirical function of $<L>$ as the value of Abs_{min}/Abs_{max}

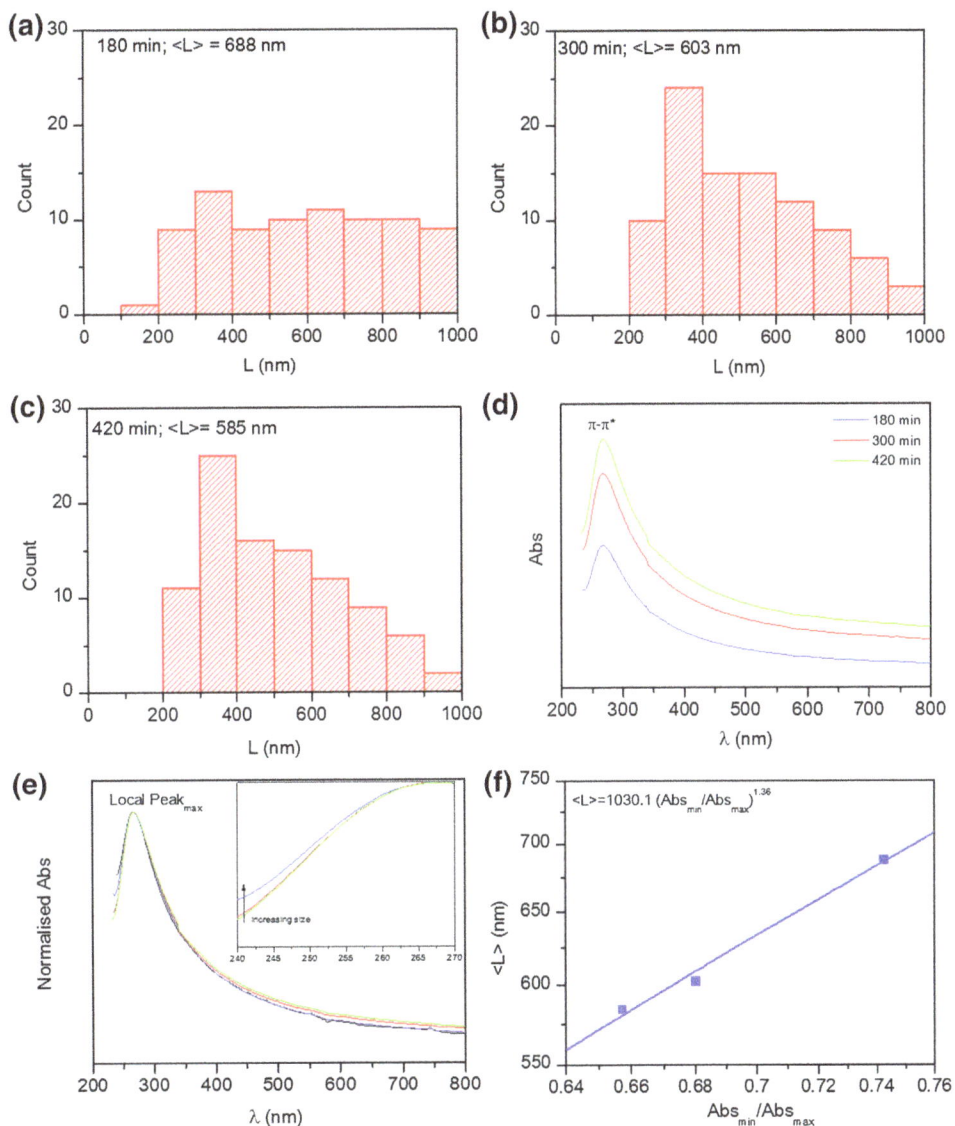

However, we would like to propose a modification of our metric, which may be suitable for prediction of sheet size smaller than 561 nm ($Abs_{min}/Abs_{max} = 0.64$) as:

$$<L> = 561\left[1 - \left(0.64 - \frac{Abs_{min}}{Abs_{max}}\right)^{1.36}\right]. \quad (7)$$

From the presented metric, the possible smallest size of graphene is computed as 255 nm, which is not far from the smallest sheet size determined by our TEM imaging. We do believe that the application of both metrics would be useful for in situ measurement of sheet size in future exfoliation stage, as these metrics allow user to evaluate the resulting $<L>$ of produced graphene using only UV spectroscopy. In spite of this advantage, we would recommend for further length investigation by TEM after each prediction for validation and confirmation of proposed $<L>$ value.

Conclusion

As conclusion, we have shown that it is possible to produce gum Arabic-modified graphene from the co-mixing of graphite and gum Arabic in a commercial kitchen mixer. Through the application of XPS and IR, we have identified the role of protein and polysaccharide role in the stabilisation of graphene in water. It is suggested that the amphiphilic protein from AGP and GP groups of gum Arabic is required for the bridging between polar polysaccharide and hydrophobic graphene in water. Through the reduction of surface tension, wetting of graphite is initiated for exfoliation and stabilisation of graphene against reaggregation. In addition, we have proposed an exfoliation model that was designed by considering the rotational motion of graphite during exfoliation to graphene. The generated model used to describe the exfoliation of graphene sheet from graphite with the magnitude of τ and θ was computed from the application of basic continuum mechanics. To accommodate these simple calculations, we must assume the geometry of our bulky graphite as an attached circular disk. For in situ evaluation of graphene sheet size after exfoliation stage, two simple metrics were developed from the combined application of UV spectroscopy and TEM imaging. However, we note that although we could estimate the sheet size from the application of both metrics, we could not be absolutely certain that the computed <L> from the metric is the actual one. Thus, we recommended the application of TEM and AFM for validation of the result after the application of these proposed metrics. In this way, while our metrics may be useful as rapid measurement tool for <L> of graphene after exfoliation, the resulting data from TEM or AFM could still be used as supporting data for later validation step.

Acknowledgements This research has received funding from Ministry of Higher Education of Malaysia and Universiti Malaysia Pahang under grant agreement of RDU 160149 and ULC 150306. In addition, we are grateful for assistance provided by Universiti Sains Malaysia (SERC) and Universiti Putra Malaysia (ITMA) during XPS and Raman characterisation. We are also very thankful to Mr. Fahmi from Universiti Malaysia Pahang (CARIFF) for technical support during the imaging of graphene by TEM.

References

1. Geim, A.K., Novoselov, K.S.: The rise of graphene. Nat. Mater. **6**(3), 183–191 (2007)
2. Lee, C., Wei, X., Kysar, J.W., Hone, J.: Measurement of the elastic properties and intrinsic strength of monolayer graphene. Science **321**(5887), 385–388 (2008)
3. Abidin, A.S.Z., Yusoh, K., Jamari, S.S., Abdullah, A.H., Ismail, Z.: Enhanced performance of alkylated graphene reinforced polybutylene succinate nanocomposite. AIP Conf. Proc. **1756**(1), 040005 (2016). doi:10.1063/1.4958766
4. Kang, M.S., Kim, K.T., Lee, J.U., Jo, W.H.: Direct exfoliation of graphite using a non-ionic polymer surfactant for fabrication of transparent and conductive graphene films. J. Mater. Chem. C **1**(9), 1870–1875 (2013)
5. Surwade, S.P., Smirnov, S.N., Vlassiouk, I.V., Unocic, R.R., Veith, G.M., Dai, S., Mahurin, S.M.: Water desalination using nanoporous single-layer graphene. Nat. Nanotechnol. **10**(5), 459–464 (2015)
6. Huang, L., Huang, Y., Liang, J., Wan, X., Chen, Y.: Graphene-based conducting inks for direct inkjet printing of flexible conductive patterns and their applications in electric circuits and chemical sensors. Nano Res. **4**(7), 675–684 (2011)
7. Andryieuski, A., Lavrinenko, A.V., Chigrin, D.N.: Graphene hyperlens for terahertz radiation. Phys. Rev. B **86**(12), 121108 (2012)
8. Yan, K., Fu, L., Peng, H., Liu, Z.: Designed CVD growth of graphene via process engineering. Acc. Chem. Res. **46**(10), 2263–2274 (2013)
9. Sutter, P.W., Flege, J.-I., Sutter, E.A.: Epitaxial graphene on ruthenium. Nat. Mater. **7**(5), 406–411 (2008)
10. Paton, K.R., Varrla, E., Backes, C., Smith, R.J., Khan, U., O'Neill, A., Boland, C., Lotya, M., Istrate, O.M., King, P., Higgins, T., Barwich, S., May, P., Puczkarski, P., Ahmed, I., Moebius, M., Pettersson, H., Long, E., Coelho, J., O'Brien, S.E., McGuire, E.K., Sanchez, B.M., Duesberg, G.S., McEvoy, N., Pennycook, T.J., Downing, C., Crossley, A., Nicolosi, V., Coleman, J.N.: Scalable production of large quantities of defect-free few-layer graphene by shear exfoliation in liquids. Nat. Mater. **13**(6), 624–630 (2014). doi:10.1038/nmat3944
11. Varrla, E., Paton, K.R., Backes, C., Harvey, A., Smith, R.J., McCauley, J., Coleman, J.N.: Turbulence-assisted shear exfoliation of graphene using household detergent and a kitchen blender. Nanoscale **6**(20), 11810–11819 (2014)
12. Yi, M., Shen, Z.: Kitchen blender for producing high-quality few-layer graphene. Carbon **78**, 622–626 (2014)
13. Pattammattel, A., Kumar, C.V.: Kitchen chemistry 101: multi-gram production of high quality biographene in a blender with edible proteins. Adv. Funct. Mater. **25**(45), 7088–7098 (2015)
14. Liu, L., Shen, Z., Yi, M., Zhang, X., Ma, S.: A green, rapid and size-controlled production of high-quality graphene sheets by hydrodynamic forces. RSC Adv. **4**(69), 36464–36470 (2014)
15. Chen, X., Dobson, J.F., Raston, C.L.: Vortex fluidic exfoliation of graphite and boron nitride. Chem. Commun. **48**(31), 3703–3705 (2012)
16. Wahid, M.H., Eroglu, E., Chen, X., Smith, S.M., Raston, C.L.: Functional multi-layer graphene–algae hybrid material formed using vortex fluidics. Green Chem. **15**(3), 650–655 (2013)
17. Shen, Z., Li, J., Yi, M., Zhang, X., Ma, S.: Preparation of graphene by jet cavitation. Nanotechnology **22**(36), 365306 (2011)
18. Yi, M., Li, J., Shen, Z., Zhang, X., Ma, S.: Morphology and structure of mono-and few-layer graphene produced by jet cavitation. Appl. Phys. Lett. **99**(12), 123112 (2011)
19. Yi, M., Shen, Z., Zhang, W., Zhu, J., Liu, L., Liang, S., Zhang, X., Ma, S.: Hydrodynamics-assisted scalable production of boron nitride nanosheets and their application in improving oxygen-atom erosion resistance of polymeric composites. Nanoscale **5**(21), 10660–10667 (2013)
20. Ismail, Z., Abdullah, A.H., Zainal Abidin, A.S., Yusoh, K.: Application of graphene from exfoliation in kitchen mixer allows mechanical reinforcement of PVA/graphene film. Appl. Nanosci. (2017). doi:10.1007/s13204-017-0574-y
21. Chabot, V., Kim, B., Sloper, B., Tzoganakis, C., Yu, A.: High yield production and purification of few layer graphene by Gum Arabic assisted physical sonication. Sci. Rep. **3**, 1378 (2013)
22. Fan, J., Shi, Z., Ge, Y., Wang, J., Wang, Y., Yin, J.: Gum arabic assisted exfoliation and fabrication of Ag–graphene-based hybrids. J. Mater. Chem. **22**(27), 13764–13772 (2012)
23. Paton, K.R., Coleman, J.N.: Relating the optical absorption coefficient of nanosheet dispersions to the intrinsic monolayer absorption. Carbon **107**, 733–738 (2016). doi:10.1016/j.carbon.2016.06.043
24. Su, R., Lin, S.F., Chen, D.Q., Chen, G.H.: Study on the absorption coefficient of reduced graphene oxide dispersion. J. Phys. Chem. C **118**(23), 12520–12525 (2014). doi:10.1021/jp500499d
25. Gómez-Navarro, C., Weitz, R.T., Bittner, A.M., Scolari, M., Mews, A., Burghard, M., Kern, K.: Electronic transport properties of individual chemically reduced graphene oxide sheets. Nano Lett. **7**(11), 3499–3503 (2007). doi:10.1021/nl072090c

26. Backes, C., Paton, K.R., Hanlon, D., Yuan, S., Katsnelson, M.I., Houston, J., Smith, R.J., McCloskey, D., Donegan, J.F., Coleman, J.N.: Spectroscopic metrics allow in situ measurement of mean size and thickness of liquid-exfoliated few-layer graphene nanosheets. Nanoscale **8**(7), 4311–4323 (2016)

27. Stobinski, L., Lesiak, B., Malolepszy, A., Mazurkiewicz, M., Mierzwa, B., Zemek, J., Jiricek, P., Bieloshapka, I.: Graphene oxide and reduced graphene oxide studied by the XRD, TEM and electron spectroscopy methods. J. Electron Spectrosc. Relat. Phenom. **195**, 145–154 (2014)

28. Daoub, R.M.A., Elmubarak, A.H., Misran, M., Hassan, E.A., Osman, M.E.: Characterization and functional properties of some natural Acacia gums. J. Saudi Soc. Agric. Sci. (2016). doi:10.1016/j.jssas.2016.05.002

29. Unalan, I.U., Wan, C., Trabattoni, S., Piergiovanni, L., Farris, S.: Polysaccharide-assisted rapid exfoliation of graphite platelets into high quality water-dispersible graphene sheets. RSC Adv. **5**(34), 26482–26490 (2015)

30. An, X., Simmons, T., Shah, R., Wolfe, C., Lewis, K.M., Washington, M., Nayak, S.K., Talapatra, S., Kar, S.: Stable aqueous dispersions of noncovalently functionalized graphene from graphite and their multifunctional high-performance applications. Nano Lett. **10**(11), 4295–4301 (2010). doi:10.1021/nl903557p

31. Xie, L., Rielly, C.D., Özcan-Taşkin, G.: Break-up of nanoparticle agglomerates by hydrodynamically limited processes. J. Dispers. Sci. Technol. **29**(4), 573–579 (2008)

32. Lyklema, J.: The surface tension of pure liquids. Colloids Surf. A **156**(1), 413–421 (1999). doi:10.1016/S0927-7757(99)00100-4

33. Cao, C., Zhang, L., Zhang, X.-X., Du, F.-P.: Effect of gum arabic on the surface tension and surface dilational rheology of trisiloxane surfactant. Food Hydrocoll. **30**(1), 456–462 (2013). doi:10.1016/j.foodhyd.2012.07.006

34. Birdi, K.: Handbook of surface and colloid chemistry. CRC Press, Boca Raton (2015)

35. Renard, D., Lepvrier, E., Garnier, C., Roblin, P., Nigen, M., Sanchez, C.: Structure of glycoproteins from Acacia gum: an assembly of ring-like glycoproteins modules. Carbohydr. Polym. **99**, 736–747 (2014). doi:10.1016/j.carbpol.2013.08.090

36. Beer, F.P., Johnston, E.R., DeWolf, J.T., Mazurek, D.F.: Mechanics of materials. McGraw-Hill Education, New York (2015)

37. Liu, Z., Zhang, S.-M., Yang, J.-R., Liu, J.Z., Yang, Y.-L., Zheng, Q.-S.: Interlayer shear strength of single crystalline graphite. Acta Mech. Sinica **28**(4), 978–982 (2012). doi:10.1007/s10409-012-0137-0

38. Lancaster, J.: A review of the influence of environmental humidity and water on friction, lubrication and wear. Tribol. Int. **23**(6), 371–389 (1990)

39. Liu, X., Metcalf, T.H., Robinson, J.T., Houston, B.H., Scarpa, F.: Shear modulus of monolayer graphene prepared by chemical vapor deposition. Nano Lett. **12**(2), 1013–1017 (2012). doi:10.1021/nl204196v

40. Ismail, Z., Yusoh, K.: Facile method for liquid-exfoliated graphene size prediction by UV-visible spectroscopy. AIP Conf. Proc. **1756**(1), 070002 (2016). doi:10.1063/1.4958778

41. Khan, U., O'Neill, A., Porwal, H., May, P., Nawaz, K., Coleman, J.N.: Size selection of dispersed, exfoliated graphene flakes by controlled centrifugation. Carbon **50**(2), 470–475 (2012)

5

N$_2$O interaction with the pristine and 1Ca- and 2Ca-doped beryllium oxide nanotube: a computational study

Mahdi Rezaei-Sameti[1] · **Negin Hemmati**[1]

Abstract In this study, the electrical and structural parameters of pristine and 1Ca- and 2Ca-doped beryllium oxide nanotubes (BeONTs) before and after N$_2$O adsorption are studied using density function theory (DFT). In the first step, we selected 15 models for the adsorption of N$_2$O gas on the exterior and interior surfaces of nanotube and then the considered models are optimized using the B3LYP/6-31G(d, p) level of theory. The results indicate that the adsorption processes in all the models are physisorption and are endothermic. A strong interaction between N$_2$O and 1Ca-, 2Ca-doped BeONTs increases the conductivity of nanotube, which acts a good candidate for make sensor for N$_2$O gas. The ESP analysis shows that the nanotube is relatively electron rich in N$_2$O/BeONTs complex, and the N$_2$O is relatively electron poor. With 1Ca and 2Ca doping, stabilization energy (E^2) and charge density of three oxygen atoms around the dopant decrease and the dipole moment of nanotube increases significantly from original values.

Keywords BeONTs · 1Ca- and 2Ca-doped · Adsorption of N$_2$O · DFT · ESP · DOS

Introduction

Nitrous oxide (N$_2$O) gas is a colorless, non-flammable gas, with a slightly sweet odor and taste. It is known as "laughing gas" due to the euphoric effects of inhaling it [1]. N$_2$O gas has been generated as a byproduct of nitric and adipic acids [2–4]. It is used in surgery, dentistry, as a powerful oxidizer in rocket propellants, in motor racing, and as a good solvent for many organic compounds [5–7]. N$_2$O gas extremely reacts with ozone in the stratosphere and acts as a regulator of stratospheric ozone and it is known also as air pollutant and a major greenhouse gas [8–10]. Therefore, extensive research has been carried out to control and monitor N$_2$O gas to overcome the human health and environmental concerns caused by it. Recent investigations indicated that CuO, BeO, In$_2$O$_3$, SnO$_2$, WO$_3$, and ZnCdO can be employed for monitoring and adsorbing N$_2$O gas [10, 11]. The interaction of N$_2$O gas with alkaline earth oxides [12–17], TiO$_2$ [18], molecular zeolite and metals [19–26], isolated Cu$^+$ ion [27, 28], BNNTs surfaces, Al-doped (6,0) zigzag SWCNTs, AlNNTs, and AlPNTs, Co-doped MgONTs, BeONTs, and ZnONTs have been extensively investigated in many fields [29–34]. After the discovery of carbon nanotube and studying the structural parameters and applications of it, many efforts have been focused to find other nanotubes. One of them was beryllium oxide nanotube (BeONTs). After the discovery of BeONTs by Continenza et al. in 1990 [35], Baumeier et al. and Wu et al. demonstrated that the fluorinated and hydrogenated BeO nanosheets behave as semiconductors [36–38]. The other computational studies showed that B, C, and N dopant impurity atoms decreased the energy gap of BeONTs and effectively improved the electronic structure, optical properties, magnetism, and other applications of nanotube [39–41].

✉ Mahdi Rezaei-Sameti
mrsameti@gmail.com; mrsameti@malayeru.ac.ir

[1] Department of Applied Chemistry, Faculty of Science, Malayer University, Malayer 65174, Iran

In the recent years, interaction and adsorption of H_2, CO_2, H_2S, and N_2O gas with pristine BeONTs were investigated, and these results indicated BeONTs as a good candidate for adsorbing gas and making a gas sensor [40–43]. For this aim, in the current research, the effects of 1Ca- and 2Ca-doped BeONTs on the adsorption of N_2O on both exterior and interior surfaces of nanotube are studied using density functional theory and the B3LYP/(d, p) level of theory, in order to reveal some clues for chemical sensor design. These results can provide valuable information about its interaction and reactivity of N_2O gas with pristine and Ca-doped BeO nanotube.

Computational methods

For finding stable adsorption structures, we consider many different configurations and after optimizing all configurations with small basis set, fifteen favorable configurations are selected for this study and are named AI, AII, AIII, BI, BII, BIII, CI, CII, CIII, DI, DII, DIII, EI, EII, and EIII models (see Fig. 1). Here, the A and B models are used to identify the vertical adsorption of N_2O gas from the O side and N side on the outer surface of nanotube, respectively, whereas the C and D models are used to show the horizontal adsorption of N_2O gas on the forward and backward surfaces of nanotube, respectively. The E model is utilized for the adsorption of N_2O gas on the inner surface of nanotube. In all models, indexes I, II, and III are used to depict the pristine and 1Ca- and 2Ca-doped BeONTs, respectively. Moreover, all the models are optimized by B3LYP/6-31G (d, p) basis set using the GAMESS program package [44].

The adsorption energy (E_{ads}) for N_2O gas on the surface of the pristine and 1Ca- and 2Ca-doped BeONTs can be computed using the following equation:

$$E_{ads} = E_{BeONTs-N_2O} - (E_{BeONTs} + E_{N_2O}) + BSSE, \quad (1)$$

where $E_{BeONTsN_2O}$ is the total energy of the complex consisting of N_2O gas and BeONTs, while E_{BeONTs} and E_{N_2O} are the total energies of BeONTs and N_2O gas, respectively, and BSSE is the basis set superposition error for the adsorption energy. Here, the negative and positive values of the E_{ads} indicate exothermic and endothermic reaction, respectively.

The gap energy (E_g) between HOMO and LUMO orbital and Fermi level (E_{FL}) are calculated using the following two equations:

$$E_g = E_{LUMO} - E_{HOMO} \quad (2)$$

$$E_{FL} = \frac{E_{HOMO} + E_{LUMO}}{2}. \quad (3)$$

According to Parr et al. [45, 46], the chemical potential (μ) and electronegativity (χ) are defined as follows:

$$\mu = -(I + A)/2 \quad (4)$$

$$\chi = -\mu. \quad (5)$$

According to Koopmans theorem [47], global hardness (η), the fractional number of electrons transfer (ΔN), global softness (S), and electrophilicity index (ω) can be approximated using the following equations:

$$\eta = (I - A)/2 \quad (6)$$

$$S = 1/2\eta \quad (7)$$

$$\omega = \mu^2/2\eta \quad (8)$$

$$\Delta N = -\frac{\mu}{\eta}, \quad (9)$$

where E_{HOMO} is the energy of the highest occupied molecular orbital and E_{LUMO} is the energy of the lowest unoccupied molecular orbital of the considered structure. I ($-E_{HOMO}$) is the ionization potential and A ($-E_{LUMO}$) is the electron affinity of the molecule.

The chemical shielding isotropic (CSI) and chemical shielding anisotropic (CSA) of the sites of ^7Be, ^{17}O nuclei are calculated using the following equations [48–50]:

$$CSI(ppm) = \frac{1}{3}(\sigma_{11} + \sigma_{22} + \sigma_{33}) \quad (10)$$

$$CSA(ppm) = \sigma_{33} - (\sigma_{11} + \sigma_{22})/2 \qquad \sigma_{33} > \sigma_{22} > \sigma_{11}. \quad (11)$$

Results and discussion

Structural geometry of N_2O adsorption on the pristine and Ca-doped BeONTs

The 15 optimized models (AI–EIII models) designed for the N_2O gas adsorption on the outer and inner surfaces of pristine and 1Ca- and 2Ca-doped BeONTs are depicted in Fig. 1. The length and the diameter of the pristine (4, 4) armchair BeONTs are calculated to be about 9.99 and 6.32 Å, respectively. The optimized geometries including bond lengths and bond angles for all the adsorption models are listed in Table 1. The average bond length and bond angle for the pristine BeONTs are 1.57 Å and 119°, respectively, and it is in agreement with other research studies [39, 40]. The average bond length and bond angle of Be–O and Ca–O in the AII, AIII, BI, BII, CII, CIII, DII, DIII, EII, and EIII are 1.54 and 2.23 Å, respectively (Table 1). In both cases, the Ca dopant atom protrudes a little out of the nanotube surface to occupy more space due to its relatively larger atomic radius than that of Be atom. It is also found that the adsorption persuades little local structural deformation on both the N_2O molecule and the BeONTs.

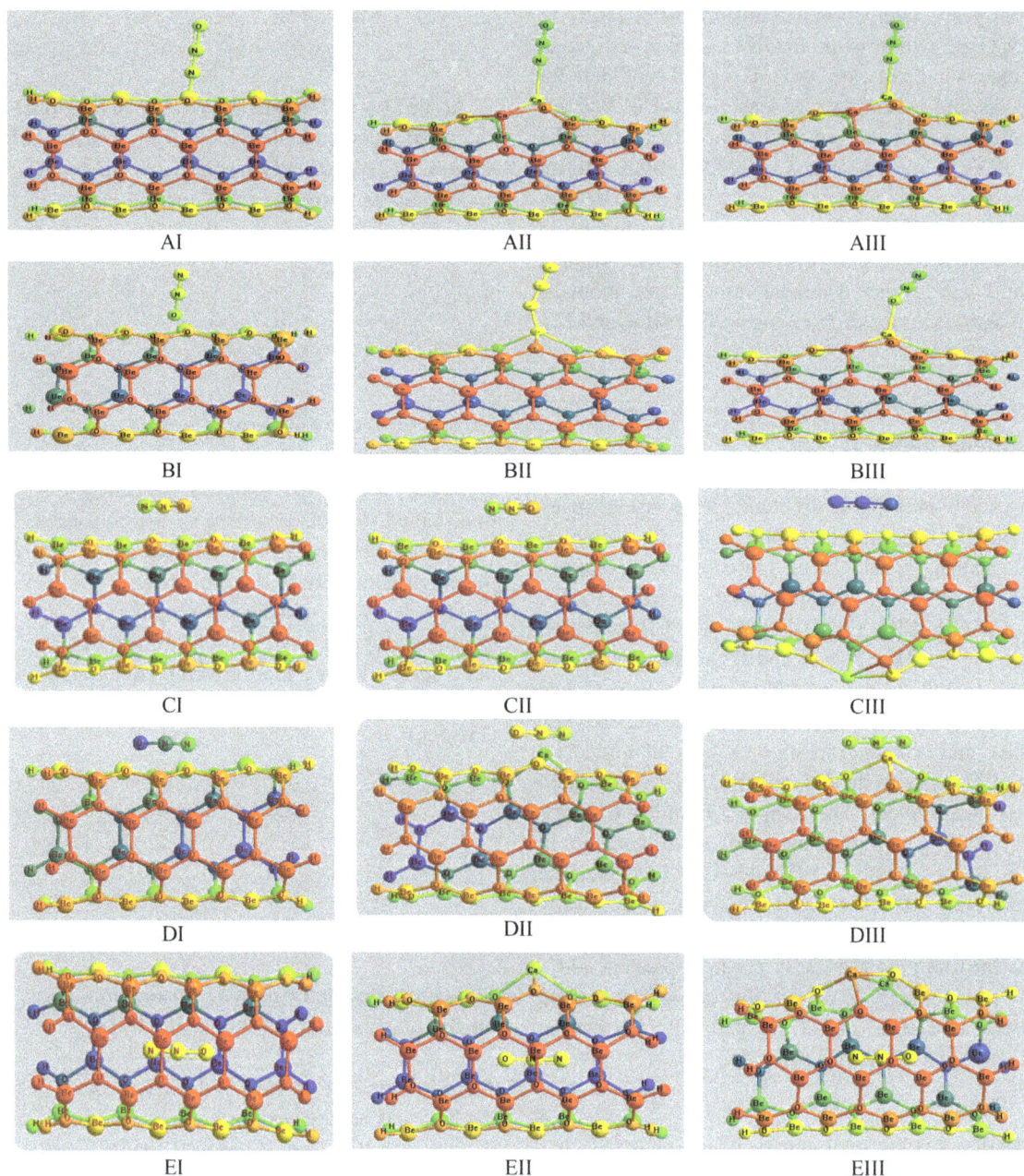

Fig. 1 2D views of the optimized structures of N_2O adsorption on the surface of pristine and 1Ca- and 2Ca-doped BeONTs (models AI–EIII)

The bond length and bond angle of the N_2O gas differ slightly between the adsorption models.

The calculated adsorption energies (E_{ads}), dipole moment (μ_D), NBO charge transfer of N_2O (Qe), and distance between N_2O and nanotube (d) for the AI to EIII models are listed in Table 2. The E_{ads} for all models is in the range of 37.73–61.91 kcal/mol; the distance between N_2O and nanotube (d) is in the range of 2.42–3.20 Å; and NBO charge transfer of N_2O (Qe) for all the models is in the range 0.04–0.08e. The BSSE values for all the models are in the range of 0.002–0.005. The positive values of E_{ads} and the small transferred charge from N_2O to the nanotube indicate that the weak adsorption of N_2O gas on the surface of BeONTs which is endothermic and also the process is called physical adsorption or physisorption. On the other hand, comparing our results with adsorbing energy of N_2O gas on the surface of CNTs [30], AlNNTs [31], and TiO_2 [33] show that the adsorption process on the surface of BeONTs is not favorable in thermodynamic approach. This result is in agreement with other research study [34].

Table 1 Bond length and bond angle of adsorption models (See Fig. 1)

Properties	AI	AII	AIII	BI	BII	BIII	CI	CII	CIII	DI	DII	DIII	EI	EII	EIII
Bond length (Å)															
Be51/Ca-O52	1.54	2.26	2.25	1.56	2.60	2.25	1.56	2.26	2.25	1.56	2.24	2.24	1.57	2.25	2.25
Be42/Ca-O32	1.54	1.59	2.28	1.55	1.59	2.28	1.55	1.59	2.28	1.56	1.59	2.28	1.55	1.59	2.29
Be51/Ca-O41	1.54	2.23	2.25	1.55	2.23	2.25	1.55	2.23	2.25	1.55	2.22	2.24	1.56	2.22	2.24
Be31/Ca-O41	1.54	1.52	1.54	1.54	1.52	1.54	1.54	1.52	1.54	1.54	1.52	1.54	1.57	1.58	1.54
Be42/Ca-O42	1.54	1.58	2.22	1.56	1.58	2.22	1.56	1.58	2.22	1.56	1.57	2.22	1.57	1.55	2.23
Be62-O62	1.54	1.57	1.62	1.56	1.57	1.62	1.56	1.57	1.62	1.56	1.54	1.55	1.57	1.55	1.55
Be31/Ca-O32	1.54	1.59	1.57	1.56	1.59	1.57	1.56	1.59	1.57	1.56	1.54	1.54	1.57	1.55	1.54
Bond angle															
<Be22-O32-Be42/Ca	119	117	105	120	117	105	119	117	105	120	117	105	120	117	105
<O52-Be51/Ca-O41	119	96	109	118	96	109	119	96	109	118	96	109	118	96	109
<O52-Be42/Ca-O32	120	123	103	121	123	103	120	123	103	121	123	103	121	123	103
<O32-Be31/Ca-O41	118	121	124	117	121	124	118	121	124	117	121	124	117	121	124
<O61-Be51/Ca-O52	118	97	101	117	97	101	118	97	101	117	97	101	117	97	101
<O42-Be42/Ca-O52	119	119	103	118	119	103	119	119	103	117	115	100	118	119	103
<O42-Be42/Ca-O32	118	115	100	117	115	100	118	115	100	118	119	103	117	115	100
<Be51/Ca-O41-Be41	114	123	119	115	123	119	114	123	119	123	115	119	123	115	119
<Be51/Ca-O41-Be31	120	113	118	121	113	118	120	113	118	121	113	118	121	113	118

Table 2 Adsorption energy (Kcal/mol), dipole moment (Debye), NBO charge of N_2O (Q_{N2O}), and three O atoms around the Ca dopant (Q_{O2}) and distance between N_2O and nanotube (d) in AI to EIII models (See Fig. 1)

Model	E_{ads}	μ_D	d (Å)	Q_{N2O}/e	Q_{O2}/e
AI	48.23	0.64	2.61	0.04	−3.59
AII	37.79	10.32	2.42	0.05	−3.98
AIII	37.73	12.43	2.46	0.04	−4.03
BI	61.91	0.66	2.46	0.06	−3.67
BII	59.84	9.22	2.58	0.06	−3.91
BIII	56.53	12.02	2.60	0.06	−4.02
CI	47.23	1.88	2.21	0.08	−3.63
CII	37.79	10.32	2.47	0.06	−3.90
CIII	37.73	12.42	2.47	0.07	−4.11
DI	49.68	0.67	2.67	0.06	−3.63
DII	48.11	5.50	2.61	0.08	−3.93
DIII	38.74	13.31	2.18	0.08	−4.06
EI	61.91	0.40	3.13	0.07	−3.69
EII	59.84	6.70	3.20	0.07	−3.97
EIII	56.53	10.28	2.93	0.07	−4.08

HOMO and LUMO orbital energy

In order to study the adsorption properties of N_2O gas on the surface of the pristine and 1Ca- and 2Ca-doped BeONTs, the electronic energies, the highest occupied molecular orbital (HOMO), and the lowest unoccupied molecular orbital (LUMO) for all the AI–EIII adsorption modes are calculated, the results of which are shown in Fig. 2. As we can see in the AI–EIII models, HOMO orbitals are uniformly distributed inward of the nanotube and LUMO orbitals are localized around N_2O gas. The positive values of NBO charge around N_2O gas is also in agreement with the density of LUMO orbital around N_2O. To study the chemical activity of the nanotube, the HOMO and LUMO energy, gap energy (E_g), and the other quantum descriptors involving the chemical potential (μ), electronegativity (χ), chemical hardness (η), chemical softness (S), and electrophilicity index (ω) are calculated and shown in Table 3 and Table S2. As seen in Table S2, the E_g of pristine and 1Ca- and 2Ca-doped BeONTs are 7.12, 4.84, and 4.81 eV, respectively, and it is notable that with 1Ca and 2Ca doping, the gap energy decreases significantly from original value and thereby increasing the conductivity of nanotube. Inspection of calculated results in Table 3 indicates that in all models, the E_g values are in the range of 3.74–6.53 eV and slightly lower than original values, as the HOMO and LUMO levels move to higher energies after N_2O adsorption.

To facilitate a more detailed study on the change in gap energy, the reduction in percent for E_g (%ΔE_g) is calculated using the following:

$$\%\Delta E_g = \frac{E_{g(Nano)} - E_{g(Nano/N_2O)}}{E_{g(Nano)}} \times 100 \qquad (12)$$

The %ΔE_g values in all the adsorption models decrease in the following order: EI > DI > BII > BIII > AII = DII > BI = CI > AIII = DIII > AI > EII > EIII > CIII > CII (see Table 3). Comparison results indicate that the %ΔE_g values in CII and CIII models (0.40 and 0.56 %, respectively) are lower than those in the other models, whereas %ΔE_g values in DI and EI (22.75 and 23.31 %, respectively) are more than those in the other models.

In order to gain a better understanding about the adsorption process, the density of state (DOS) plots and total charge density around the nanotube are determined from the output of HOMO–LUMO, and the results are shown in Fig. 3. From Fig. 3, it is found that the total charge densities of AI, BI, CI, DI, and EI models are uniformly distributed on the inner layer of the nanotube, but with 1Ca and 2Ca doping in AII, AIII, BII, BIII, CII, CIII, DII, DIII, EII, and EIII models, total charge density gets distributed into the various inner surfaces of the nanotube. Meanwhile, the total electron density and scheme of DOS plots depended on the orientations of N_2O gas on the surface of the nanotube. Comparison of the DOS plots exhibits that the electronic properties of nanotube after doping with 1Ca and 2Ca atoms and also N_2O adsorption changed significantly. It can be clearly observed that in the AII, BII, CII, DII, EII, AIII, BIII, CIII, DIII and EIII adsorption models one and two picks are generated in the gap region due to 1Ca and 2Ca dopants, and so the gap energy reduced remarkably from original values, thereby increasing the electrical conductivity of nanotube. From these results, it can be proposed that the 1Ca- and 2Ca-doped BeONTs are good candidates for making sensor for N_2O gas.

To further study the adsorption process, the quantum parameters involving global hardness, the chemical potential, and electronegativity are calculated using Eqs. (3−5) and the results are shown in Table 3 and Fig S4. As is evident from Fig S4, upon adsorbing N_2O gas, the global hardness of pristine BeONTs decreased from 3.56 to 3.26 eV and that of 1Ca- and 2Ca-doped BeONTs decreased from 2.42 to 1.95 and 2.35 to 1.99 eV, respectively. This decrease in global hardness leads to decrease in stability and increase in reactivity of the species. On the other hand, the calculated results of all adsorption models show that the electronic chemical potentials and electrophilicity of BeONTs will be decreased.

The positive values of the fractional number of electron transfer (ΔN) indicate that when N_2O gas is adsorbed on the outer and inner surfaces of the BeONTs, the charge transfer occurred from N_2O gas toward nanotube, which suggests that their electronic transport properties could be altered upon adsorptions of N_2O. According to

Fig. 2 2D views of the HOMO–LUMO structures of N$_2$O adsorption on the surface of pristine and 1Ca- and 2Ca-doped BeONTs (models AI–EIII)

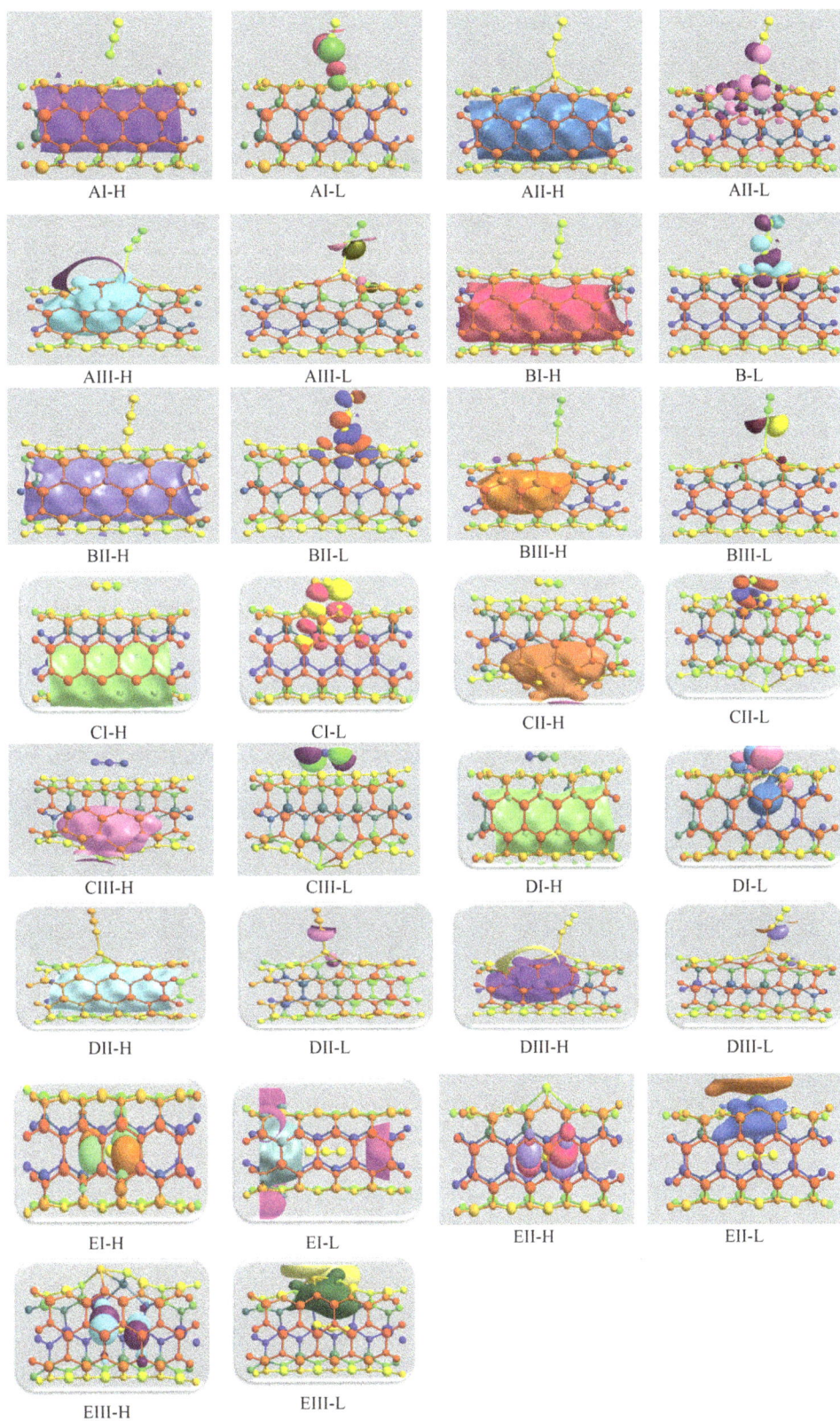

thermodynamic approach, the direction of electron flow will occur from higher chemical potential to the lower electronic chemical potential, until the electronic chemical potentials become equal. The chemical potential of the nanotube is lower than N$_2$O gas, and so the electron flow occurs from N$_2$O gas toward the nanotube. The

Table 3 Quantum parameters of AI–EIII adsorption models (See Fig. 1)

Properties/ eV	AI	AII	AIII	BI	BII	BIII	CI	CII	CIII	DI	DII	DIII	EI	EII	EIII
EHOMO	−7.69	−7.15	−7.00	−7.49	−7.17	−6.99	−7.49	−7.26	−7.01	−7.43	−7.15	−7	−7.50	−7.23	−7.02
ELUMO	−1.15	−3.25	−3.01	−1.64	−3.43	−3.23	−1.64	−2.36	−2.33	−1.92	−3.25	−3.01	−2.04	−2.41	−2.36
Eg	6.53	3.90	3.98	5.84	3.74	3.76	5.84	4.86	4.68	5.50	3.90	3.98	5.46	4.48	4.66
%ΔEg	8.28	19.42	17.25	17.97	22.72	21.82	17.97	0.40	0.56	22.75	19.42	17.25	23.31	7.43	3.11
E_{Fermi}	−4.42	−5.20	−5.01	−4.56	−5.30	−5.11	−4.56	−4.49	−4.67	−4.67	−5.20	−5.01	−4.47	−4.48	−4.69
A	1.15	3.25	3.01	1.64	3.43	3.23	1.64	2.36	2.33	1.92	3.25	3.01	2.04	2.41	2.36
I	7.69	7.15	7.00	7.49	7.17	6.99	7.49	7.22	7.01	7.43	7.15	7.00	7.50	7.23	7.02
η	3.26	1.95	1.99	2.92	1.87	1.88	2.94	2.43	2.34	2.75	1.95	1.99	2.73	2.40	2.33
S	0.15	0.26	0.25	0.17	0.27	0.27	0.17	0.21	0.21	0.18	0.26	0.25	0.18	0.21	0.21
μ	−4.42	−5.20	−5.01	−4.56	−5.30	−5.11	−4.56	−4.79	−4.67	−4.67	−5.20	−5.00	−4.77	−4.82	−4.69
$\Delta\phi$	−3.26	−1.95	−1.99	−2.92	−1.87	−1.88	−3.56	−4.72	−4.66	−2.75	−1.95	−1.99	−2.73	−2.40	−2.33
ΔN	1.35	2.66	2.51	1.56	2.83	2.72	1.60	1.97	1.99	1.7	2.66	2.51	1.74	2.00	2.01
W	2.99	6.93	6.30	3.56	7.50	6.92	4.56	4.79	4.67	3.97	6.93	6.30	4.77	4.82	4.72
χ	4.42	5.20	5.01	−7.49	−7.17	−6.99	4.56	4.79	4.67	4.67	5.20	5.00	4.77	4.82	4.69

electrophilicity index determines the maximum flow of electron from donor to acceptor species and supplies data connected to structural stability and reactivity of chemical species. It is notable that the electrophilicity index of 1Ca- (4.79 eV) and 2Ca (4.81 eV)-doped BeONTs is more than pristine (2.23 eV) due to the donor electron effect of dopant atoms. On the other hand, when N_2O gas adsorbed on the surface of BeONTs in the AI–EIII models the electrophilicity index increased significantly from the original values, indicating that the direction of the charge flow occurred from N_2O gas toward the nanotube. Meanwhile, this result emphasizes that the adsorption of N_2O gas on the surface of pristine and 1Ca- and Ca-doped BeONTs significantly alters their electronic properties and demonstrates that only a weak interaction exists between them. Here, we also investigated the changes of Fermi energy (E_{Fermi}) for adsorbing N_2O gas on the pristine and 1Ca- and 2Ca-doped BeONTs. Inspection of the calculated result in Fig S4 and Tables S2 and S3 reveals that the E_{Fermi} values of nanotube after interaction with N_2O gas reduced from −3.94 eV to −4.42, −4.56, −4.56, −4.67, and −4.47 eV in the AI, BI,CI, DI, and EI models, respectively. For 1Ca– doped ones, the E_{Fermi} values of nanotube decreased from −4.81 eV to −5.20, −5.30, and −5.20 eV in the AII, BII, and DII models, respectively. For 2Ca-doped ones, the E_{Fermi} values of nanotube reduced from −4.81 eV to −5.01, −5.11, and −5.01 eV in the AIII, BIII, and DIII models, respectively. It is notable that the E_{Fermi} values of nanotube in the CII, CIII, EII, and EIII increased slightly from the original values. The changes in the Fermi level energy demonstrate that a remarkable number of electrons transfer during the interaction between nanotube and N_2O gas; therefore, the electronic structure of the adsorption

system changed and the electrical conductance of the material thus is significantly altered.

NMR parameters

The NMR parameters involving the CSI and CAS of ^{9}Be and ^{17}O nuclei are calculated using Eqs. 10 and 11, and the results are shown in Tables S3–S14 and Fig S5. The CSI values of ^{9}Be and ^{17}O nuclei in all AI–EIII models are in the range of 108–112 and 230–253 ppm, respectively. Comparison of the results indicates that with adsorption of N_2O gas, the CSI values of nanotube increase significantly from unabsorbed models due to the donor electron effect of N_2O gas. To further study the electronic structures of nanotube/N_2O complex, the shielding density plots for XX and YY tensors are displayed in Fig S6. The results indicate that in the AI–EIII adsorption models, the electrical charges are distributed continuously around XX direction, whereas the charge density is distributed discontinuously around YY direction due to the electrical effects of N_2O gas on the surface of nanotube.

The NMR plots of O and Be atoms for all adsorption models are shown in Fig S7. According to the obtained NMR results, it can be clearly observed that the shielding parameters for O atoms in the [AI and EI] and the [BI and CI] models are in the range of 140–248 and 100–252 ppm, respectively, and for DI model it is in the range of 56–252 ppm. On the other hand, in the AII, DII, and EII models, the shielding parameters are in the range of 132–258 ppm, whereas in the BII and CII models, they are in the range of 100–258 and 88–260 ppm, respectively. And also the shielding parameters in AIII, DIII, and EIII models are in the range of 132–280 ppm, while in BIII and

Fig. 3 2D views of the DOS plot and total density electron of N₂O adsorption on the surface of pristine and 1Ca- and 2Ca-doped BeONTs (models AI–EIII)

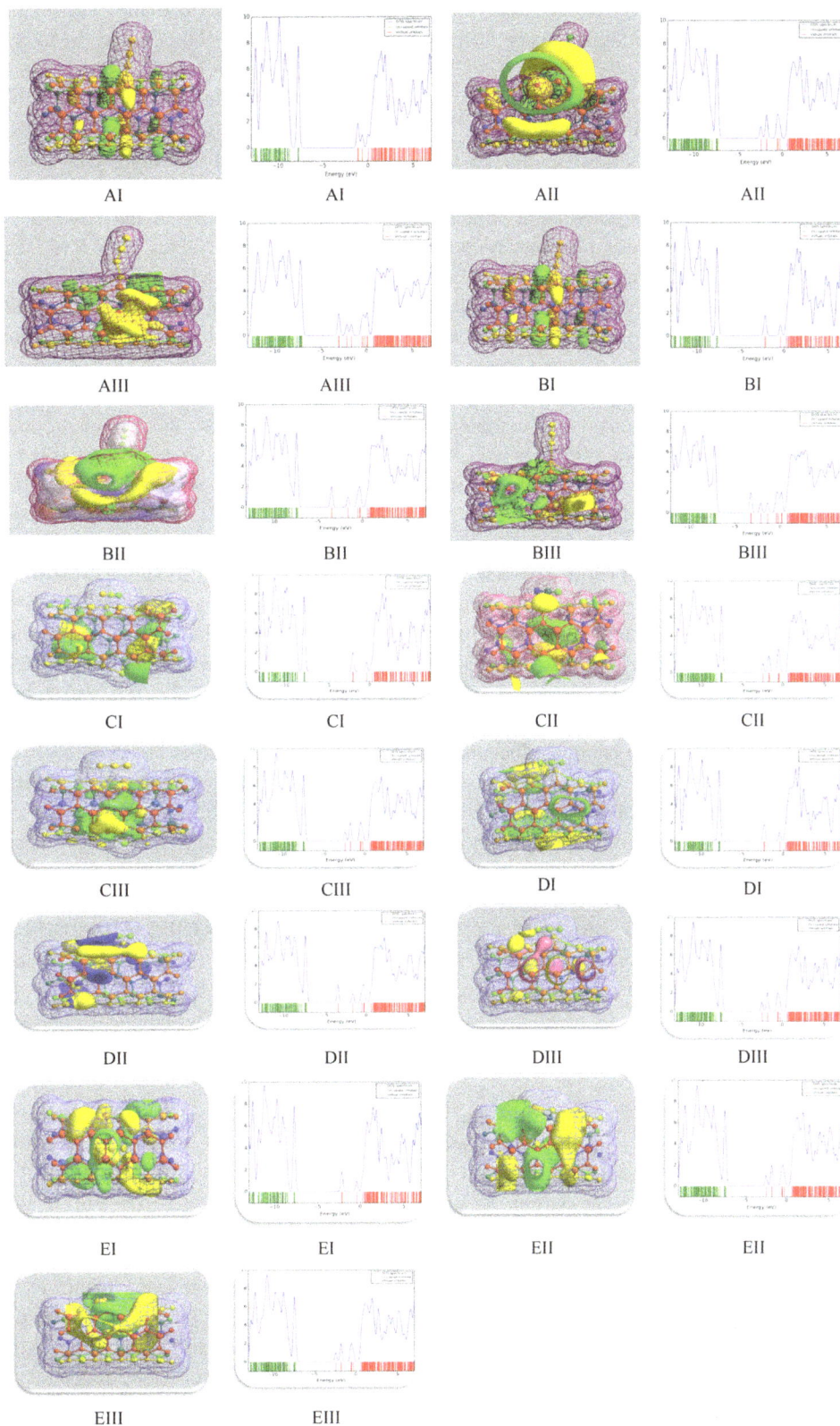

AI   AI   AII   AII

AIII   AIII   BI   BI

BII   BII   BIII   BIII

CI   CI   CII   CII

CIII   CIII   DI   DI

DII   DII   DIII   DIII

EI   EI   EII   EII

EIII   EIII

CIII models they are in the range of 100–282 and 84–282 ppm, respectively. Comparing results reveal that with doping 1Ca and 2Ca the shielding parameters values for O atoms increase significantly from original values due to donor electron effect of Ca doped. However, the shielding parameters for Be atoms in all adsorption models

(A, B, C, D, and E models) are in the range of 108.5–111 ppm, which slightly vary from original values with Ca doping.

Electrostatic potential maps

To further elucidate the adsorption process, we have calculated the electrostatic potential at selected points on the 0.0002 isodensity surfaces. Electrostatic potential maps are very useful three-dimensional diagrams of molecules. They enable us to visualize the charge distributions of molecules and charge-related properties of molecules. They also allow us to visualize the size and shape of molecules. In organic chemistry, electrostatic potential maps are invaluable in predicting the behavior of complex molecules.

Here, different colors are used to identify different potential of matter. The most negative and positive potentials are colored in red and blue, respectively. Intermediate potentials are assigned colors according to the color spectrum: Red < Orange < Yellow < Green < Blue.

Based on this scheme, one can usually identify red regions of a map as being the most electron-rich regions of a molecule and blue regions of a map as being the most electron-poor regions of a molecule. To accurately analyze the charge distribution of a molecule, a very large quantity of electrostatic potential energy values must be calculated. The best way to convey the data is to visually represent it, as in an electrostatic potential map [51–54]. Electrostatic potential (ESP) maps of all adsorption models (AI–EII) are shown in Fig. 4. Inspections of the results indicate that red regions are found on the surface of nanotube and blue regions are found on the surface of N_2O gas. This means that the nanotube is relatively electron rich in this molecule, and the N_2O is relatively electron poor.

NBO analysis

Natural bond orbital (NBO) analysis is an important technique for investigating covalent and hybridization effects in polyatomic wave functions, which is based on local block Eigen vector of the one-particle density matrix. NBO analysis provides information about interaction between both filled and virtual orbital spaces, which in turn could provide information about intra- and intermolecular interactions. For each donor (i) and acceptor (j), the stabilization energy $E^{(2)}$ is estimated from the second-order perturbation approach [55–60] as given below:

$$\Delta E_{i-j}^{(2)} = q_i \frac{\left\langle \sigma_i \left| F \right| \sigma_j \right\rangle^2}{\varepsilon_i - \varepsilon_j}, \quad (13)$$

where q_i is the donor orbital occupancy ε_i and ε_j are the diagonal elements, and F(i, j) is the off-diagonal NBO Fock matrix element. These calculations allow us to analyze the probable charge transfers and the intermolecular bond paths [56, 57]. The larger the $E^{(2)}$ value, the more intensive is the interaction, i.e., the more donation tendency from electron donors to electron acceptors and the greater the extent of conjugation in the whole system [58]. In this study, NBO analysis and second-order perturbation theory analysis of the AI–EIII models are presented in Table 4 and Tables S15–S19. The intramolecular interaction is formed by the orbital overlap between σBe62–O52 and σ*Be42–O52 bond orbital, σBe31–O41 → σ*Be41–O41, which results in intramolecular charge transfer causing stabilization of the system. The most interactions in the AI–EIII models occur in the range of 0.67–5.67 kcal/mol. A very strong interaction has been observed due to the electron density transfer from the σO42–Be52 to σ*Be32–O42 with stabilization energy of 6.58 kcal/mol for CI model. This indicates that the strongest charge transfer is responsible for interaction between N_2O gas and nanotube in this model. Furthermore, for σBe62–O52 to σ*Be42–O52 in the AIII and EIII models, due to the smaller $E^{(2)}$ of less than about 0.67 kcal/mol, weaker interaction occurred between them. These facts may be the probable reasons behind the relative stability of the axial and equatorial adsorptions of N_2O gas on the outer and inner surfaces of nanotube based on energetic data and NBO interpretation. The obtained $E^{(2)}$ values for the interaction between σBe62–O52 and σ*Be42–O52 of the studied AI, BI, CI, DI, and EI models decreased in the following order: AI > BI > CI > EI > DI; and for AII, BII, CII, DII, and EII models the values decreased in the following order: EII > CII > AII = DII > BII; moreover, for AIII, BIII, CIII, DIII, and EIII models, the values decreased in the order as follows: CIII > AIII = EIII = DIII > BIII.

It can be clearly observed that $E^{(2)}$ values in adsorption of N_2O gas on the pristine and 1Ca and 2Ca dopants decrease in the following order: pristine (I) > 1Ca-doped (II) > 2Ca-doped.

We found that with 1Ca and 2Ca atom doping, the charge transfer between σBe62–O52 and σ*Be42–O52 decreases significantly from original values; on the other hand, the orientation of N_2O adsorbed significantly influences the charge transfer. Inspection of the dipole moment (μ_D) results at Table 2 reveal that $E^{(2)}$ has the linear relation with the dipole moment inversion. On the other hand with doping one and two Ca atoms the dipole moment of systems increase significantly from original values. Therefore, dopants 1Ca and 2Ca atoms bollix the structures of nanotube thus increasing the dipole moment and decreasing the charge transfer in the σBe62–O52 and σ*Be42–O52 bond orbital. In addition, from Table 2, it can be inferred that with 1Ca and 2Ca atom doping, the NBO charge on the oxygen atoms decreases

Fig. 4 ESP plots of N$_2$O adsorption on the surface of pristine and 1Ca- and 2Ca-doped BeONTs (models AI–EIII)

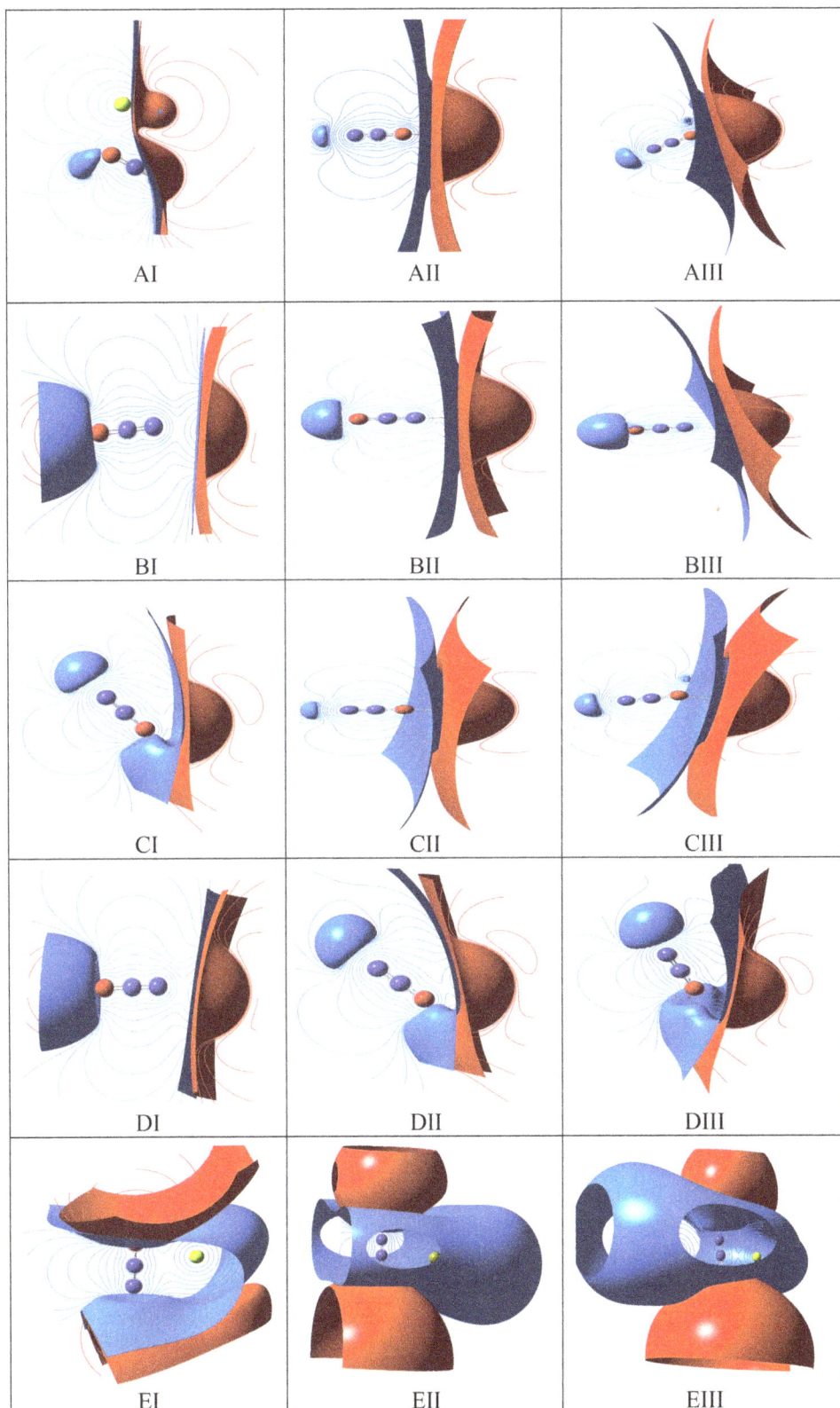

significantly due to the donor electron effect of the dopant atom, and so the Be–O bonds are highly polarized and are responsible for the large shift of charge from metal to its neighbors. It is found that there is small back-bonding interaction for Ca which account for the fact that it has large charge distribution and smaller covalent contribution, this also supports the high adsorption energy of Ca.

Table 4 NBO analysis of some important orbital interactions of studied complexes BeONTs/N2O

Structure	Donor (i)		Acceptor (j)	E^2 (Kcal/mol)	$E_j - E_i$	F(i,j)
AI	σ Be62-O52	→	σ* Be42-O52	5.62	1.17	0.072
	σ Be62-O72	→	σ* O82-Be82	1.15	0.09	0.030
AII	σ Be62-O52	→	σ* Be42-O52	4.39	1.25	0.066
	σ Be62-O72	→	σ* O82-Be82	0.93	0.98	0.027
AIII	σ Be62-O52	→	σ* Be42-O52	0.67	0.82	0.021
	σ Be62-O72	→	σ* O82-Be82	1.20	0.95	0.031
BI	σ Be62-O52	→	σ* Be42-O52	5.27	1.15	0.070
	σ Be31-O41	→	σ* Be41-O41	5.03	1.10	0.066
BII	σ Be62-O52	→	σ* Be42-O52	4.29	1.25	0.065
	σ Be31-O41	→	σ* Be41-O41	2.59	1.21	0.050
BIII	σ Be62-O52	→	σ* Be42-O52	0.51	0.81	0.018
	σ Be31-O41	→	σ* Be41-O41	3.25	1.21	0.056
CI	σ Be62-O52	→	σ* Be42-O52	5.24	1.15	0.069
	σ O42-Be52	→	σ* Be32-O42	6.58	1.17	0.079
CII	σ Be62-O52	→	σ* Be42-O52	4.65	1.24	0.068
	σ O21-Be31	→	σ* O11-Be21	3.52	1.06	0.052
CIII	σ Be62-O52	→	σ* Be42-O52	0.87	1.33	0.010
	σ O21-Be31	→	σ* O11-Be21	3.25	1.05	0.052
DI	σ Be31-O41	→	σ* Be41-O51	2.70	1.15	0.050
	σO62-Be52	→	σ* Be42-O52	4.96	1.09	0.066
DII	σ Be31-O41	→	σ* Be41-O51	3.13	1.19	0.055
	σ Be62-O52	→	σ* Be42-O52	4.39	1.25	0.066
DIII	σ Be31-O41	→	σ* Be41-O51	3.62	1.19	0.058
	σ Be62-O52	→	σ* Be42-O52	0.67	0.82	0.021
EI	σ Be62-O52	→	σ* Be42-O52	5.46	1.13	0.070
	σ Be62-O72	→	σ* O82-Be82	1.03	1.00	0.029
EII	σ Be62-O52	→	σ* Be42-O52	4.95	1.24	0.070
	σ O42-Be52	→	σ* Be32-O42	5.45	1.16	0.071
EIII	σ Be62-O52	→	σ* Be42-O52	0.67	0.81	0.021
	σ O42-Be52	→	σ* Be32-O42	4.98	1.25	0.070

Conclusions

In this research, the electrical and structural parameters of pristine and 1Ca- and 2Ca-doped BeONTs, both on the exterior and interior surfaces, before and after N_2O adsorption are investigated using density function theory. The computational results established that the adsorption process in all models is endothermic. The decrease in percentage of gap energy ($\%\Delta E_g$) confirms a strong interaction between N_2O and BeONTs, and an increase in conductivity of nanotube can propose that the 1Ca- and 2Ca-doped BeONTs are good candidates for making sensor for N_2O gas. Upon N_2O gas adsorption, the HOMO and LUMO levels moved to higher energies and the energy gap and other electrical properties of nanotube changed. Comparing results of dipole moment and the stabilization energy show that with increasing $E^{(2)}$ the dipole moment decrease and at the all adsorption models with doping 1Ca and 2Ca atoms the dipole moment increases significantly from original values.

Acknowledgments The author thanks the computational information center of Malayer University for providing the necessary facilities to carry out the research.

References

1. Overview of greenhouse gases—nitrous oxide, US EPA, 164 (2014)
2. Iwamoto, M., Hamada, H.: Removal of nitrogen monoxide from exhaust gases through novel catalytic processes. Catal. Today **10**, 57–71 (1991)
3. Kaptein, F., Rodriguez-Mirasol, J., Moulijn, J.A.: Heterogeneous catalytic decomposition of nitrous oxide. Appl. Catal. B Environ. **9**, 25–64 (1996)
4. Kondratenko, E.V., Pérez-Ramirez, J.: Micro-kinetic analysis of direct N2O decomposition over steam-activated. Fe-silicalite

from transient experiments in the TAP reactor. Catal. Today **121**, 197–203 (2007)

5. U.S. Greenhouse gas inventory report: 1990–2013, US EPA. 60 (2015)

6. CFR Part 98—revisions to the greenhouse gas reporting rule and final confidentiality US EPA. Environmental Protection Agency. (2014)

7. Steinfeld, H., Gerber, P., Wassenaar, T., Castel, V., Rosales, M., de Haan, C.: Livestock's long shadow—Environmental issues and options. (2006)

8. Nitrous oxide emissions. US Environmental Protection Agency (2016)

9. Sources of greenhouse gases. IPCC TAR WG1 2001 (2012)

10. Jeong, T.S., Yu, J.H., Mo, H.S., Kim, T.S., Youn, C.J., Hong, K.J.: Detection mechanism and characteristics of ZnO-based N2O sensors operating with photons. J. Korean Phys. Soc. **63**, 2012–2017 (2013)

11. Kanazawa, E., Sakai, G., Shimanoe, K., Kanmura, Y., Teroaka, Y., Miura, N., Yamazoe, N.: Metal oxide semiconductor N2O sensor for medical use. Sens. Actuators B **77**, 72–77 (2001)

12. Kantorovich, L.N., Gillan, M.J.: The energetics of N_2O dissociation on CaO(001). Surf. Sci. **376**, 169–176 (1997)

13. Karlsen, E.J., Nygren, M.A., Pettersson, L.G.M.: Theoretical study on the decomposition of N_2O over rocksalt metal-oxides MgO-BaO. J. Phys. Chem. A **106**, 7868–7875 (2002)

14. Xu, Y.J., Li, J.Q., Zhang, Y.F.: Conversion of N_2O to N_2 on MgO (001) surface with vacancy: a DFT study. Chin. J. Chem. **21**, 1123–1135 (2003)

15. Lu, X., Xu, X., Wang, N.Q., Zhang, Q.E.: N2O decomposition on MgO and Li/MgO catalysts: a quantum chemical study. J. Phys. Chem. B **103**, 3373–3379 (1999)

16. Karlsen, L.G., Pettersson, M.: N2O decomposition over BaO, including effects of coverage. J. Phys. Chem. B **106**, 5719–5721 (2002)

17. Snis, A., Stromberg, D., Panas, I.: N_2O adsorption and decomposition at a CaO(100) surface, studied by means of theory. Surf. Sci. **292**, 317–324 (1993)

18. Ovideo, J., Sanz, J.F.: N2O decomposition on TiO2 (110) from dynamic first-principles calculations. J. Phys. Chem. B **109**, 16223–16226 (2005)

19. Solans-Monfort, X., Sodupe, M., Branchadell, V.: Spin-forbidden N_2O dissociation in Cu–ZSM-5. Chem. Phys. Lett. **368**, 242–246 (2003)

20. Andelman, B.J., Beutel, T., Lei, G.D., Sachtler, W.M.H.: On the mechanism of selective NO_x reduction with alkanes over Cu/ZSM-5 Appl. Catal. B Environ. **11**, 1–9 (1996)

21. Chen, H.J., Matsuoka, M., Zhang, J.L., Anpo, M.: Investigations on the effect of Mn ions on the local structure and photocatalytic activity of Cu(I)-ZSM-5 catalysts. J. Phys. Chem. B **110**, 4263–4269 (2006)

22. Zhanpeisov, N.U., Ju, W.S., Matsuoka, M., Anpo, M.: Quantum chemical calculations on the structure and adsorption properties of NO and N2O on Ag+ and Cu+ ion-exchanged zeolites. Struct. Chem. **14**(3), 247–255 (2003)

23. Kachurovskaya, N.A., Zhidomirov, G.M., Hensen, E.J.M., Santen, R.A.v.: Cluster model DFT Study of the intermediates of benzene to phenol oxidation by N_2O on FeZSM-5 zeolites. Catal. Lett. **86**, 25–31 (2003)

24. Kaucky, D., Sobalik, Z., Schwarze, M., Vondrova, A., Wichterlova, B.: Effect of FeH-zeolite structure and Al-Lewis sites on N_2O decomposition and NO/NO2-assisted reaction. J. Catal. **238**(2), 293–300 (2006)

25. Heyden, A., Peters, B., Bell, A.T., Keil, F.J.: J Comprehensive DFT study of nitrous oxide decomposition over Fe-ZSM-5. Condens. Matter Mater. Surf. Interfaces Biophys. **109**(10), 4801–4804 (2005)

26. Kokalj, A., Kobal, I., Matsushima, T.: A DFT study of the structures of N2O adsorbed on the Pd (110) surface. J. Phys. Chem. B **107**(12), 2741–2747 (2003)

27. Delabie, A., Pierloot, K.: The reaction of Cu(I) (1S and 3D) with N2O: an ab initio study. J. Phys. Chem. A **106**, 5679–5685 (2002)

28. Wang, G.J., Jin, X., Chen, M.H., Zhou, M.F.: Matrix isolation infrared spectroscopic and theoretical study of the copper (I) and silver (I)–nitrous oxide complexes. Chem. Phys. Lett. **420**, 130–134 (2006)

29. Baei, M.T., Soltani, A.R., Varasteh Moradi, A., Tazikeh Lemeskic, E.: Adsorption properties of N_2O on (6,0), (7,0), and (8,0) zigzag single-walled boron nitride nanotubes: a computational study. Comput. Theor. Chem. **970**, 30–35 (2011)

30. Baei, M.T., Soltani, A.R., Varasteh Moradi, A., Moghimi, M.: Adsorption properties of N_2O on (6,0), (7,0), (8,0), and Al-doped (6,0) zigzag single-walled carbon nanotubes: a density functional study. Monatshefte Chem. **142**, 573–578 (2011)

31. Soltani, A., Ramezani Taghartapeh, M., Tazikeh Lemeski, E., Abroudi, M., Mighani, H.: A theoretical study of the adsorption behavior of N_2O on single-walled AlN and AlP nanotubes. Superlattices Microstruct. **58**, 178–190 (2013)

32. Stelmachowski, P., Zasada, F., Piskorz, W., Kotarba, A., Paul, J.F., Sojka, Z.: Experimental and DFT studies of N_2O decomposition over bare and Co-doped magnesium oxide—insights into the role of active sites topology in dry and wet conditions. Catal. Today **137**, 423–428 (2008)

33. Wanbayor, R., De, P., Frauenheim, T., Ruangpornvisuti, V.: First-principles investigation of adsorption of N2O on the anatase TiO2 (101) and the CO pre-adsorbed TiO2 surfaces. Comput. Mater. Sci. **58**, 24–30 (2012)

34. Nayebzadeh, M., Soleymanabadi, H., Bagher, Z.: Adsorption and dissociation of nitrous oxide on pristine and defective BeO and ZnO nanotubes: DFT studies. Monatshefte Chem. **145**, 1745–1752 (2014)

35. Continenza, A., Wentzcovitch, R.M., Freeman, A.J.: Theoretical investigation of graphitic BeO. Phys. Rev. B **41**, 3540–3544 (1990)

36. Baumeier, B., Kruger, P., Pollmann, J.: Structural, elastic, and electronic properties of SiC, BN, and BeO nanotubes. Phys. Rev. B **76**, 085407(1)–085407(10) (2007)

37. Wu, W., Lu, P., Zhang, Z., Guo, W.: Electronic and magnetic properties and structural stability of BeO sheet and nanoribbons. ACS Appl. Mater. Interfaces **3**, 4787–4795 (2011)

38. Wang, Y.Y.: Ding electronic structure of fluorinated and hydrogenated beryllium monoxide nanostructures. Phys. Status Solidi Rapid Res. Lett. **6**, 83–85 (2012)

39. Gorbunova, M.A., Shein, I.R., Makurin, Y.N., Ivanovskaya, V.V., Kijko, V.S., Ivanovskii, A.L.: Electronic structure and magnetism in BeO nanotubes induced by boron, carbon and nitrogen doping, and beryllium and oxygen vacancies inside tube. Phys. E **41**, 164–168 (2008)

40. Seif, A., Zahedi, E.: A DFT studies of structural and quadrupole coupling constants properties in C-doped BeO nanotubes. Superlattices Microstruct. **50**, 539–548 (2011)

41. Fathalian, A., Kanjouri, F., Jalilian, J.: BeO nanotube bundle as a gas sensor. Superlattices Microstruct. **60**, 291–299 (2013)

42. Ahmadi Peyghan, A., Yourdkhani, S.: Capture of carbon dioxide by a nanosized tube of BeO: a DFT study. Struct. Chem. **25**, 419–426 (2014)

43. Ahmadaghaei, N., Noei, M.: Density functional study on the sensing properties of nano-sized BeO tube toward H2S. J. Iran. Chem. Soc. **11**, 725–731 (2014)

44. Schmidt, M.W., Baldridge, K.K., Boatz, J.A., Elbert, S.T., Gordon, M.S., Jensen, J.H., Koseki, S., Matsunaga, N., Nguyen,

K.A., Su, S.J., Windus, T.L., Dupuis, M., Montgomery, J.A.: General atomic and molecular electronic structure system. J. Comput. Chem. **14**, 1347–1363 (1993)

45. Parr, R.G., Donnelly, R.A., Levy, M., Palke, W.E.: Electronegativity the density functional viewpoint. J. Chem. Phys. **68**, 3801–3807 (1978)

46. Parr, R.G., Szentpaly, L., Liu, S.: Electrophilicity index. J. Am. Chem. Soc. **121**, 1921–1924 (1999)

47. Koopmans, T.: Über die Zuordnung von Wellenfunktionen und Eigenwerten zuden Einzelnen Elektronen Eines Atoms. Physica **1**, 104 (1933)

48. Rezaei-Sameti, M., Samadi Jamil, E.: The adsorption of CO molecule on pristine, As, B, BAs doped (4,4) armchair AlNNTs: a computational study. J. Nanostruct. Chem. **6**, 197–205 (2016)

49. Rezaei Sameti, M., Kazmi, A.: A computational study on the interaction between O$_2$ and pristine and Ge-doped aluminum phosphide nanotubes. Turk. J. Phys. **39**, 128–136 (2015)

50. Rezaei-Sameti, M., Yaghoobi, S.: Theoretical study of adsorption of CO gas on pristine and AsGa-doped (4, 4) armchair models of BPNTs. Comput. Condens. Matter **3**, 21–29 (2015)

51. Scrocco, E., Tomasi, J.: Electronic molecular structure, reactivity and intermolecular forces: an euristic interpretation by means of electrostatic molecular potentials. Adv. Quantum Chem. **103**, 115–193 (1978)

52. Luque, F.J., Lopez, J.M., Orozco, M.: Perspective on "Electrostatic interactions of a solute with a continuum. A direct utilization of ab initio molecular potentials for the prevision of solvent effects". Theor. Chem. Acc. **103**, 343–345 (2000)

53. Scrocco, E., Tomasi, J.: The electrostatic molecular potential as a tool for the interpretation of molecular properties, the series Topics in Current Chemistry Fortschritte der Chemischen. Forschung **42**, 95–170 (2005)

54. Li, Y., Liu, Y., Wang, H., Xiong, X., Wei, P., Li, F.: Synthesis, crystal structure, vibration spectral, and DFT studies of 4-aminoantipyrine and its derivatives. Molecules **18**, 877–893 (2013)

55. Foster, J.P., Weinhold, F.: Natural hybrid orbitals. J. Am. Chem. Soc. **102**, 7211–7218 (1980)

56. James, C., Amal Raj, A., Raghunathan, R., Hubert Joe, I., Jayakumar, V.S.: Structural conformation and vibrational spectroscopic studies of 2,6-bis(p-N, N-dimethyl benzylidene) cyclohexanone using density functional theory". J. Raman Spectrosc. **12**, 1381–1392 (2006)

57. Politzer, P., Truhlar, D.G.: Chemical applications of atoms and molecular electrostatic potential. Plessum Press, New York (1981)

58. Subash chandrabose, S., Krishnana, A.R., Saleem, H., Paramashwari, R., Sundaraganesan, N., Thanikachalam, V., Maikandan, G.: Vibrational spectroscopic study and NBO analysis on bis (4-amino-5-mercapto-1, 2, 4-triazol-3-yl) methane using DFT method. Spectrochim. Acta **77A**, 877–884 (2010)

59. Arjunan, V., Raj, A., Ravindran, P., Mohan, S.: Structure–activity relations of 2-(methylthio) benzimidazole by FTIR, FT-Raman, NMR, DFT and conceptual DFT methods. Mol. Biomol. Spectrosc. **118**, 951–965 (2014)

60. Liu, J.N., Chen, Z.R., Yuan, S.F., Zhejiang, J.: Study on the prediction of visible absorption maxima of azobenzene compounds. J. Zhejiang Univ. Sci. B **6**, 584–589 (2005)

6

The effect of C, Si, N, and P impurities on structural and electronic properties of armchair boron nanotube

Farzad Molani[1]

Abstract The structural and electronic properties of non-metal atoms (X = C, N, Si, P) doped (6,0) boron nanotube (BNT) have been considered in a systematic study by performing periodic spin polarized density functional theory (DFT) calculations. The studies showed that cylindrical shape of the nanotube is changed by doping, except for C substitution. Notably, all the substitution processes are endothermic and the C-substituted becomes energetically more stable than the other dopants. It is revealed that the C-doped BNT is semi-metal, whereas the nanotube in the presence of the other dopants remains semiconductor. Hence, the substitution is an effective way in narrowing band gap of the nanotube. Doping by N atom changes the band gap from indirect to direct, which can be suitable for optical applications. Thus, electronic structure of the tube has been controlled by type of the dopant atoms. Our study predicted that the BNTs in the presence of the dopants are promising candidate as interconnects for nano-devices as well as field emission devices.

Keywords Density functional theory · Boron nanotube · Impurity · Electronic structure

Introduction

Boron atom is one of the appealing and challenging elements to create novel structures [1]. Boron nanostructures (BNS) have been received special attention as a result of their innovative uses in medicine [2], superconductivity [3], hydrogen storage [4–8], and electrical interconnects [9, 10]. Up to date, many experimental efforts have been performed aiming to synthesize the BNSs [11–15]. Boron nanotube (BNT) was originally predicted for the first time by Boustani et al. [16] and then the BNTs have been investigated in many experimental and theoretical studies [3–5, 10–13, 17–21].

The previous studies confirmed that, hexagonal and triangular patterns were more stable than a buckled triangular or hexagonal boron sheet, which two and three center bonds of these sheets (nominated as α-sheet) are main reasons for their stability [17, 18]. Recently, tetragonal, hexagonal and dodecagonal cycles for carbon, boron nitride (BNNT) and aluminum nitride nanotubes have been proposed [22]. Theoretical studies also showed that a series of the BNTs may be obtained by rolling up the stable α-sheet [17, 18]. In all these studies, researchers have found out that the surface buckling is a common phenomenon for the BNTs. Unlike carbon nanotube (CNT) [23], silicon carbide nanotube (SiCNT) [24], C_3N nanotube [25, 26], and g-C_3N_4 nanotube [27], the BNTs based on the α-sheet are predicted to be metallic [20, 21, 28–30] and semiconductor for small diameters (diameter <17 Å) [1, 17–20].

The structural and electronic properties of structures can be modified by chemical doping. Many experimental and theoretical investigations on substitution of non-metal impurity in CNT [31], boron nitride nanotube [32], SiCNT [33], phosphorene [34], TiO_2 [35], and BNT [36–38], encouraged us to explore the properties of the BNT in the presence of non-metal impurity atoms. In the BNT case, it is noted that the previous studies [36–38] have been carried out only on hexagonal form (as an unstable form). Our main concerns on the stable form are listed as follows. (1) What is the effect of C, Si, N, and P atoms on the electronic

✉ Farzad Molani
f.molani@iausdj.ac.ir; fmolani@dena.kntu.ac.ir

[1] Department of Chemistry, Sanandaj Branch, Islamic Azad University, P. O. Box 618, Sanandaj, Iran

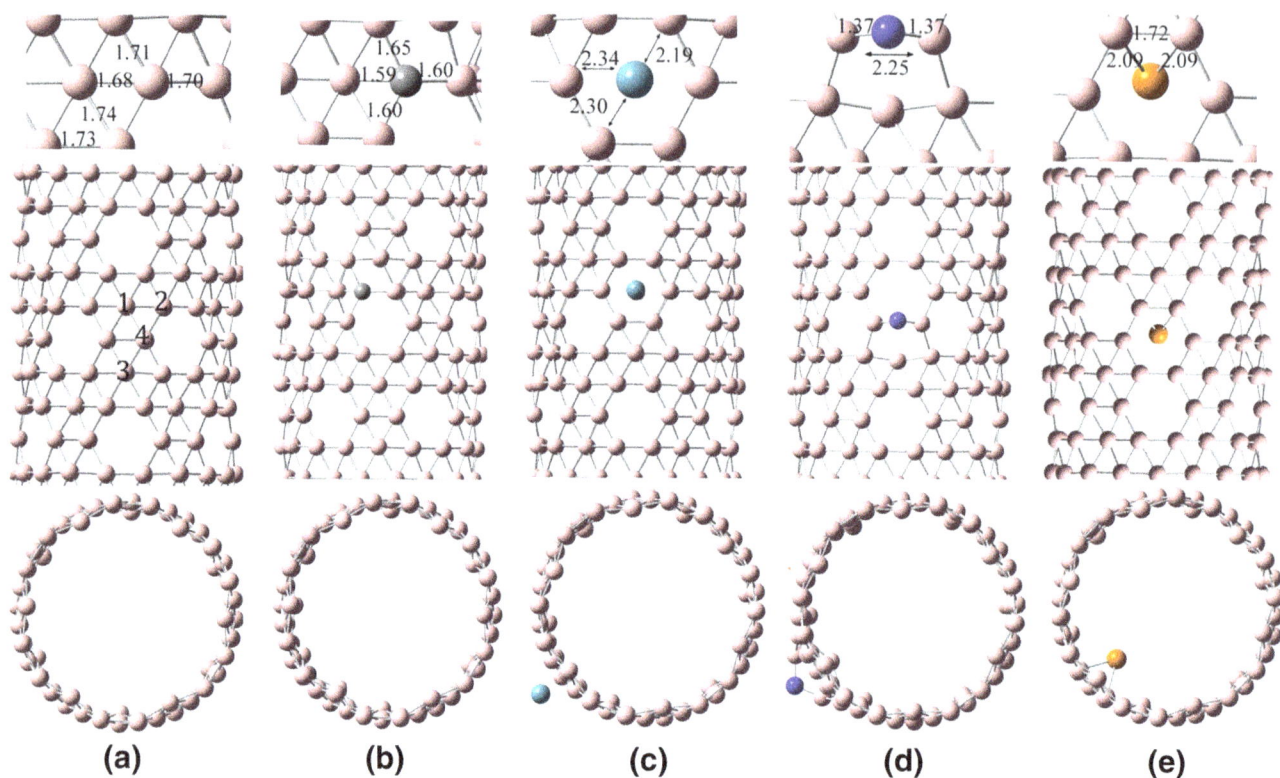

Fig. 1 Optimized structures of **a** perfect BNT, **b** C-doped BNT, **c** Si-doped BNT, **d** N-doped BNT, and **e** P-doped BNT. The tubes have been shown in different views. *Pink, gray, cyan, blue,* and *orange colors* represent B, C, Si, N, and P atoms, respectively

properties of the BNT? (2) Do these atoms change the structural properties of the BNT? (3) What is the best site for doping and which substitution has the minimum formation energy (E_{Form})?

Computational methods

The spin-polarized DFT calculations were performed by means of the Quantum ESPRESSO package [39] on (6,0) armchair BNT. A (p,0) BNT corresponds to a (n,n) (n = p) CNT and a (p,p) BNT corresponds to a (n,o) (n = 3p) CNT [17]. The system is represented by a tetragonal supercell of 25 Å × 25 Å × 17.79 Å, containing 192 atoms for the perfect BNT, which the distance of the impurity adjacent images in *z*-axis is suitable. The distance between the outer walls of neighbor periodic images of the tube is at least 25 Å to negligible the interaction between the tubes. The Monkhorst–Pack meshes [40] of 1 × 1 × 7 were used for sampling the Brillouin zone and the tetrahedron method [41] was also used for the calculation of density of state (DOS) to achieve high accuracy. A plane-wave basis set with a cutoff value of 600 eV was used in combination with ultrasoft pseudopotentials. All structures were fully relaxed until the forces on each atom were less than

0.01 eV Å^{-1}. To further analyze the electronic charge population analysis, we have performed the analysis with the Löwdin method [42]. Losing and gaining charge density are represented by negative and positive values for charge transfer (CT), respectively. Further, work function (Φ) is defined as the minimum energy required to remove an electron from the surface to a point in the vacuum [43].

Results and discussion

Full geometry optimization without any constraints on the (6,0) BNT was performed. The optimized structures are depicted in Fig. 1. The average calculated B–B bond length and diameter of the nanotube are about 1.71 and 9.5 Å, respectively. These values are in good agreement with the previous reports [17, 18]. As shown in Fig. 1a, buckling pattern is observed for the nanotube: the two atoms in the triangular regions become inequivalent with one moving radially outward (B atom labeled B1) and the other moving radially inward (B atom labeled B3). Based on Löwdin charge analysis [42], we found that the BNT is an amphoteric molecule, containing acidic sites [B atoms which lose charge like B1 (+0.19 |e|), B3 (+0.20 |e|), and B4 (+0.03 |e|)] and basic sites [B atoms which gain charge

Table 1 Calculated structural and electronic properties of the nanotubes

	Pure BNT	C-doped BNT	Si-doped BNT	N-doped BNT	P-doped BNT
d_{min} (Å)	–	1.6	2.19	1.37	2.09
E_{form} (eV)	–	+0.28	+1.85	+2.00	+2.86
ΔQ_X (e)*	–	−0.17	+0.61	−0.24	+0.53
Prot (Å)	–	0	+1.3	+2.0	−1.2
E_g (eV)	0.36-indirect	Semi-metal	0.27-indirect	0.28-direct	0.33-indirect
μ_B (Bohr mag/cell)	0.00	0.1	0.3	0.0	0.0
Φ (eV)	2.854	2.504	2.787	2.913	2.839

The d_{min}, E_{form}, ΔQ_X, Prot, E_g, μ_B, and Φ are the distance between impurity atom with the nearest B atom, the formation energy of the nanotubes, the variation of Lowdin charge on impurity atom, the protrusion of the impurity atom with respect to the surface of the tube, the band gap of the nanotube, the total magnetization, and the work function of the nanotube

* ΔQ_X (e) = Q_X (in atomic form) − Q_X (after doping)

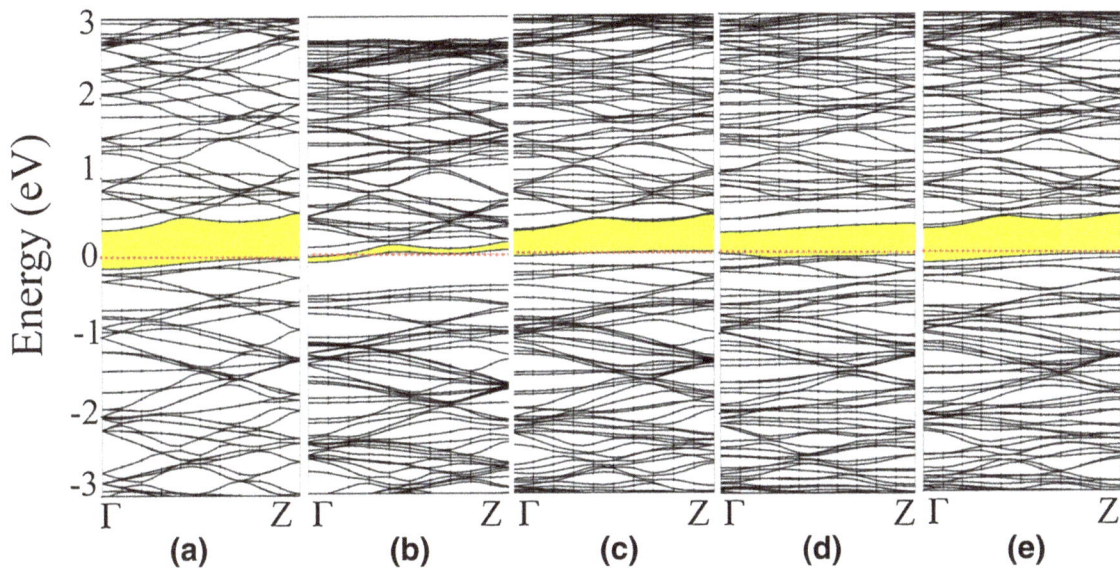

Fig. 2 Band structures of **a** perfect BNT, **b** C-doped BNT, **c** Si-doped BNT, **d** N-doped BNT, and **e** P-doped BNT. The band gaps have been assigned by *green color*

like B2 (−0.03 |e|)]. As a result of dissimilarity in the B atoms, four various positions for further analysis were considered.

Let us consider first the structural properties of the doped BNT. For the doped BNT, one B atom is substituted by another atom in the supercell. As shown in Fig. 1, C atom retains on the surface of the tube, while the other dopants will leave the surface. As shown in Table 1, the minimum distance (d_{min}) between C, Si, N, and P with the nearest B atom is found to be 1.60, 2.19, 1.37, and 2.09 Å, respectively. Due to large atomic radius of Si and P atoms, the long Si–B and P–B bond length are expected. Furthermore, Si, N, and P in the doped BNT are moving radially outward, outward, and inward, respectively. The protrusions of the dopant atom with respect to the surface of the tube are 2.0, 1.30, and −1.20 Å, for N, Si, and P atoms, respectively (the minus value is represented for

inward moving). The drastic changes in the geometric structure of BNT can be seen in Fig. 1d where the N atom, as smaller study atoms, can locate among two B atoms and bonds with pyramidal like configurations, with bond angles close to 99.4°. The N-B bond length is quite smaller than the B–B bond.

To probe the relative stability of the X-doped BNT, the computational methods of the doped systems were estimated. The enhancement of energetic stability is supported by the E_{form}. The E_{form} of the most stable configurations is presented in Table 1. Through the total energy calculation, the E_{Form} are computed using:

$$E_{Form} = E_{[impurity\ tube]} - E_{[pure\ tube]} + \mu_B - \mu_X, \quad (1)$$

where $E_{[impurity\ tube]}$ and $E_{[pure\ tube]}$ are the total energy of the doped tube with the impurity atoms and the undoped case, respectively. Further, the μ_B and μ_X are the chemical potentials for B and X (C, Si, N, and P) atoms, respectively. The μ_B, μ_C, μ_{Si}, μ_N, and μ_P are calculated as the total energy per atom in the rhombohedral structure of α-boron, pure graphene, Si bulk, gaseous N_2 molecule, and P_4 molecule, respectively. The positive value indicates that the process is endothermic. Due to unique structure of the BNT (hexagonal and triangular pattern), endothermic process can be expected. Endothermic processes for doping with non-metal atoms in different structures were obtained [44–47]. The most favorite site for C, Si, N, and P atoms is 2, 1, 1, and 3 sites, respectively. As a result, the E_{form} depends on the occupation sites. It is noted that the doping with C atom is energetically more favorable the other dopants. The results showed that the doping with P atom is energetically more expensive than the doping with Si and N atoms which need to a high external energy. Our investigations showed that four parameters may be affected on stability of the tubes. These parameters are the number of electrons in the dopants, structural distortion of the tubes, the magnetic moment of the structures, and the amount of charge variation between the guest and host atoms. C and Si atoms have an extra valence electron with respect to B atom, while the other dopants have two more electros. It seems that the BNT does not have a good ability to accept two more electrons in the system. A small distortion, as another criterion for stability, has been observed in C- and Si-doped BNT systems. An important aspect of non-metal-doped BNT is the magnetic behavior of the system. The results of magnetic moments are also presented in Table 1. From the results, we can see that no magnetism is observed when N and P atoms are doped in the tube, while magnetic moment is observed after doping with C and Si atoms. This evidence can be explained by the number of electrons in the system (odd numbers of electrons in the N- and P-doped BNT). Unlike Si atom, C atom located at the surface of the nanotube; thus, C atom can hybridize with their neighboring atoms, showing small diamagnetic ordering. Interestingly, the tubes with large magnetic moment are favorable than the other systems. As an important result, the numbers of valence electrons, atomic radius of the dopants, the amounts of structural distortion, and total magnetic moment have synergic effects on the E_{form}.

The electronic properties of small gap semiconducting (6,0) pure and impurity BNTs were also studied, as summarized in Table 1. Based on Löwdin analysis, the elements of the second (third) row of the periodic table gain (lose) charges. This charge variation can be explained by Pauling electronegativity. Although Si atom has higher electronegativity with respect to B atom, the B atoms in the BNT have more tendencies toward electrons respect than B in atomic form.

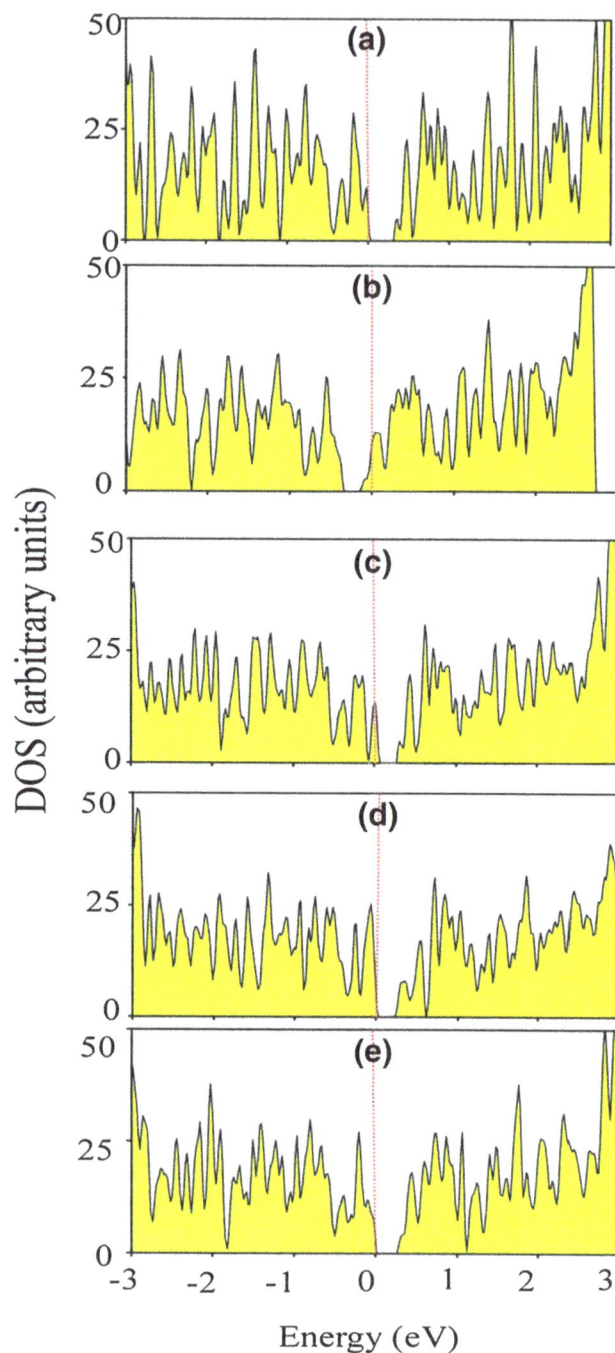

Fig. 3 Density of states of a perfect BNT, b C-doped BNT, c Si-doped BNT, d N-doped BNT, and e P-doped BNT. The Fermi level is shown by *dashed red line*. The surfaces of the valence and conduction regions are shown by *green color*

In addition, the electronic band structures and DOSs of the structures were also studied, as shown in Figs. 2 and 3. After doping with the atoms, several impurity states below the Fermi level (FL) will appear which can easily donate electron to conduction band minimum (CBM). The BNT with a donor impurity is known as n-type semiconductor. In the case of Si-, N- and, P-doped systems, the FL moves to valence band maximum (VBM) and some impurity states are introduced between the VBM and the CBM. These impurity states are filled (in the case of N and P) or present unpaired electrons (in the case of C and Si). As a result of these new states, charge carrier would become easier to transport from the VBM to the CBM. These transfers make a hole in the states under the FL. In other words, the conductivity of the BNT is enhanced by the impurity atoms.

The results of Table 1 revealed that the C, Si, and P can increase the field emission current of BNT by reducing the work function (Φ). In semiconductors, the field emission strongly depends on the Φ. By definition, the Φ (computed here as the difference between the energies of FL and the vacuum region) is the minimum energy needed by one electron to escape the surface. The inverse relation between Φ and electron emission from the semiconductor surface in the presence of an electric field can be observed [43]. We believe that the electronic and chemical properties of semiconductor BNT can alter in the presence of the doped atoms; therefore, it could be potentially used in several applications.

Conclusion

In summary, a first-principles method has been applied to investigate the influence of non-metal (C, Si, N, and P atoms) impurities on the stability and electronic properties of the BNT. The obtained results showed that the number of electrons in the dopants, structural distortion, the amount of charge variation between the guest and the host atoms, and the magnetic moment of the structures have impact role in the E_{form}. Doping with C atom is energetically more favorable than the other dopants. The results showed that the doping with Si and N atoms needs to a high external energy and doping with P atom is more expensive energetically. The calculations indicated that the performance of BNT in field emission devices should be affected by non-metal impurities, and the magnetic moment is very sensitive to the number of electrons of guest atom. The calculated results showed that the substitution of B atom by the non-metal may induce change in the electronic structure, where the band gaps are narrowed for all the doped systems. Our study proposes that the BNT in the presence of the doped atom is potentially useful for spintronic applications and the development of magnetic nanostructures.

References

1. Lau, K.C., Pandey, R.: Ab initio study of single-walled boron nanotubes. In: Saxena, S. (ed.) Handbook of Boron Nanostructures, p. 49. Pan Stanford Publishing, Danvers (2016)
2. Sethia, K., Roy, I.: Applications of boron nanostructures in medicine. In: Saxena, S. (ed.) Handbook of Boron Nanostructures, p. 101. Pan Stanford Publishing, Danvers (2016)
3. Zhao, Y., Zeng, S., Ni, J.: Superconductivity in two-dimensional boron allotropes. Phys. Rev. B **93**(1), 014502 (2016)
4. Li, M., Li, Y., Zhou, Z., Shen, P., Chen, Z.: Ca-coated boron fullerenes and nanotubes as superior hydrogen storage materials. Nano Lett. **9**(5), 1944–1948 (2009)
5. An, H., Liu, C.-S., Zeng, Z.: Radial deformation-induced high-capacity hydrogen storage in Li-coated zigzag boron nanotubes. Phys. Rev. B **83**(11), 115456 (2011)
6. Li, F., Wei, W., Sun, Q., Yu, L., Huang, B., Dai, Y.: Prediction of single-wall boron nanotube structures and the effects of hydrogenation. J. Phys. Chem. C **121**(10), 5841–5847 (2017)
7. Wang, J., Du, Y., Sun, L.: Ca-decorated novel boron sheet: a potential hydrogen storage medium. Int. J. Hydrog. Energy **41**(10), 5276–5283 (2016)
8. Molani, F., Jalili, S., Schofield, J.: Al-doped B$_{80}$ fullerene as a suitable candidate for H$_2$, CH$_4$, and CO$_2$ adsorption for clean energy applications. J. Saudi Chem. Soc. (2017). doi:10.1016/j.jscs.2017.07.001
9. Lin, B., Dong, H., Du, C., Hou, T., Lin, H., Li, Y.: B$_{40}$ fullerene as a highly sensitive molecular device for NH$_3$ detection at low bias: a first-principles study. Nanotechnology **27**(7), 075501 (2016)
10. Baňacký, P., Noga, J., Szöcs, V.: Electronic structure of boron nanotubes: perspective material for nanotechnologies. Quantum Matter **4**(4), 367–372 (2015)
11. Liu, F., Shen, C., Su, Z., Ding, X., Deng, S., Chen, J., Xu, N., Gao, H.: Metal-like single crystalline boron nanotubes: synthesis and in situ study on electric transport and field emission properties. J. Mater. Chem. **20**(11), 2197–2205 (2010)
12. Ciuparu, D., Klie, R.F., Zhu, Y., Pfefferle, L.: Synthesis of pure boron single-wall nanotubes. J. Phys. Chem. B **108**(13), 3967–3969 (2004)
13. Patel, R.B., Chou, T., Iqbal, Z.: Synthesis of boron nanowires, nanotubes, and nanosheets. J. Nanomater. **2015**, 14 (2015)
14. Mannix, A.J., Zhou, X.-F., Kiraly, B., Wood, J.D., Alducin, D., Myers, B.D., Liu, X., Fisher, B.L., Santiago, U., Guest, J.R.: Synthesis of borophenes: anisotropic, two-dimensional boron polymorphs. Science **350**(6267), 1513–1516 (2015)
15. Zhai, H.-J., Zhao, Y.-F., Li, W.-L., Chen, Q., Bai, H., Hu, H.-S., Piazza, Z.A., Tian, W.-J., Lu, H.-G., Wu, Y.-B.: Observation of an all-boron fullerene. Nat. Chem. **6**(8), 727–731 (2014)
16. Boustani, I., Quandt, A.: Nanotubules of bare boron clusters: Ab initio and density functional study. EPL **39**(5), 527 (1997)
17. Yang, X., Ding, Y., Ni, J.: Ab initio prediction of stable boron sheets and boron nanotubes: structure, stability, and electronic properties. Phys. Rev. B **77**(4), 041402 (2008)
18. Tang, H., Ismail-Beigi, S.: First-principles study of boron sheets and nanotubes. Phys. Rev. B **82**(11), 115412 (2010)
19. Singh, A.K., Sadrzadeh, A., Yakobson, B.I.: Probing properties of boron α-tubes by ab initio calculations. Nano Lett. **8**(5), 1314–1317 (2008)
20. Bezugly, V., Kunstmann, J., Grundkötter-Stock, B., Frauenheim, T., Niehaus, T., Cuniberti, G.: Highly conductive boron nanotubes: transport properties, work functions, and structural stabilities. ACS Nano **5**(6), 4997–5005 (2011)

21. Gonzalez Szwacki, N., Tymczak, C.: The symmetry of the boron buckyball and a related boron nanotube. Chem. Phys. Lett. **494**(1), 80–83 (2010)

22. BabaeiPour, M., Safari, E.K., Shokri, A.: First-principles study of nanotubes within the tetragonal, hexagonal and dodecagonal cycle structures. Phys. E **86**, 129–135 (2017)

23. Odom, T.W., Huang, J.-L., Kim, P., Lieber, C.M.: Structure and electronic properties of carbon nanotubes. J. Phys. Chem. B **104**(13), 2794–2809 (2000)

24. Molani, F., Jalili, S., Schofield, J.: A computational study of platinum adsorption on defective and non-defective silicon carbide nanotubes. Monatsh. Chem. **146**(6), 883–890 (2015)

25. Jalili, S., Molani, F., Akhavan, M., Schofield, J.: Role of defects on structural and electronic properties of zigzag C_3N nanotubes: a first-principle study. Phys. E **56**, 48–54 (2014)

26. Molani, F., Jalili, S., Schofield, J.: A novel candidate for hydrogen storage: Ca-decorated zigzag C_3N nanotube. Int. J. Hydrog. Energy **41**(18), 7431–7437 (2016)

27. Jalili, S., Molani, F., Schofield, J.: First principles study on energetic, structural, and electronic properties of defective g-C_3N_4-zz3 nanotubes. J. Theor. Comput. Chem. **13**(04), 1450021 (2014)

28. Lau, K.C., Pati, R., Pandey, R., Pineda, A.C.: First-principles study of the stability and electronic properties of sheets and nanotubes of elemental boron. Chem. Phys. Lett. **418**(4), 549–554 (2006)

29. Kunstmann, J., Quandt, A.: Broad boron sheets and boron nanotubes: an ab initio study of structural, electronic, and mechanical properties. Phys. Rev. B **74**(3), 035413 (2006)

30. Boustani, I., Quandt, A., Hernández, E., Rubio, A.: New boron based nanostructured materials. J. Chem. Phys. **110**(6), 3176–3185 (1999)

31. Ye, J., Shao, Q., Wang, X., Wang, T.: Effects of B, N, P and B/N, B/P pair into zigzag single-walled carbon nanotubes: a first-principle study. Chem. Phys. Lett. **646**, 95–101 (2016)

32. Li, K., Ye, J., Zhang, J., Wang, X., Shao, Q.: The stability and electronic structures of Si/O/Al/P atom doped (5, 0) boron nitrogen nanotubes with Stone-Wales defects: density functional theory studies. Phys. E **87**, 112–117 (2017)

33. Gali, A.: Ab initio study of nitrogen and boron substitutional impurities in single-wall SiC nanotubes. Phys. Rev. B **73**(24), 245415 (2006)

34. Boukhvalov, D.W.: Atomic and electronic structure of nitrogen- and boron-doped phosphorene. Phys. Chem. Chem. Phys. **17**(40), 27210–27216 (2015)

35. Di Valentin, C., Pacchioni, G.: Trends in non-metal doping of anatase TiO_2: B, C, N and F. Catal. Today **206**, 12–18 (2013)

36. Jain, S.K., Srivastava, P.: Structural stability of nitrogen-doped ultrathin single-walled boron nanotubes: an Ab initio study. Appl. Nanosci. **2**(3), 345–349 (2012)

37. Jain, S.K., Srivastava, P.: Electronic properties of C-substituted boron nanotubes: a first principles study. EPJ B **86**(7), 1–7 (2013)

38. Jain, S.K., Srivastava, P.: Effect of nitrogen impurity on electronic properties of boron nanotubes. Adv. Cond. Matter Phys. **2014**, 706218 (2014)

39. Giannozzi, P., Baroni, S., Bonini, N., Calandra, M., Car, R., Cavazzoni, C., Ceresoli, D., Chiarotti, G.L., Cococcioni, M., Dabo, I.: QUANTUM ESPRESSO: a modular and open-source software project for quantum simulations of materials. J. Phys. Condens. Matter **21**(39), 395502 (2009)

40. Monkhorst, H.J., Pack, J.D.: Special points for Brillouin-zone integrations. Phys. Rev. B **13**(12), 5188–5192 (1976)

41. Blöchl, P.E., Jepsen, O., Andersen, O.K.: Improved tetrahedron method for Brillouin-zone integrations. Phys. Rev. B **49**(23), 16223 (1994)

42. Löwdin, P.O.: On the non-orthogonality problem connected with the use of atomic wave functions in the theory of molecules and crystals. J. Chem. Phys. **18**(3), 365–375 (1950)

43. Kittel, C.: Introduction to Solid State Physics, 8th edn. Wiley, Hoboken (2005)

44. Zhou, Z., Gao, X., Yan, J., Song, D., Morinaga, M.: A first-principles study of lithium absorption in boron-or nitrogen-doped single-walled carbon nanotubes. Carbon **42**(12), 2677–2682 (2004)

45. Cruz-Silva, E., Lopez-Urias, F., Munoz-Sandoval, E., Sumpter, B.G., Terrones, H., Charlier, J.-C., Meunier, V., Terrones, M.: Electronic transport and mechanical properties of phosphorus- and phosphorus-nitrogen-doped carbon nanotubes. ACS Nano **3**(7), 1913–1921 (2009)

46. Audiffred, M., Elías, A.L., Gutiérrez, H.R., López-Urías, F., Terrones, H., Merino, G., Terrones, M.: Nitrogen–silicon heterodoping of carbon nanotubes. J. Phys. Chem. C **117**(16), 8481–8490 (2013)

47. Yuan, N., Bai, H., Ma, Y., Ji, Y.: First-principle simulations on silicon-doped armchair single-walled carbon nanotubes of various diameters. Phys. E **64**, 195–203 (2014)

7

Theoretical identification of structural heterogeneities of divalent nickel active sites in NiMCM-41 nanoporous catalysts

Mahboobeh Balar[1] · Zahra Azizi[1] · Mohammad Ghashghaee[2]

Abstract This paper deals with the theoretical identification of the digrafted Ni species exchanged into the defect sites of MCM-41 using hybrid density functional theory. The nickel–siloxane clusters included seven 2T–6T rings. The 2MR and 5MR structures were found to be the least and most favorable sites to form thermodynamically. The Ni–O distances ranged from 1.69 to 1.79 Å with the highest asymmetry found in 5MR. The 4MR and 5MR clusters showed also interesting intertwined nickel configurations. Overall, the QTAIM calculations revealed the transient electrostatic nature of the Ni–O bonds.

Keywords Nickel · MCM-41 · DFT · Thermodynamics · Cluster modeling · Silica

Introduction

Nickel ions supported on different silica, silica–alumina, natural clay, and zeolite-type porous materials are well-known catalysts for the selective dimerization [1–9] and oligomerization [1, 5, 10–19] of olefins in both gas and liquid phases. The application of silica-supported nickel catalysts for the ethylene dimerization dates back to 1980s [20–22]. Later, the large and well-ordered cavities of the Ni-exchanged MCM-41 proved highly favorable for the oligomerization of olefins [23]. For instance, high productivities of oligomers were obtained over Ni-incorporated MCM-41 catalysts [17] being higher than those reported previously with silica–alumina supports [19]. The strong interactions of Ni^{2+} cations residing in the mesoporous cavities with the support framework made the reduction of the nickel ions difficult [17]. Further electron spin resonance (ESR) and Fourier transform infrared (FTIR) spectroscopic data [11, 24] served as evidence to support the role of Ni^{2+} cations as active sites in ethylene dimerization.

The nickel ions incorporated into the ordered mesoporous materials (OMMs) have also been suggested as efficient catalysts for the gas-phase transformation of ethylene to propylene [25–34]. For instance, the performance and durability of Ni/MCM-41 catalysts prepared from a nickel citrate precursor was ranked as auspicious with a promising potential for closing the gap between the propylene demand and supply [35]. Ikeda et al. [36] proposed that layered nickel-silicate species were the main players in the conversion of ethylene on NiMCM-41 catalysts. In a similar study [4], the threefold coordinated Ni^{2+} ions situated in the 5-membered rings of the phyllosilicate pore walls of NiMCM-41 were taken as the active sites of the reaction.

Computational studies of Ni^{2+} binding in FER [37], MFI [38, 39], AFI [40], and silica [4, 41] have been reported earlier. Analogous reports have addressed neutral atoms [42, 43]. The present study aims at a systematic theoretical modeling of the locations of Ni^{2+} species in a model MCM-41 material at the fundamental level. To our knowledge, the most relevant publication to this target is that by Neiman [41] who considered an $O_3(SiO)_2Ni$ structure unit cell with a formal charge of +2 for the nickel ion. The author did not investigate all of the defect sites available for the Ni^{2+} siting. This work is then the first

✉ Zahra Azizi
zahra.azizi@kiau.ac.ir; zahraazizi@yahoo.com

[1] Department of Chemistry, Karaj Branch, Islamic Azad University, P.O. Box 31485-313, Karaj, Iran

[2] Faculty of Petrochemicals, Iran Polymer and Petrochemical Institute, P.O. Box 14975-112, Tehran, Iran

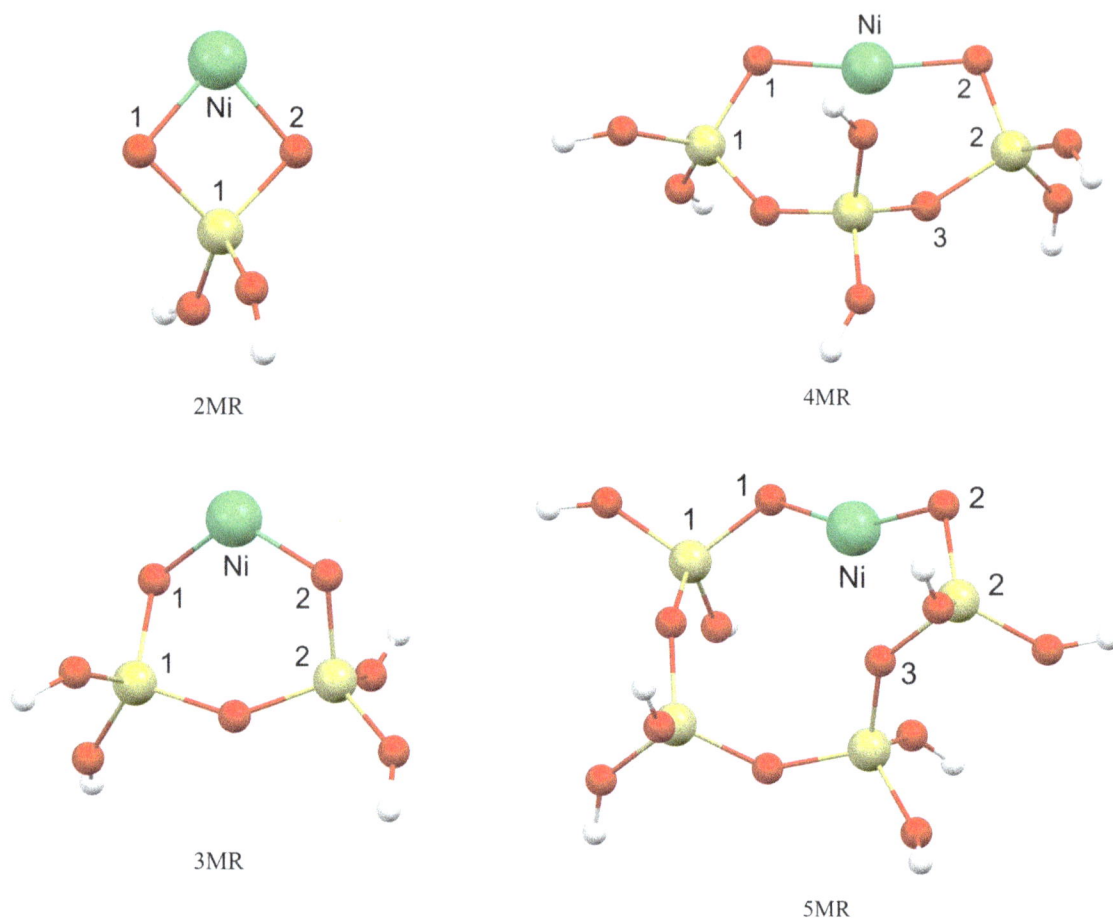

Fig. 1 The optimized geometries of the NiMCM-41 clusters where the *darker atoms* refer to the framework oxygen, the *plain bigger balls* are silicon atoms and the *small white balls* are terminal hydrogen atoms. The *largest atoms* represent nickel

fundamental report on the molecular-level heterogeneities of the NiMCM-41 catalysts.

Computational method

The cluster model approach was employed to simulate the active sites of a nanoporous NiMCM-41 catalyst through exploring the available defect sites of MCM-41 within an extended unit cell. As a common approach [44, 45], the model nanoclusters were terminated by H atoms frozen in agreement with the geometries obtained from the crystallographic data [46–51]. The divalent nickel cation and the immediate neighbors including the O atoms from two defect-site hydroxyls were allowed to relax during the optimizations. As adopted earlier [52], the remaining part of the cluster was held fixed to mimic the mechanical restrictions of the matrix.

As an approximation, the structure of cristobalite is normally taken as a good representative for the amorphous silica materials in terms of type and density of the surface

hydroxyl groups [51, 53–56]. Many references [42, 43, 57–60] have then applied this model to ordered silica mesoporous materials such as MCM-41. The optimizations and single-point computations were implemented using hybrid functional M06 [61] and the Def2-TZVP basis set [62–64]. Moreover, the natural bond orbital (NBO) population [65] as well as the quantum theory of atoms in molecules (QTAIM) [66–71] analyses were carried out at the same level of theory.

The ion exchange energies (ΔE_{ex}) were calculated for the following reaction:

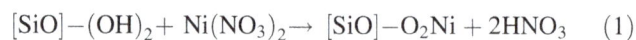

$$[SiO]-(OH)_2 + Ni(NO_3)_2 \rightarrow [SiO]-O_2Ni + 2HNO_3 \quad (1)$$

Moreover, the binding energies (ΔE_b) for the nickel ions at the defect sites of MCM-41 were assessed for the following reaction:

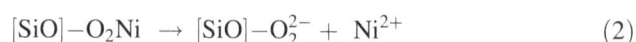

$$[SiO]-O_2Ni \rightarrow [SiO]-O_2^{2-} + Ni^{2+} \quad (2)$$

NWChem 6.5 [72] and Multiwfn 3.3.8 [73] were used for the computations. Finally, the graphical outputs were generated by Mercury 3.3 [74–77].

6MR-1

6MR-3

6MR-2

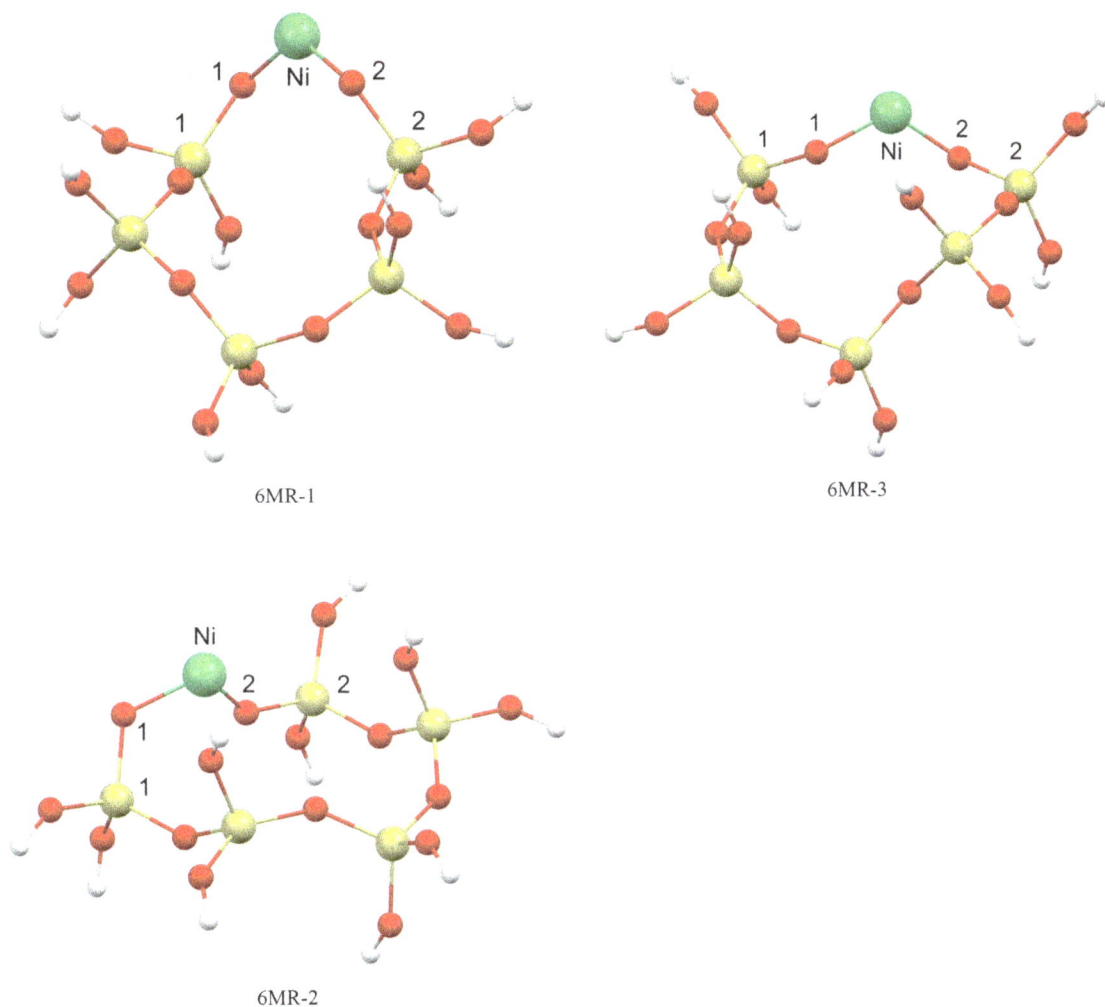

Fig. 1 continued

Results and discussion

The viable defect sites of MCM-41 for the binding of the divalent nickel ions were determined from exploring the available pairs of vicinal and close non-vicinal silanol groups. Different NiMCM-41 clusters were considered in terms of the number of T-atoms present in the final rings and the interatomic distances between the next-nearest-neighbor Si atoms with respect to the Ni cation. The optimized geometries are shown in Fig. 1. As can be seen, different oxide rings can be formed varying from a simple 2T (a digrafted nickel species stabilized with an interaction with two geminal-type Si–O groups) to tri, tetra, penta, and hexa membered ones. To our knowledge, this heterogeneity and its consequences are discussed theoretically for the first time. As such, the simplest cluster was a 2MR ring followed by a 3MR structure with a nickel cation chemisorbed on two vicinal OH groups located at the two ends of a siloxane bridge. The other clusters represent relatively more complex types of Ni binding with proximate non-

vicinal OH groups at two defect sites of the surface linked together via two or more consecutive siloxane bridges (Fig. 1). For the comparisons made here, it will be assumed [78] that the dispersion of the Ni^{2+} species at the defect sites of MCM-41 is controlled by their thermodynamic stability and not by any kinetic phenomenon in the preparation method.

Table 1 reports the NBO charges of the selected atoms of the NiMCM-41 clusters. As can be seen, the partial charge of the Ni cation ranged from 0.889 e for 2MR to 1.077 e for 4MR, being almost half the formal charge of $+2$ for the Ni ions. Except 2MR and 3MR, the bridging O atoms of the O_2Ni species did not bear identical charges, falling into the range of -1.157 e to -1.020 e being almost half the formal charge of -2 for the O atoms. Although not linearly correlated, there was a general proportional trend between $q(Ni)$ and $-q(O)$. Such an interconnection was not confirmed on the Mulliken atomic charges (see Table 1). Moreover, the average Mulliken charges of Ni and O atoms were generally 0.60 and 0.46 times those of the NBO

Table 1 Calculated charges of selected atoms of NiMCM-41 for different cluster models at M06/Def2-TZVP level of theory

Cluster	Ni	O_1	O_2	Si_1	Si_2
NBO partial charges					
2MR	0.889	−1.021	−1.020	2.353	–
3MR	0.952	−1.071	−1.069	2.442	2.434
4MR	1.077	−1.157	−1.110	2.428	2.410
5MR	1.061	−1.149	−1.115	2.442	2.402
6MR-1	0.951	−1.073	−1.089	2.464	2.460
6MR-2	1.016	−1.022	−1.150	2.445	2.451
6MR-3	0.939	−1.076	−1.069	2.445	2.444
Mulliken atomic charges					
2MR	0.590	−0.522	−0.521	0.918	–
3MR	0.589	−0.535	−0.537	0.855	0.864
4MR	0.566	−0.486	−0.492	0.782	0.822
5MR	0.565	−0.492	−0.500	0.770	0.812
6MR-1	0.603	−0.505	−0.509	0.758	0.755
6MR-2	0.582	−0.467	−0.437	0.770	0.764
6MR-3	0.568	−0.479	−0.498	0.795	0.791

Table 4 QTAIM data for different optimized clusters at M06/Def2-TZVP level of theory

Cluster	BCP	ρ	λ_1	λ_2	λ_3	$\nabla^2\rho$
2MR	Ni–O_1	0.152	−0.245	−0.225	1.111	0.641
	Ni–O_2	0.152	−0.245	−0.226	1.112	0.642
3MR	Ni–O_1	0.154	−0.264	−0.251	1.241	0.725
	Ni–O_2	0.153	−0.261	−0.247	1.217	0.709
4MR	Ni–O_1	0.132	−0.172	−0.155	1.066	0.739
	Ni–O_2	0.134	−0.153	−0.144	1.015	0.719
	Ni–O_3	0.032	−0.028	−0.012	0.167	0.127
5MR	Ni–O_1	0.144	−0.268	−0.217	1.264	0.779
	Ni–O_2	0.130	−0.151	−0.137	0.988	0.700
	Ni–O_3	0.043	−0.045	−0.036	0.276	0.195
6MR–1	Ni–O_1	0.172	−0.312	−0.297	1.493	0.884
	Ni–O_2	0.169	−0.303	−0.290	1.462	0.869
6MR–2	Ni–O_1	0.165	−0.277	−0.276	1.347	0.794
	Ni–O_2	0.156	−0.251	−0.233	1.306	0.822
6MR–3	Ni–O_1	0.172	−0.316	−0.299	1.512	0.897
	Ni–O_2	0.172	−0.313	−0.296	1.490	0.881

Table 2 The Gibbs free energy (kcal/mol), enthalpy (kcal/mol), and entropy (cal/mol/K) of the exchange reaction and the binding energy (kcal/mol) of the digrafted Ni ions in NiMCM-41 at M06/Def2-TZVP level of theory (please see the corresponding reactions in the text)

Cluster	ΔH_{ex}	ΔS_{ex}	ΔG_{ex}	ΔE_b
2MR	84.90	0.04	72.72	831.5
3MR	77.78	0.04	66.47	783.4
4MR	71.35	0.03	61.72	780.6
5MR	63.01	0.03	53.99	793.8
6MR-1	72.52	0.04	60.51	813.4
6MR-2	72.54	0.04	61.67	813.3
6MR-3	75.60	0.05	60.22	803.6

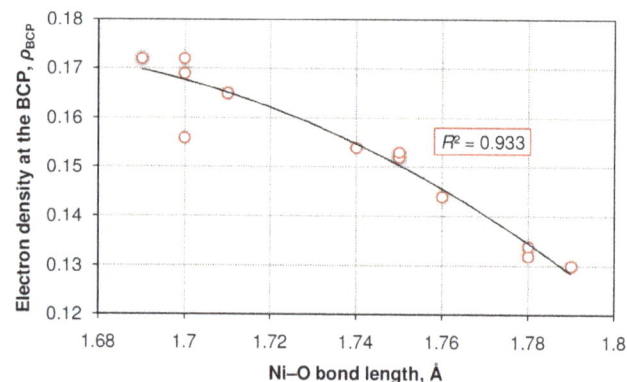

Fig. 2 The correlation found between the topological properties and the Ni–O distances in the NiMCM-41 catalysts

Table 3 Nickel–oxygen bond length (Å) and interbond angle (in degree) for different optimized cluster models at M06/Def2-TZVP level of theory

Cluster	Bond lengths		Angles
	Ni–O_1	Ni–O_2	O_1–Ni–O_2
2MR	1.75	1.75	85.6
3MR	1.74	1.75	119.4
4MR	1.78	1.78	168.9
5MR	1.76	1.79	154.6
6MR-1	1.69	1.70	118.9
6MR-2	1.71	1.72	105.2
6MR-3	1.69	1.70	120.9

Table 5 Calculated HOMO and LUMO and HOMO–LUMO energy gaps ($\Delta E_{HOMO-LUMO}$) for the investigated clusters at M06/Def2-TZVP level of theory

Cluster	E_{HOMO} (eV)	E_{LUMO} (eV)	$\Delta E_{HOMO-LUMO}$ (eV)
2MR	−7.19	−3.04	4.15
3MR	−7.46	−3.86	3.61
4MR	−7.30	−4.29	3.01
5MR	−7.11	−3.85	3.26
6MR-1	−7.11	−3.32	3.78
6MR-2	−7.16	−3.27	3.89
6MR-3	−7.42	−3.70	3.72

calculations and then even more distant from the nominal charges.

Table 2 contains the enthalpy, entropy, and Gibbs free energy of the exchange reaction for the NiMCM-41clusters and the binding energies of Ni at the investigated sites following Eqs. (1) and (2), respectively. As evident, the grafting reaction considered here is non-spontaneous at 298 K and atmospheric pressure on all of the clusters, connoting the necessity for devising more severe (e.g., hydrothermal) preparation conditions to overcome the exchange non-spontaneity. In any event, the thermodynamic favorability for grafting of the Ni^{2+} cations onto the available defect sites of MCM-41 followed the sequence of 2MR < 3MR < 4MR < 6MR-3 < 6MR-1 < 6MR-2 < 5MR (53.99–72.72 kcal/mol). This is in accordance with the previous ideas on the favorability of 5MR structures mentioned previously. Moreover, the enthalpy of the exchange reaction followed the order of 5MR < 4MR < 6MR-1 < 6MR-2 < 6MR-3 < 3MR < 2MR, all being endothermic (63.09–84.10 kcal/mol). However, the binding energy for this reaction varied in the order of 4MR < 3MR < 5MR < 6MR-3 < 6MR-2 < 6MR-1 < 2MR (780.6–831.5 kcal/mol). As a result, the 2MR site is more demanding to form despite its largest binding energy.

In Table 3, the nickel–oxygen bond lengths and the interbond angles of Ni have been listed for comparison. The bridging (Ni–O) bond lengths fell into the range of 1.69–1.79 Å with the highest asymmetry [a difference of 0.03 Å between $r(Ni–O1)$ and $r(Ni–O2)$] found in the case of 5MR. The O–Ni–O angles changed over a markedly wider range here with the minimum angle (85.6°) on 2MR and the largest one (168.9°) on 4MR. The data reported here indicated that neither the Ni–O distances nor the O1–Ni–O2 angles correlated with the size of the siloxane rings as analogously expressed in our previous study of Cr–silicalite-2 [79].

The characterizing features [71, 80, 81] of QTAIM were employed to determine whether the nickel–oxide solid interactions were of shared or closed-shell nature. The calculated topological properties are shown in Table 4. The tabulated data represent moderate electron densities (ρ_{BCP}). Moreover, positive $\nabla^2\rho_{BCP}$ values were obtained as anticipated from a common character of metal–oxygen interactions [80]. The interactions of nickel with the siloxane rings of MCM-41 were all found to be rather polar than purely covalent. More interestingly, however, the 4MR and 5MR structures revealed third Ni–O interactions of electrostatic nature that made the nickel ion interlocked three-coordinated in the plane of the ring. Potential correlations between the properties of the NiMCM-41 active sites were probed. Figure 2 shows an obvious correlation

between $r(Ni–O)$ and ρ_{BCP} which was also observed in our previous works on MEL structure [79, 82]. No further correlation with an acceptably high coefficient of determination was found between any other two parameters.

The energy levels of electrons and their gaps provide useful data for implications on the reactivity of the active sites [79, 82, 83] according to the frontier molecular orbital (FMO) theory [84]. Table 5 reports the HOMO and LUMO levels and HOMO–LUMO energy gaps for the optimized clusters shown in Fig. 1. The HOMO–LUMO energy gaps increased in the order of 4MR < 5MR < 3MR < 6MR-3 < 6MR-1 < 6MR-2 < 2MR ranging from 3.01 to 4.15 eV. Considering the chemical hardness as $\eta = (E_{LUMO} - E_{HOMO})/2$ [85–89], one can state that the chemical hardness of the digrafted nickel ion was maximized at 2MR and minimized at 4MR and 5MR sites.

Conclusion

This paper investigated the diversity of the cluster models of NiMCM-41 in a systematic computational framework. Total of seven active sites (2T–6T rings) were found at the defect sites of an MCM-41 silica model. The NBO partial charge of the nickel cation was less positive on 2MR and largest on 4MR. The thermodynamic favorability of the NiMCM-41 clusters followed the order of 2MR < 3MR < 4MR < 6MR-3 < 6MR-1 < 6MR-2 < 5MR. The optimized structures indicated the Ni–O distances in the range of 1.69–1.79 Å with the highest asymmetry observed in 5MR. The highest reactivity was observed in the case of the digrafted nickel ions at 4MR and 5MR sites and the lowest one at 2MR. The 4MR and 5MR clusters showed also some intertwining features for nickel hosting. The QTAIM calculations revealed intermediate polar Ni–O bonds. Moreover, the electron densities at the BCP correlated with the Ni–O distances.

References

1. Andrei, R.D., Popa, M.I., Fajula, F., Hulea, V.: Heterogeneous oligomerization of ethylene over highly active and stable Ni-AlSBA-15 mesoporous catalysts. J. Catal. **323**, 76–84 (2015)
2. Andrei, R.D., Mureseanu, M., Popa, M.I., Cammarano, C., Fajula, F., Hulea, V.: Ni-exchanged AlSBA-15 mesoporous materials as outstanding catalysts for ethylene oligomerization. Eur. Phys. J. Special Topics **224**(9), 1831–1841 (2015)

3. Zhang, H., Li, X., Zhang, Y., Lin, S., Li, G., Chen, L., Fang, Y., Xin, H., Li, X.: Ethylene oligomerization over heterogeneous catalysts. Energy Environ. Focus **3**(3), 246–256 (2014)
4. Tanaka, M., Itadani, A., Kuroda, Y., Iwamoto, M.: Effect of pore size and nickel content of Ni-MCM-41 on catalytic activity for ethene dimerization and local structures of nickel ions. J. Phys. Chem. C **116**(9), 5664–5672 (2012)
5. Iwamoto, M.: One step formation of propene from ethene or ethanol through metathesis on nickel ion-loaded silica. Molecules **16**(9), 7844–7863 (2011)
6. Choo, H., Kevan, L.: Catalytic study of ethylene dimerization on Ni(II)-exchanged clinoptilolite. J. Phys. Chem. B **105**(27), 6353–6360 (2001)
7. Nkosi, B., Ng, F.T.T., Rempel, G.L.: The oligomerization of butenes with partially alkali exchanged NiNaY zeolite catalysts. Appl. Catal. A-Gen. **158**(1), 225–241 (1997)
8. Sohn, J.R., Park, J.H.: Characterization of dealuminated NiY zeolite and effect of dealumination on catalytic activity for ethylene dimerization. Appl. Catal. A-Gen. **218**(1–2), 229–234 (2001)
9. Ng, F.T.T., Creaser, D.C.: Ethylene dimerization over modified nickel exchanged Y-zeolite. Appl. Catal. A-Gen. **119**(2), 327–339 (1994)
10. Moussa, S., Arribas, M.A., Concepción, P., Martínez, A.: Heterogeneous oligomerization of ethylene to liquids on bifunctional Ni-based catalysts: the influence of support properties on nickel speciation and catalytic performance. Catal. Today (2016). doi:10.1016/j.cattod.2015.11.032
11. Martínez, A., Arribas, M.A., Concepción, P., Moussa, S.: New bifunctional Ni–H-Beta catalysts for the heterogeneous oligomerization of ethylene. Appl. Catal. A-Gen. **467**, 509–518 (2013)
12. Lallemand, M., Finiels, A., Fajula, F., Hulea, V.: Continuous stirred tank reactor for ethylene oligomerization catalyzed by NiMCM-41. Chem. Eng. J. **172**(2), 1078–1082 (2011)
13. Brückner, A., Bentrup, U., Zanthoff, H., Maschmeyer, D.: The role of different Ni sites in supported nickel catalysts for butene dimerization under industry-like conditions. J. Catal. **266**(1), 120–128 (2009)
14. Lallemand, M., Finiels, A., Fajula, F., Hulea, V.: Catalytic oligomerization of ethylene over Ni-containing dealuminated Y zeolites. Appl. Catal. A-Gen. **301**(2), 196–201 (2006)
15. Hulea, V., Lallemand, M., Finiels, A., Fajula, F.: Catalytic oligomerization of ethylene over Ni-containing MCM-22, MCM-41 and USY. In: J. Čejka, N.Ž., Nachtigall, P. (eds.) Studies in Surface Science and Catalysis, Part B, vol. 158. pp. 1621–1628. Elsevier, Amsterdam (2005)
16. Lin, S., Shi, L., Zhang, H., Zhang, N., Yi, X., Zheng, A., Li, X.: Tuning the pore structure of plug-containing Al-SBA-15 by post-treatment and its selectivity for C_{16} olefin in ethylene oligomerization. Microporus and Mesoporus Mat. **184**, 151–161 (2014)
17. Hulea, V., Fajula, F.: Ni-exchanged AlMCM-41—an efficient bifunctional catalyst for ethylene oligomerization. J. Catal. **225**(1), 213–222 (2004)
18. Nicolaides, C.P., Scurrell, M.S., Semano, P.M.: Nickel silica-alumina catalysts for ethene oligomerization—control of the selectivity to 1-alkene products. Appl. Catal. A-Gen. **245**(1), 43–53 (2003)
19. Heydenrych, M.D., Nicolaides, C.P., Scurrell, M.S.: Oligomerization of ethene in a slurry reactor using a nickel(II)-exchanged silica–alumina catalyst. J. Catal. **197**(1), 49–57 (2001)
20. Kimura, K., Hideo, A.-I., Ozaki, A.: Tracer study of ethylene dimerization over nickel oxide-silica catalyst. J. Catal. **18**(3), 271–280 (1970)
21. Sohn, J.R., Ozaki, A.: Structure of NiO-SiO$_2$ catalyst for ethylene dimerization as observed by infrared absorption. J. Catal. **59**(2), 303–310 (1979)

22. Sohn, J.R., Ozaki, A.: Acidity of nickel silicate and its bearing on the catalytic activity for ethylene dimerization and butene isomerization. J. Catal. **61**(1), 29–38 (1980)
23. Finiels, A., Fajula, F., Hulea, V.: Nickel-based solid catalysts for ethylene oligomerization—a review. Catal. Sci. Technol. **4**(8), 2412–2426 (2014)
24. Martínez, A., Arribas, M.A., Moussa, S.: Development of bifunctional Ni-based catalysts for the heterogeneous oligomerization of ethylene to liquids. In: Kanellopoulos, N. (ed.) Small-Scale Gas to Liquid Fuel Synthesis. pp. 377–400. CRC Press, Boca Raton (2015)
25. Frey, A.S., Hinrichsen, O.: Comparison of differently synthesized Ni(Al)MCM-48 catalysts in the ethene to propene reaction. Microporus and Mesoporus Mat. **164**, 164–171 (2012)
26. Alvarado Perea, L., Wolff, T., Veit, P., Hilfert, L., Edelmann, F.T., Hamel, C., Seidel-Morgenstern, A.: Alumino-mesostructured Ni catalysts for the direct conversion of ethene to propene. J. Catal. **305**, 154–168 (2013)
27. Alvarado Perea, L., Wolff, T., Hamel, C., Seidel-Morgenstern, A.: Direct conversion of ethene to propene on Ni/AlMCM-41–study of the reaction mechanism. In: Jahrestreffen Deutscher Katalytiker, Weimar, Germany (2014)
28. Alvarado Perea, L.: Direct conversion of ethene to propene on Ni-alumino-mesostructured catalysts: synthesis, characterization and catalytic testing. PhD dissertation, Otto-von-Guericke Universität (2014)
29. Iwamoto, M., Kosugi, Y.: Highly selective conversion of ethene to propene and butenes on nickel ion-loaded mesoporous silica catalysts. J. Phys. Chem. C **111**(1), 13–15 (2007)
30. Iwamoto, M.: One step formation of propene from ethene or ethanol through metathesis on nickel ion-loaded silica. Molecules **16**, 7844–7863 (2011)
31. Taoufik, M., Le Roux, E., Thivolle-Cazat, J., Basset, J.-M.: Direct transformation of ethylene into propylene catalyzed by a tungsten hydride supported on alumina: trifunctional single-site catalysis. Angew. Chem. **119**, 7340–7343 (2007)
32. Taoufik, M., Le Roux, E., Thivolle-Cazat, J., Basset, J.-M.: Direct transformation of ethylene into propylene catalyzed by a tungsten hydride supported on alumina: trifunctional single-site catalysis. Angew. Chem. Int. Ed. **46**, 7202–7205 (2007)
33. Iwamoto, M.: Conversion of ethene to propene on nickel ion-loaded mesoporous silica prepared by the template ion exchange method. Catal. Surv. Asia **12**(1), 28–37 (2008)
34. Lehmann, T., Wolff, T., Zahn, V.M., Veit, P., Hamel, C., Seidel-Morgenstern, A.: Preparation of Ni-MCM-41 by equilibrium adsorption—Catalytic evaluation for the direct conversion of ethene to propene. Catal. Commun. **12**(5), 368–374 (2011)
35. Zahn, V.M., Wolff, T., Lehmann, T., Hamel, C., Seidel-Morgenstern, A.: Direct synthesis of propene using supported bifunctional nickel catalysts: preparation and potential. In: ISCRE 21-21st International Symposium on Chemical Reaction Engineering (2010)
36. Ikeda, K., Kawamura, Y., Yamamoto, T., Iwamoto, M.: Effectiveness of the template-ion exchange method for appearance of catalytic activity of Ni–MCM-41 for the ethene to propene reaction. Catal. Commun. **9**(1), 106–110 (2008)
37. Šponer, J.E., Sobalík, Z., Leszczynski, J., Wichterlová, B.: Effect of metal coordination on the charge distribution over the cation binding sites of zeolites. A combined experimental and theoretical study. J. Phys. Chem. B **105**(35), 8285–8290 (2001)
38. Rice, M.J., Chakraborty, A.K., Bell, A.T.: Site availability and competitive siting of divalent metal cations in ZSM-5. J. Catal. **194**(2), 278–285 (2000)
39. Rice, M.J., Chakraborty, A.K., Bell, A.T.: Theoretical studies of the coordination and stability of divalent cations in ZSM-5. J. Phys. Chem. B **104**(43), 9987–9992 (2000)

40. Brogaard, R.Y., Olsbye, U.: Ethene oligomerization in Ni-containing zeolites: theoretical discrimination of reaction mechanisms. ACS Catal. **6**(2), 1205–1214 (2016)

41. Neiman, M.L.: Interaction of cobalt and nickel with the hydroxylated silanol silica surface: a theoretical study. MSc thesis, Lehigh University (2002)

42. Ma, Q., Klier, K., Cheng, H., Mitchell, J.W., Hayes, K.S.: Interaction between catalyst and support. 1. Low coverage of Co and Ni at the silica surface. J. Phys. Chem. B **104**(45), 10618–10626 (2000)

43. Ma, Q., Klier, K., Cheng, H., Mitchell, J.W., Hayes, K.S.: Interaction between catalyst and support. 3. Metal agglomeration on the silica surface. J. Phys. Chem. B **105**(38), 9230–9238 (2001)

44. Pietrzyk, P.: Spectroscopy and computations of supported metal adducts. 1. DFT study of CO and NO adsorption and coadsorption on Cu/SiO$_2$. J. Phys. Chem. B **109**(20), 10291–10303 (2005)

45. Ferullo, R.M., Garda, G.R., Belelli, P.G., Branda, M.M., Castellani, N.J.: Deposition of small Cu, Ag and Au particles on reduced SiO2. J. Mol. Struct. **769**(1–3), 217–223 (2006)

46. Downs, R.T., Palmer, D.C.: The pressure behavior of α cristobalite. Am. Mineral. **79**(1–2), 9–14 (1994)

47. Wright, A.F., Leadbetter, A.J.: The structures of the β-cristobalite phases of SiO$_2$ and AlPO$_4$. Philos. Mag. **31**(6), 1391–1401 (1975)

48. Hatch, D.M., Ghose, S.: The α-β phase transition in cristobalite, SiO$_2$. Phys. Chem. Miner. **17**(6), 554–562 (1991)

49. Leadbetter, A.J., Smith, T.W., Wright, A.F.: Structure of high cristobalite. Nature **244**, 125–126 (1973)

50. Peacor, D.R.: High-temperature single-crystal study of the cristobalite inversion. zkri **138**(1–4), 274–298 (1973)

51. Pophal, C., Fuess, H.: Investigation of the medium range order of polyhedra forming the walls of MCM-41—An X-ray diffraction study. Microporus and Mesoporus Mat. **33**(1–3), 241–247 (1999)

52. Lopez, N., Illas, F., Pacchioni, G.: Adsorption of Cu, Pd, and Cs atoms on regular and defect sites of the SiO$_2$ surface. J. Am. Chem. Soc. **121**(4), 813–821 (1999)

53. Chuang, I.S., Maciel, G.E.: A detailed model of local structure and silanol hydrogen bonding of silica gel surfaces. J. Phys. Chem. B **101**(16), 3052–3064 (1997)

54. Maciel, G.E.: Probing hydrogen bonding and the local environment of silanols on silica surfaces via nuclear spin cross polarization dynamics. J. Am. Chem. Soc. **118**(2), 401–406 (1996)

55. Chuang, I.S., Kinney, D.R., Bronnimann, C.E., Zeigler, R.C., Maciel, G.E.: Effects of proton-proton spin exchange in the silicon-29 CP-MAS NMR spectra of the silica surface. J. Phys. Chem. **96**(10), 4027–4034 (1992)

56. Sindorf, D.W., Maciel, G.E.: Silicon-29 NMR study of dehydrated/rehydrated silica gel using cross polarization and magic-angle spinning. J. Am. Chem. Soc. **105**(6), 1487–1493 (1983)

57. Lillehaug, S.: A theoretical study of Cr/oxide catalysts for dehydrogenation of short alkanes. PhD dissertation, University of Bergen (2006)

58. Handzlik, J., Kurleto, K.: Assessment of density functional methods for thermochemistry of chromium oxo compounds and their application in a study of chromia–silica system. Chem. Phys. Lett. **561–562**, 87–91 (2013)

59. Lillehaug, S., Jensen, V.R., Børve, K.J.: Catalytic dehydrogenation of ethane over mononuclear Cr(III)–silica surface sites. Part 2: C—H activation by oxidative addition. J. Phys. Org. Chem. **19**, 25–33 (2006)

60. Chempath, S., Zhang, Y., Bell, A.T.: DFT studies of the structure and vibrational spectra of isolated molybdena species supported on silica. J. Phys. Chem. C **111**(3), 1291–1298 (2006)

61. Zhao, Y., Truhlar, D.: The M06 suite of density functionals for main group thermochemistry, thermochemical kinetics, noncovalent interactions, excited states, and transition elements: two new functionals and systematic testing of four M06-class functionals and 12 other functionals. Theor. Chem. Account **120**(1–3), 215–241 (2008)

62. Weigend, F.: Accurate Coulomb-fitting basis sets for H to Rn. Phys. Chem. Chem. Phys. **8**(9), 1057–1065 (2006)

63. Weigend, F., Ahlrichs, R.: Balanced basis sets of split valence, triple zeta valence and quadruple zeta valence quality for H to Rn: design and assessment of accuracy. Phys. Chem. Chem. Phys. **7**(18), 3297–3305 (2005)

64. Feller, D.: The role of databases in support of computational chemistry calculations. J. Comput. Chem. **17**(13), 1571–1586 (1996)

65. Glendening, E., Badenhoop, J., Reed, A., Carpenter, J., Weinhold, F.: NBO 3.1. Theoretical Chemistry Institute, University of Wisconsin, Madison (1996)

66. Rodríguez, J.I., Bader, R.F.W., Ayers, P.W., Michel, C., Götz, A.W., Bo, C.: A high performance grid-based algorithm for computing QTAIM properties. Chem. Phys. Lett. **472**(1–3), 149–152 (2009)

67. Bader, R.F.W.: A quantum theory of molecular structure and its applications. Chem. Rev. **91**(5), 893–928 (1991)

68. Bader, R.F.W.: The quantum mechanical basis of conceptual chemistry. Monatsh. Chem. **136**(6), 819–854 (2005)

69. Bader, R.F.W.: Molecular fragments or chemical bonds. Accounts Chem. Res. **8**(1), 34–40 (1975)

70. Bader, R.F.W.: Atoms in Molecules: A Quantum Theory, vol. 22. International series of monographs on chemistry. Oxford University Press, USA (1994)

71. Matta, C.F., Boyd, R.J.: The quantum theory of atoms in molecules: from solid state to DNA and drug design. Wiley, New York (2007)

72. Valiev, M., Bylaska, E.J., Govind, N., Kowalski, K., Straatsma, T.P., Van Dam, H.J.J., Wang, D., Nieplocha, J., Apra, E., Windus, T.L., de Jong, W.A.: NWChem: a comprehensive and scalable open-source solution for large scale molecular simulations. Comput. Phys. Commun. **181**(9), 1477–1489 (2010)

73. Lu, T., Chen, F.: Multiwfn: a multifunctional wavefunction analyzer. J. Comput. Chem. **33**(5), 580–592 (2012)

74. Bruno, I.J., Cole, J.C., Edgington, P.R., Kessler, M., Macrae, C.F., McCabe, P., Pearson, J., Taylor, R.: New software for searching the Cambridge structural database and visualizing crystal structures. Acta Crystallogr. B **58**(3 Part 1), 389–397 (2002)

75. Macrae, C.F., Bruno, I.J., Chisholm, J.A., Edgington, P.R., McCabe, P., Pidcock, E., Rodriguez-Monge, L., Taylor, R., van de Streek, J., Wood, P.A.: Mercury CSD 2.0—new features for the visualization and investigation of crystal structures. J. Appl. Crystallogr. **41**(2), 466–470 (2008)

76. Macrae, C.F., Edgington, P.R., McCabe, P., Pidcock, E., Shields, G.P., Taylor, R., Towler, M., van de Streek, J.: Mercury: visualization and analysis of crystal structures. J. Appl. Crystallogr. **39**(3), 453–457 (2006)

77. Taylor, R., Macrae, C.F.: Rules governing the crystal packing of mono- and dialcohols. Acta Crystallogr. B **57**(6), 815–827 (2001)

78. Handzlik, J., Grybos, R., Tielens, F.: Structure of monomeric chromium(VI) oxide species supported on silica: periodic and cluster DFT studies. J. Phys. Chem. C **117**(16), 8138–8149 (2013)

79. Ghambarian, M., Azizi, Z., Ghashghaee, M.: Diversity of monomeric dioxo chromium species in Cr/silicalite-2 catalysts: a hybrid density functional study. Comp. Mater. Sci. **118**, 147–154 (2016)

80. Sierraalta, A., Añez, R., Brussin, M.-R.: Theoretical study of NO$_2$ adsorption on a transition-metal zeolite model. J. Catal. **205**(1), 107–114 (2002)

81. Gatti, C.: Chemical bonding in crystals: new directions. zkri **220**, 399–457 (2005)

82. Ghashghaee, M., Ghambarian, M., Azizi, Z.: Characterization of extraframework Zn^{2+} cationic sites in silicalite-2: a computational study. Struct. Chem. **27**(2), 467–475 (2016)

83. Ghambarian, M., Ghashghaee, M., Azizi, Z., Balar, M.: Effect of cluster size in computational modeling of Cu^+/ZSM-11 catalysts. In: 3rd International Congress of Chemistry and Chemical Engineering, pp. 292–299 (2016)

84. Fukui, K., Yonezawa, T., Shingu, H.: A molecular orbital theory of reactivity in aromatic hydrocarbons. J. Chem. Phys. **20**(4), 722–725 (1952)

85. Parthasarathi, R., Subramanian, V., Chattaraj, P.K.: Effect of electric field on the global and local reactivity indices. Chem. Phys. Lett. **382**(1–2), 48–56 (2003)

86. Parr, R.G., Szentpály, Lv, Liu, S.: Electrophilicity Index. J. Am. Chem. Soc. **121**(9), 1922–1924 (1999)

87. Parr, R.G., Pearson, R.G.: Absolute hardness: companion parameter to absolute electronegativity. J. Am. Chem. Soc. **105**(26), 7512–7516 (1983)

88. Datta, D.: On Pearson's HSAB Principle. Inorg. Chem. **31**(13), 2797–2800 (1992)

89. Parr, R.G., Weitao, Y.: Density-functional theory of atoms and molecules. International Series of Monographs on Chemistry. Oxford University Press, New York (1989)

8

Synthesis of palladium–carbon nanotube–metal organic framework composite and its application as electrocatalyst for hydrogen production

Zahra Ghiamaty[1] · Ali Ghaffarinejad[1,2] · Mojtaba Faryadras[3] · Abbas Abdolmaleki[4] · Hojjat Kazemi[5]

Abstract There are very rare reports on using metal–organic framework (MOF) catalysts for electrochemical hydrogen production. In this study, a composite of palladium, single-walled carbon nanotube (SWCNT) and MOF-199 (Pd/SWCNTs@MOF-199) was synthesized by hydrothermal method, and its application as electrocatalyst in carbon paste electrode (CPE) structure for hydrogen production was studied. Scanning electron microscopy, X-ray diffraction, Brunauer–Emmett–Teller and thermal gravimetric analysis were used to characterize Pd/SWCNTs@MOF-199 catalyst. The performance of the proposed modified CPE for electrochemical hydrogen production was studied by cyclic voltammetry, linear sweep voltammetry, electrochemical impedance spectroscopy and chronoamperometry techniques. The effect of solution pH and the amount of binder and catalyst in the paste composition were investigated. The results showed that the CPE modified with Pd/SWCNTs@MOF-199 reveals better catalytic characteristics such as highest catalytic activity and lowest onset potential compared to CPE and CPE modified with MOF-199 for hydrogen production in aqueous solution.

Keywords Metal organic framework composite · Electrochemical hydrogen production · Carbon paste electrode · Electrocatalyst

Introduction

Today, fossil fuels with disadvantages such as limited resources, warming up the earth and the environmental pollution are known as the main energy sources. Many efforts are being made to use other energy sources instead of fossil fuels which do not have disadvantages mentioned above. Nuclear energy, which its sources are limited, and working with it requires training of skilled manpower and use of advanced systems for protection against radioactive waste, will not be able to supply the required energy of the world.

Nowadays, hydrogen as a raw material is used in different industries [1]. Hydrogen produces high amount of energy with almost no pollution. It is a renewable energy carrier, and its resource is infinite [2]. An energy carrier can change energy to the forms which are usable to consumers. Although, renewable energy sources such as sun and wind cannot provide energy all the time, they could produce electric energy and hydrogen, which can be stored and transported until they are needed.

Several ways such as hydrolysis [3], thermal catalysis and thermochemical [4, 5], photocatalysis [6, 7], photoelectrocatalysis [8], steam reforming [9], gasification [10]

✉ Ali Ghaffarinejad
Ghaffarinejad@iust.ac.ir

[1] Research Laboratory of Real Samples Analysis, Faculty of Chemistry, Iran University of Science and Technology, Tehran 1684613114, Iran

[2] Electroanalytical Chemistry Research Centre, Iran University of Science and Technology, Tehran 1684613114, Iran

[3] Faculty of Chemistry, Iran University of Science and Technology, Tehran 1684613114, Iran

[4] Department of Chemistry, Malek-Ashtar University of Technology, Tehran, P.O. Box 16765-34543, Iran

[5] Research Institute of Petroleum Industry, Tehran 1485733111, Iran

and electrolysis [5, 11] are used to produce hydrogen. Although in the electrochemical studies platinum (Pt) is known as an excellent electrode for hydrogen generation, but because of limitation of its resource and its expensive price we should look for a suitable alternative to Pt [12]. There are several reports which have attempted to reduce the amount of loaded Pt in the electrode body [13] or replace it with another catalyst [14–16].

MOFs are synthesis by linking inorganic and organic units by strong bonds that lead to combined properties of organic and inorganic porous materials [17–19]. Among the wide range of available catalysts, MOFs have recently received noticeable attention owing to their characteristics such as porosity, specific surface area and adjustable pore size [20, 21]. Related to hydrogen energy, MOFs have been mostly used for hydrogen storage [22–25] and there are some reports in hydrogen generation area with photocatalytic method [26–28], and a few reports for electrochemical hydrogen generation [29–31].

In this study, a composite of MOF-199 with Pd and single-walled carbon nanotube (SWCNT) was prepared (Pd/SWCNTs@ MOF-199), and was applied for electrochemical hydrogen generation. Structure and morphology of the composite were characterized by various techniques, and its electrochemical hydrogen generation performance was compared with MOF-199 in CPE.

Experimental

Apparatus and software

The electrochemical impedance measurements were performed with an Autolab potentiostat/galvanostat (PGSTAT30) equipped with FRA board, also all other electrochemical studies were done with a μ-Autolab type II potentiostat/galvanostat that the software for both of them was Nova version 1.7.8. A three electrode cell containing, Pt rod as the counter electrode, bare or modified CPE as working electrode and Ag/AgCl (3 M KCl) as the reference electrode was used. In this paper, for easier comparison with other reports all potential values are reported vs. reversible hydrogen electrode (RHE). Teflon cylinder (3.0 mm i.d.) that was firmly packed with carbon paste was used as the body of the working electrode. To creation electrical connect a stainless steel rod inserted into the carbon paste. All electrochemical experiments were done at room temperature, also before every experiment the electrolyte solution was deaerated with high purity nitrogen gas for at least 10 min.

The MOF structure was investigated between $5° \leq 2\theta \leq 90°$ with X-ray diffraction (XRD) instrument model of PHILIPS 1830 with a Cu (Kα) radiation source ($\lambda = 1.5418$ Å).

The catalyst morphology and its elemental analysis was investigated with a scanning electron microscopy (SEM) (VEGA\\TESCAN-XMU) equipped with energy dispersive X-ray analysis (EDX).

Thermogravimetric analysis (TGA) was used to determine the thermal behaviour of MOF-199 and Pd/SWCNT@MOF-199 using SHIMADZU, TG-50/DTA-50. The measurements were conducted at 10 °C min⁻¹ from room temperature.

Nitrogen adsorption–desorption isotherms were measured at 77 K by a BET instrument model of ASAP™ micromeritics 2020.

Materials and solutions

Graphite fine powder (extra pure with P.S. <50 μm), paraffin oil (spectroscopic grade, Uvasol®), H_3PO_4 (85 %), KCl (99.5 %), H_2SO_4 (98 %), $Cu(NO_3)_2 \cdot 3H_2O$ (99.5 %), benzene-1,3,5-tricarboxylic acid (H_3BTC, 95 %), $PdCl_2$ (59 % Pd basis) and ethanol (96 %) were obtained from Merck and were used without furture purifications. The single walled carbon nanotube (>95 %) was purchased from Sigma and was treated with nitric acid (5 M) for 15 min. After treatment, the SWCNTs were filtered and washed with deionised water to remove any remained nitric acid and impurities. All aqueous solutions were prepared with deionized water.

MOF-199 synthesis

MOF-199 was synthesized as previously reported [32]. In brief, 2.3268 g of $Cu(NO_3)_2 \cdot 3H_2O$ was dissolved in 24 mL deionized water, and 1.4140 g H_3BTC was dissolved in 24 mL ethanol. Both solutions were mixed with magnetic stirrer and then put into a 100 mL stainless steel autoclave at 120 °C for 12 h. The product was washed with ethanol and water, and then put at 70 °C for 30 min, and after this the resultant was put at 150 °C for 30 min. The final product was MOF-199.

Pd/SWCNT synthesis

$PdCl_2$ (0.16 g) was dissolved in 5 mL ethanol, 0.10 g of treated SWCNT was added to the proposed solution, and was dispersed using ultrasonic bath. After dispersion, during stirring of the mixture, 1 mL hydrazine was added dropwise under nitrogen atmosphere. Then the pH was adjusted to 9 with 1 M NaOH solution. For complete reduction of palladium the mixture was stirred on the magnetic stirrer for 2 h at 100 °C. Finally, the precipitate was filtered and washed with deionized water and then dried in oven at 100 °C for 4 h. This product is called Pd/SWCNT.

Synthesis of Pd/SWCNTs@ MOF-199 nanocomposite

Pd/SWCNTs@MOF-199 nanocomposite was prepared according to a previously reported procedure for synthesis of SWCNT@MOF-5 [33]. For this purpose, 2.320 g of $Cu(NO_3)_2 \cdot 3H_2O$ was dissolved in 24 mL deionized water, and 1.414 g H_3BTC was dissolved in 24 mL ethanol. Both solutions were mixed with magnetic stirrer for 20 min. Then 0.100 g of Pd/SWCNTs was added to the obtained solution, and was stirred for about 24 h at room temperature. The resultant mixture was transferred to autoclave and heated at 120 °C for 12 h. The precipitate was filtered, and washed with boiling deionized water and acetone. The washed precipitate was dried in vacuum oven at 100 °C for 4 h. The final product (Pd/SWCNTs@MOF-199 nanocomposite) was characterized with BET, TGA, XRD, EDX and SEM.

Preparation of bare and modified CPEs

Paraffin oil (as binder) and graphite powder (15:85 wt%) were mixed with an agate mortar to obtain a carbon paste. The modified carbon pastes were obtained by mixing a certain amount of Pd/SWCNTs@MOF-199 or MOF-199 (as modifiers) with binder and graphite. For preparation of the bare and modified electrodes, the working electrode cavities with 3 mm diameter were filled with unmodified and modified pastes. For removing any holes and improve the conductivity the pastes were packed on a smooth surface. Before each electrochemical measurement the electrode surface was renewed, polished on a smooth weighing paper, and washed with deionized water.

Results and discussion

Morphological characterization of Pd/SWCNTs@MOF-199

Figure 1 shows the SEM images of the MOF-199 and Pd/SWCNT@MOF-199. Figure 1a shows that the synthesized MOF-199 has a regular octahedral structure. Figure 1b shows the MOF-199 after modification with Pd/SWCNT, in which the carbon nanotubes on the octahedral shaped MOF-199 is clearly observed.

For a more exact examination, the EDX analysis of Pd/SWCNT@MOF-199 was also performed. Figure 2 shows

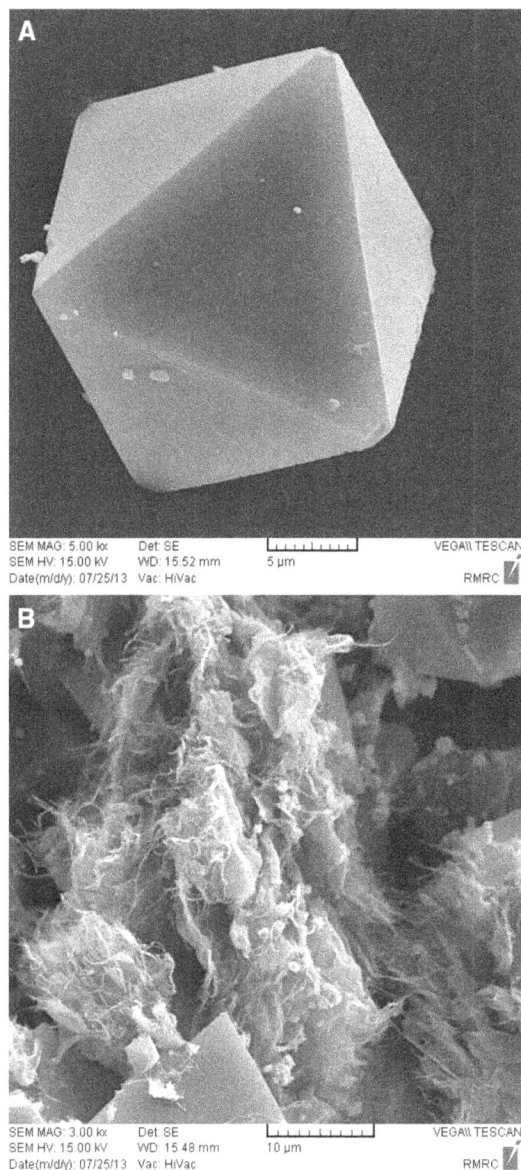

Fig. 1 SEM images of MOF-199 (**a**), and Pd/SWCNTs@MOF-199 nanocomposite (**b**)

the EDX analysis for Pd/SWCNT@MOF-199. In this experiment the presence of Cu and Pd is confirmed. As shown in this figure, the week peaks observed in the EDX spectrum between 2 and 3 keV are related to Pd L_α and Pd L_β, which reveals the loaded Pd in the composite is low. In relation to EDX analysis data the Pd and Cu wt% are 17.62 and 0.98 %, respectively. According to the Pd:Cu mole ratio in the feed (about 1:27) and according to EDX analysis (about 1:30) it could be concluded that this ratio is almost constant after composite synthesis. The Au peaks are related to gold coating for the sample preparation before taking SEM images and it is not an impurity in the composite.

Fig. 2 EDX analysis of the Pd/
SWCNTs@MOF-199
nanocomposite

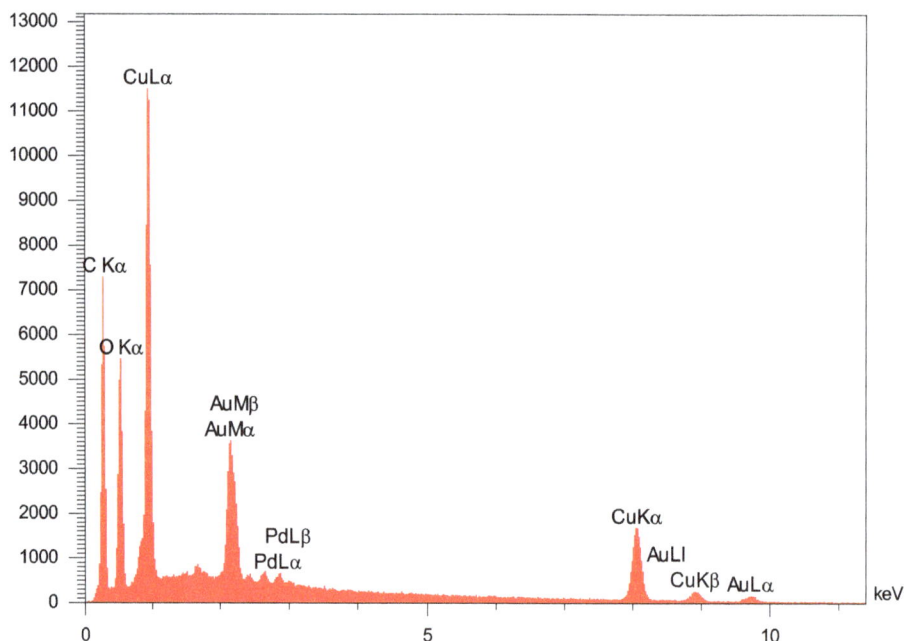

X-ray diffraction (XRD) characterization of nanocomposite

The X-ray diffraction pattern of the MOF-199 is illustrated in Fig. 3A. The peaks between 10° and 20° are related to cubic crystalline structure of MOF-199, which is consistent with previous reports [34]. The peaks at 37° and 43° are related to Cu_2O impurities, but their intensities are very low related to main peaks of the MOF-199, which confirm that the purity of the MOF is reasonable. Figure 3B shows the XRD pattern of the Pd/SWCNT@MOF-199. In this pattern the main peak of $Cu_3(BTC)_2$ at 11.68° corresponds to plane (2 2 2) reflection is not changed after modification, which indicates that the crystal structure of MOF-199 after modification has not changed. The peaks at 40.2° and 46.5° with the relative intensity of 10–15 % are related to the reflection of cubic unit cells of Pd.

Thermogravimetric analysis

Thermal stability of MOF-199 in the air atmosphere was analysed with thermal gravimetric analysis (TGA). Figure 4a illustrates that this compound is stable up to 300 °C. The weight loss is observed in two stages. The first stage (25–240 °C) is related to removal of water molecules from pores, and the solvent molecules trapped in the MOF structure during synthesis. In this stage a 22.15 % weight loss was observed. At the second stage degradation of OH groups and portion of organic frameworks at 300–350 °C occurred, which is responsible for a 46.50 % weight loss. Copper oxides are probably the final remaining products at higher temperatures. Figure 4b shows the TGA diagram for the Pd/SWCNT@MOF-199. As

illustrated in this figure the thermal stability of the composite is slightly better than MOF-199, which may be due to the presence of hydrophobic carbon nanotubes in its structure.

Determination of nitrogen adsorption–desorption isotherms

Figure 5a shows the N_2 adsorption–desorption isotherm of MOF-199, which according to IUPAC classification corresponds to type I isotherm. This isotherm indicates that the available porosity is in micrometer dimensions. Achieving to level off conditions and saturation at relatively low pressures emphasizes that the sample is microporous. At partial pressures near 1 atm some increase in adsorption is observed, which is related to accumulation of the existing porosities in the framework at relatively high pressures. Initial filling of porosities at low pressures in logarithmic scales represents that the porosities are small and there is an effective interaction between MOF and N_2. In addition, negligible hysteresis loop in adsorption/desorption isotherm shows that most of holes are in the microporous range.

Figure 5b shows the N_2 adsorption–desorption isotherm of Pd/SWCNT@MOF-199. Results show that MOF modification with Pd does not change the microporous structure of MOF, but as expected the surface area is slightly decreased. Table 1 summarizes the results of these experiments.

Electrochemical hydrogen evolution reaction on Pd/SWCNTs@MOF-199 modified CPE

Electrochemical studies showed that Pd/SWCNTs@ MOF-199 is a good electrocatalyst for HER. Therefore, for

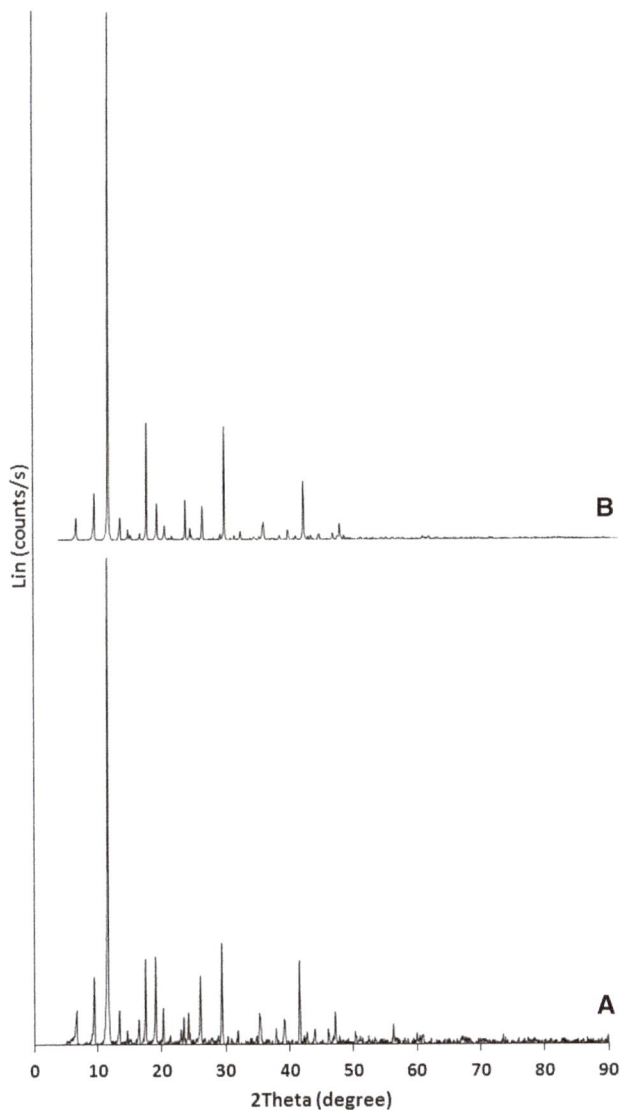

Fig. 3 XRD pattern of MOF-199 (A) and Pd/SWCNT@MOF-199 (B)

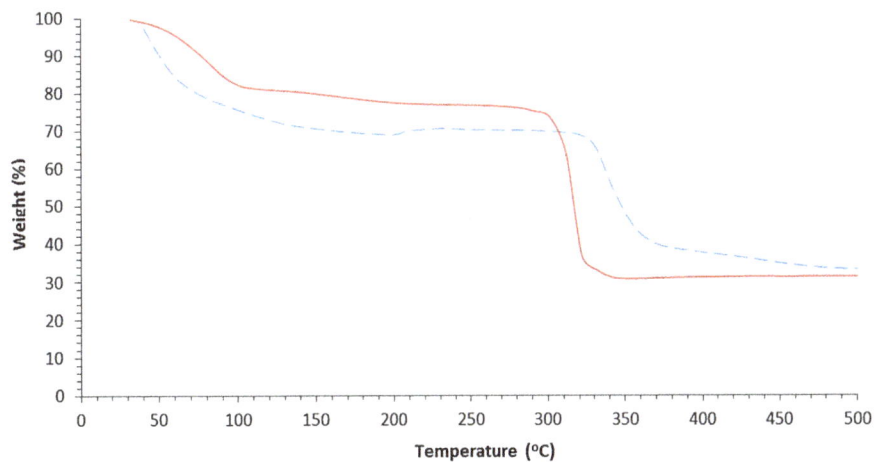

Fig. 5 Adsorption-desorption analysis at 77 K for MOF-199 (a), and Pd/SWCNT@MOF-199 (b)

Table 1 The structure data of the pores for MOF-199 and Pd/SWCNT@MOF-199 with adsorption–desorption analysis

Sample	V_p (cm^3 g^{-1})	d spacing (nm)	A_{BET} (m^2 g^{-1})
MOF-199	0.573	1.67	1370.0
Pd/SWCNTs@MOF-199	0.555	1.98	1125.8

Fig. 4 TGA thermograms of MOF-199 (solid line), and Pd/SWCNT@MOF-199 (dashed line)

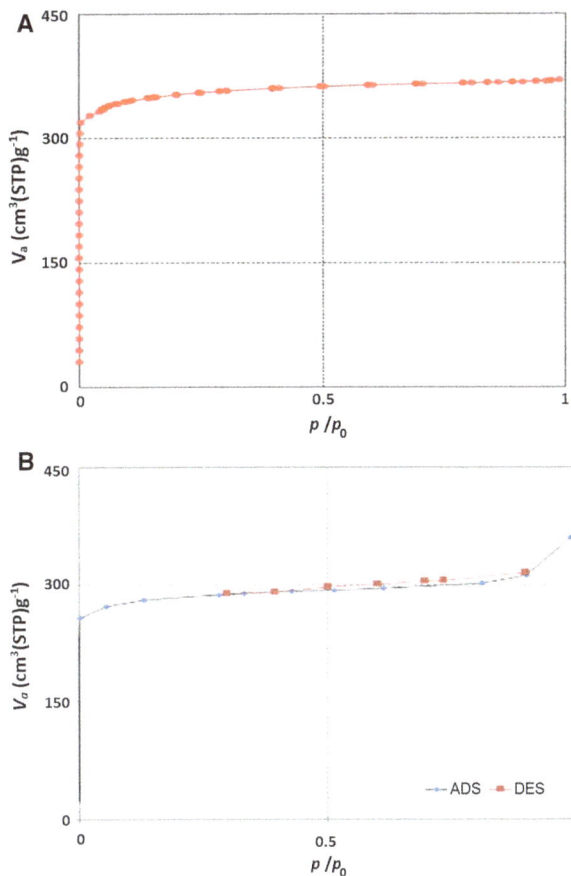

further study, some important parameters including elec-trocatalyst amounts in the electrode composition, paraffin binder and electrolyte pH were optimized.

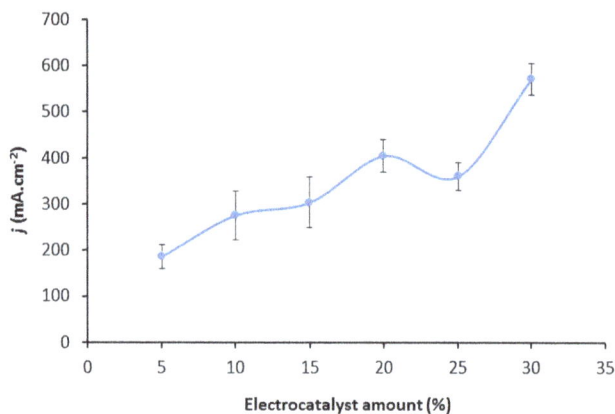

Fig. 6 The effect of catalyst amount on the HER performance. At each composition 20 successive CVs were applied and the j value at the last cycle was measured at -1 V (scan rate 100 mV s^{-1}, 15 % binder and 2 M sulphuric acid as supporting electrolyte, each point is the average of 3–5 measurements)

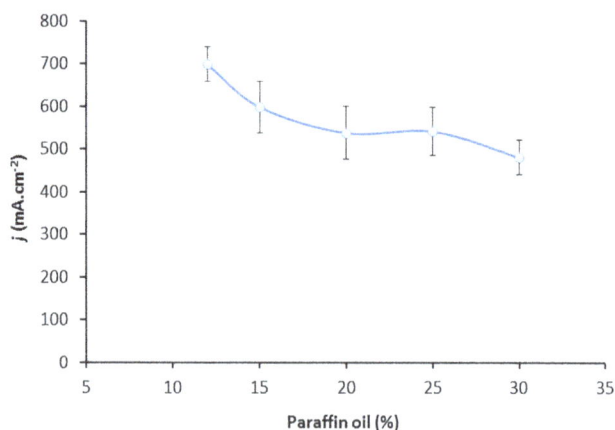

Fig. 7 The effect of oil binder amount on the HER performance. At each composition 20 successive CVs were applied and j at the last cycle was measured at -1 V (scan rate 100 mV s^{-1}, 30 % catalyst and 2 M sulphuric acid as supporting electrolyte, each point is the average of 3 or 5 measurements)

Fig. 8 The effect of supporting electrolyte pH on the HER performance for Pd/SWCNTs@MOF-199-CPE. At each solution 20 successive CVs were applied and j at the last cycle was measured at -1 V. (scan rate 100 mV s^{-1}, 30 % catalyst and 15 % Oil, each point is the average of 3 or 4 measurements)

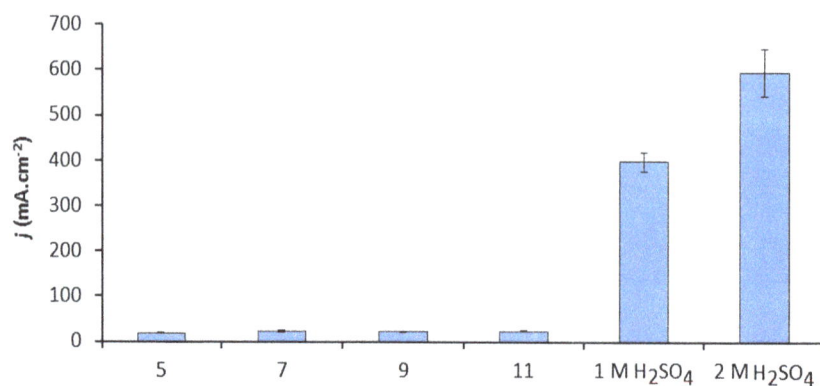

The catalyst amount

For optimization of the catalyst amount, various percentages of Pd/SWCNTs@MOF-199 were added to the paste com-position, while the paraffin percent was kept constant at 15 %. At these compositions 20 successive cyclic voltam-mograms (CVs) with the scan rate of 100 mV s^{-1} at the potential range of 1.2 to -1 V were applied on the electrodes in 2 M sulphuric acid solution. At potential of -1 V the current density of the HER (j), which could be interpreted as the hydrogen production rate, was measured. The result of this study is summarized in Fig. 6. According to Fig. 6 it is obvious that the j value increasing is a direct consequence of increasing amount of catalyst. However, for catalyst amounts greater than 30 % the paste consolidation decreased, so 30 % was used for the rest of experiments.

Optimization of the paraffin binder amount

The conductivity and active surface area of the electrode can be affected by the amount of paraffin which is a non-conductive binder. According to this fact, the amount of binder was changed, while the catalyst percentage was kept at 30 % and the electrolyte was 2 M sulphuric acid, and the HER was investigated by applying 20 successive CVs on the electrode surface with the scan rate of 100 mV s^{-1} in the potential range of 1.2 to -1 V. Figure 7 illustrates the plot of j at 1 V vs. paraffin oil percent under the mentioned conditions. Figure 7 clearly demonstrates that when the amount of binder is decreased, it favorably affected on the performance of the HER which can be related to the increase in electrode conductivity. Amounts lower than 12 % were not used, because at these compositions the paste consolidation was very poor.

pH effect

The effect of pH on HER efficiency on the Pd/SWCNTs@MOF-199-CPE was studied in 0.1 M

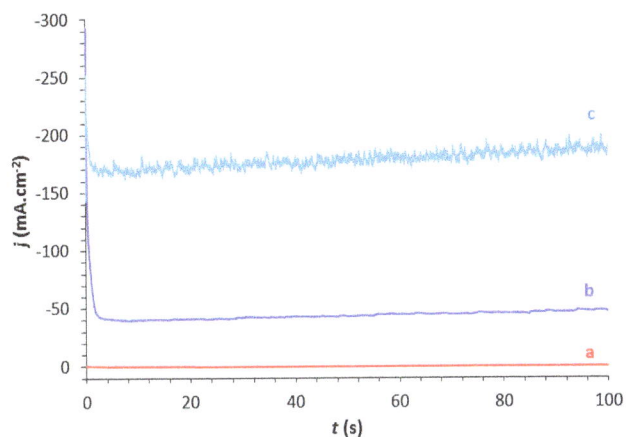

Fig. 9 Chronoamperometry on CPE (*a*), MOF-199-CPE (*b*) and Pd/SWCNTs@MOF-199-CPE (*c*) in 2 M H_2SO_4 with -0.6 V applied potential

phosphate buffer (pH = 5, 7, 9 and 11) and 1 and 2 M sulphuric acid solutions. The results of these experiments are illustrated in Fig. 8. By increasing the hydronium concentration, the current of hydrogen production increases, therefore, in this study the largest *j* was observed in 2 M sulphuric acid solution, so this solution was selected for the rest of experiments. More concentrated sulphuric acid, or other types of strong acids such as HNO_3 and HCl were not used because of their corrosive effects.

Chronoamperometry study

The electrode response stability was evaluated with chronoamperometry technique. In this experiment a constant potential (-0.6 V) was applied on CPE, MOF-199-CPE and Pd/SWCNTs@MOF-199-CPE under optimal conditions (Fig. 9). According to this figure the highest *j* for HER is observed for Pd/SWCNTs@MOF-199-CPE. Also the results of modified electrodes show that *j does* not decrease, but increases with time, and compare to CPE the current densities of HER on modified electrodes are considerably greater.

EIS studies

In this study before and after applying 20 successive CVs on the CPE, MOF-199-CPE and Pd/SWCNTs@MOF-199-CPE, the EIS test was done in 2 M H_2SO_4 in the frequency range of 5×10^{-2} to 1×10^5 Hz and at -0.25 V DC potential. Figure 10 shows that after applying CVs the wave height in the Bode plots, which is related to charge transfer resistance (R_{ct}), decreases. This observation confirms that successive CVs improve the rate of electron transfer. This improvement may be due to the activation of SWCNT and graphite surface.

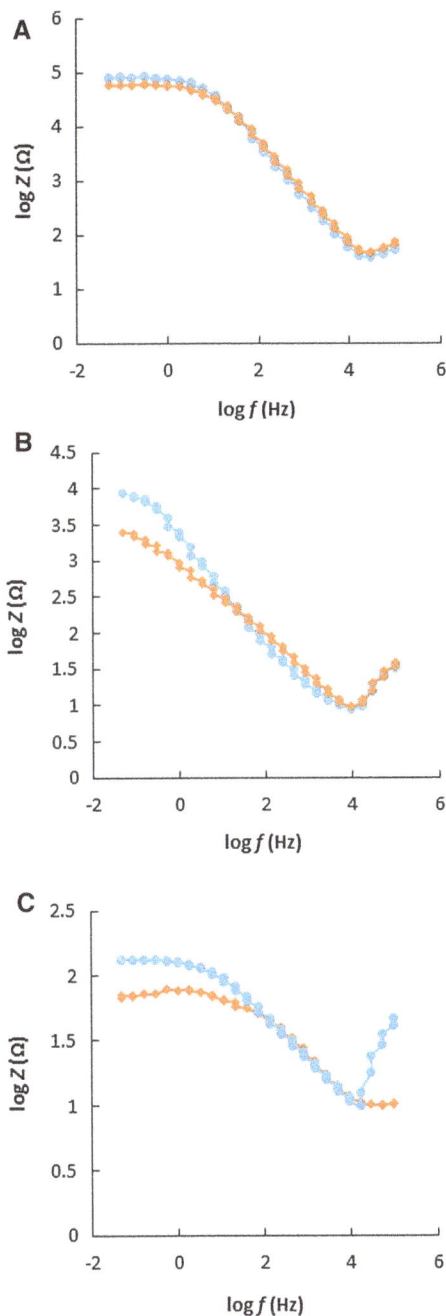

Fig. 10 *Bode plots* in 0.1 M KCl containing 10 mM $K_3Fe(CN)_6$ in the frequency range of 5×10^{-2} to 1×10^5 Hz before (**a**) and after (**b**) applying 20 successive CVs in the HER process for CPE (**a**), MOF-199-CPE (**b**), and Pd/SWCNTs@MOF-199-CPE (**c**)

Steady-state polarization curves for HER at bare and modified CPEs

To evaluate electrocatalytic activities of CPE and modified CPEs, steady-state polarization curves for HER at electrodes were measured by linear sweep voltammetry (LSV) at a very low scan rate (1 mV s^{-1}) (Fig. 11). There is a

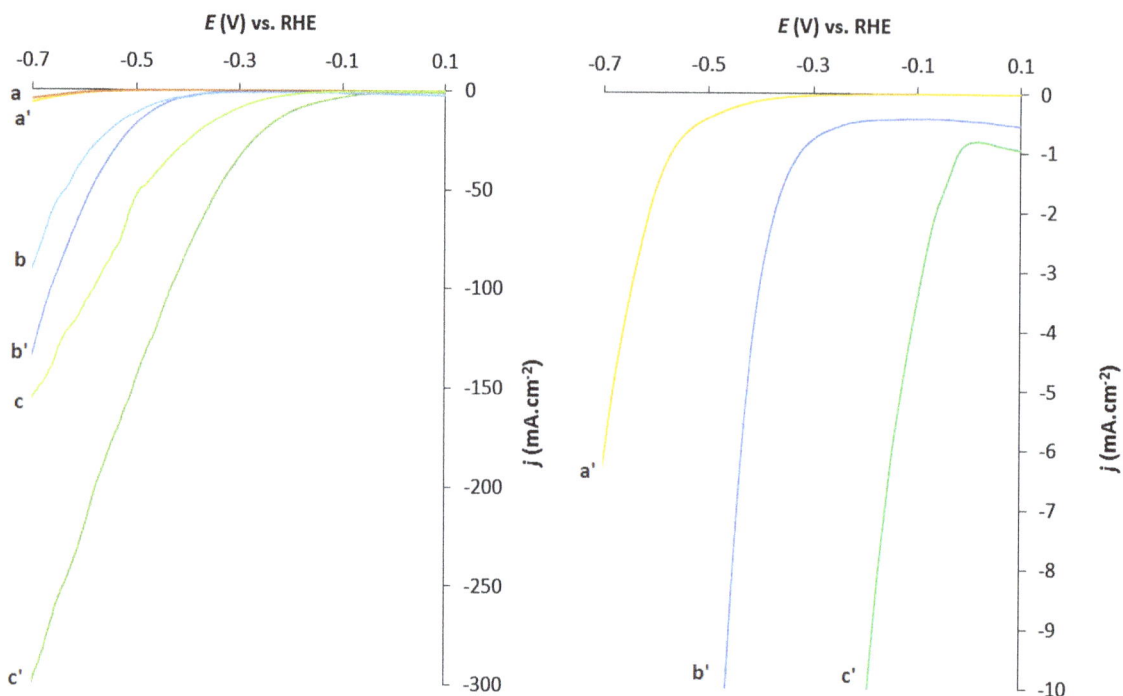

Fig. 11 LSV on CPE (a, a'), MOF-199-CPE (b, b') and Pd/SWCNTs@MOF-199-CPE (c, c') in 2 M H$_2$SO$_4$ with the scan rate of 1 mV s^{-1} before and after ($'$) applying 20 successive CVs in 2 M H$_2$SO$_4$. The *right panel* is the magnification of the LSVs after applying CVs for better vision of the HER onset potential

Table 2 Tafel slopes (b), exchange current densities (j_o) and transfer coefficients (α) for HER at CPE, MOF-199-CPE and Pd/SWCNTs@MOF-199-CPE

Electrode	b (mV decade^{-1})		j_o (mA cm^{-2})		α	
	Before	After	Before	After	Before	After
CPE	147	185	6.41×10^{-6}	6.09×10^{-5}	0.40	0.32
MOF-199-CPE	135	164	9.48×10^{-5}	5.89×10^{-4}	0.36	0.44
Pd/SWCNTs@ MOF-199-CPE	172	208	9.73×10^{-3}	1.19×10^{-1}	0.34	0.28

linear relation between η and log j under steady-state condition and large cathodic overpotential (η), (Tafel equation),

$$\eta = \frac{2.3RT}{\alpha F} \log j_o - \frac{2.3RT}{\alpha F} \log j \text{ (Tafel equation)},$$

where R is the ideal gas constant (8.314 J mol^{-1} K^{-1}), j_o is the exchange current density, T is the absolute temperature and F is the Faraday constant. For these three electrodes the values of Tafel slope (b), j_o and transfer coefficient (α) are summarized in Table 2.

The value of j_o increases for all three electrodes after applying CVs, which confirms again that the applied CVs improve the electron transfer rate (or decreases R_{ct}). The j_o values are in this order: Pd/SWCNTs@MOF-199-CPE > MOF-199-CPE > CPE. In right panel of the Fig. 11, the LSVs after applying CVs are magnified. According to these LSVs it is obvious that the lowest onset potential is for Pd/SWCNTs@MOF-199-CPE (about 0 V vs. RHE).

The results obtained from polarization and EIS studies both showed that the Pd/SWCNTs@MOF-199-CPE has the best performance related to CPE and MOF-199-CPE. The reason of this observation may be related to the presence of Pd and SWCNT in the composite structure. There are several reports which have used Pd as a catalyst for electrochemical hydrogen production [35, 36], but this metal is expensive, therefore, in this work the loaded Pd on the catalyst is low. Also there are some reports which have used carbon nanotubes in the electrode composition for electrochemical HER [37–39]. In addition, there are several reports which have applied the MOF [22–25], Pd [40, 41] and SWCNT [37, 38] as adsorbent for hydrogen, and one of the famous mechanisms for electrochemical HER in acidic media is based on hydrogen adsorption on the electrode surface in the rate determining step (Volmer reaction) [42]. Therefore, these materials by improving hydrogen adsorption can facilitate the HER progress.

In comparison with the first report for electrochemical HER with polyoxometalate-based metal organic frameworks (POMOFs) [29], although the onset potential is relatively higher, but it seems that the j value for the proposed electrode is greater. Also compared to two recently published reports for electrochemical HER on MOF-5 modified CPE [31] and Ni-based MOF modified GCE [30], the proposed modified CPE electrode has better current densities or onset potential for HER.

Conclusion

In this study, a nanocomposite of MOF-199 with SWCNT and Pd was synthesized and its characteristics was evaluated by XRD, BET, TGA and SEM. According to the rare reports for application of MOFs in electrochemical hydrogen generation, the performance of this composite as an electrocatalyst in CPE was studied, and some important parameters were optimized. The results show that this modifier has a good performance for HER, which compared to MOF-199 and bare CPE has the best performance. Also electrochemical studies showed that applying successive CVs or a constant potential on the electrodes surface does not decrease the electrode performance, but improves the electrocatalytic properties.

Acknowledgments We gratefully acknowledge the partial support of this work from the Research Council of the Iran University of Science and Technology.

References

1. Dincer, I.: Environmental and sustainability aspects of hydrogen and fuel cell systems. Int. J. Energy Res. **31**, 29–55 (2007)
2. Momirlan, M., Veziroglu, T.N.: The properties of hydrogen as fuel tomorrow in sustainable energy system for a cleaner planet. Int. J. Hydrogen Energy **30**, 795–802 (2005)
3. Liu, B., Li, Z.: A review: hydrogen generation from borohydride hydrolysis reaction. J. Power Sources **187**, 527–534 (2009)
4. Naterer, G.F., Suppiah, S., Stolberg, L., Lewis, M., Wang, Z., Daggupati, V., Gabriel, K., Dincer, I., Rosen, M.A., Spekkens, P., Lvov, S.N., Fowler, M., Tremaine, P., Mostaghimi, J., Easton, E.B., Trevani, L., Rizvi, G., Ikeda, B.M., Kaye, M.H., Lu, L., Pioro, I., Smith, W.R., Secnik, E., Jiang, J., Avsec, J.: Canada's program on nuclear hydrogen production and the thermochemical Cu–Cl cycle. Int. J. Hydrogen Energy **35**, 10905–10926 (2010)
5. Balta, M.T., Dincer, I., Hepbasli, A.: Geothermal-based hydrogen production using thermochemical and hybrid cycles: a review and analysis. Int. J. Energy Res. **34**, 757–775 (2010)
6. Shen, S., Guo, L., Chen, X., Ren, F., Mao, S.S.: Effect of Ag$_2$S on solar-driven photocatalytic hydrogen evolution of nanostructured CdS. Int. J. Hydrogen Energy **35**, 7110–7115 (2010)
7. Mangrulkar, P.A., Joshi, M.V., Kamble, S.P., Labhsetwar, N.K., Rayalu, S.S.: Hydrogen evolution by a low cost photocatalyst: bauxite residue. Int. J. Hydrogen Energy **35**, 10859–10866 (2010)
8. Zhao, W., Wang, Z., Shen, X., Li, J., Xu, C., Gan, Z.: Hydrogen generation via photoelectrocatalytic water splitting using a tungsten trioxide catalyst under visible light irradiation. Int. J. Hydrogen Energy **37**, 908–915 (2012)
9. Avasthi, K.S., Reddy, R.N., Patel, S.: Challenges in the production of hydrogen from glycerol—a biodiesel byproduct via steam reforming process. J. Proc. Eng. **51**, 423–429 (2013)
10. Khan, Z., Yusup, S., Ahmad, M.M., Rashidi, N.A.: Integrated catalytic adsorption (ICA) steam gasification system for enhanced hydrogen production using palm kernel shell. Int. J. Hydrogen Energy **39**, 3286–3293 (2014)
11. Wang, M., Wang, Z., Gong, X., Guo, Z.: The intensification technologies to water electrolysis for hydrogen production—a review. Renew. Sustain. Energy Rev. **29**, 573–588 (2014)
12. Yang, J., Cheng, F., Liang, J., Chen, J.: Hydrogen generation by hydrolysis of ammonia borane with a nanoporous cobalt–tungsten–boron–phosphorus catalyst supported on Ni foam. Int. J. Hydrogen Energy **36**, 1411–1417 (2011)
13. Raoof, J.B., Ojani, R., Esfeden, S.A., Nadimi, S.R.: Fabrication of bimetallic Cu/Pt nanoparticles modified glassy carbon electrode and its catalytic activity toward hydrogen evolution reaction. Int. J. Hydrogen Energy **35**, 3937–3944 (2011)
14. Ghaffarinejad, A., Sadeghi, N., Kazemi, H., Khajehzadeh, A., Amiri, M., Noori, A.: Effect of metal hexacyanoferrate films on hydrogen evolution reaction. J. Electroanal. Chem. **685**, 103–108 (2012)
15. Ghaffarinejad, A., Magsoudi, E., Sadeghi, N.: Hydrogen generation by shimalite Ni catalyst. Anal. Bioanal. Chem. **5**, 316–324 (2013)
16. Safavi, A., Kazemi, H., Kazemi, S.H.: In situ electrodeposition of graphene/nano-palladium on carbon cloth for electrooxidation of methanol in alkaline media. J. Power Sources **256**, 354–360 (2014)
17. Klein, N., Senkovska, I., Gedrich, K., Stoeck, U., Henschel, A., Mueller, U., Kaskel, S.: A mesoporous metal–organic framework. Angew. Chem. Int. Ed. **48**, 9954–9957 (2009)
18. O'Keeffe, M., Eddaoudi, M., Li, H., Reineke, T., Yaghi, O.M.: Frameworks for extended solids: geometrical design principles. J. Solid State Chem. **152**, 3–20 (2000)
19. Hoskins, B., Robson, R.: Design and construction of a new class of scaffolding-like materials comprising infinite polymeric frameworks of 3D-linked molecular rods. A reappraisal of the zinc cyanide and cadmium cyanide structures and the synthesis and structure of the diamond-related frameworks [N(CH$_3$)$_4$][Cu^{I}Zn^{II}(CN)$_4$] and Cu^I [4, 4′, 4″, 4‴-tetracyanotetraphenyl-methane] BF$_4$·xC$_6$H$_5$N. J. Am. Chem. Soc. **112**, 1546–1554 (1990)
20. Oh, M., Mirkin, C.A.: Chemically tailorable colloidal particles from infinite coordination polymers. Nature **438**, 651–654 (2005)
21. Lee, J., Farha, O.K., Roberts, J., Scheidt, K.A., Nguyen, S.T., Hupp, J.T.: Metal–organic framework materials as catalysts. Chem. Soc. Rev. **38**, 1450–1459 (2009)
22. Rowsell, J.L., Yaghi, O.M.: Strategies for hydrogen storage in metal–organic frameworks. Angew. Chem. Int. Ed. **44**, 4670–4679 (2005)
23. Murray, L.J., Dincă, M., Long, J.R.: Hydrogen storage in metal–organic frameworks. Chem. Soc. Rev. **38**, 1294–1314 (2009)
24. Moellmer, J., Moeller, A., Dreisbach, F., Glaeser, R., Staudt, R.: High pressure adsorption of hydrogen, nitrogen, carbon dioxide and methane on the metal–organic framework HKUST-1. J. Microporous Mesoporous Mat. **138**, 140–148 (2011)
25. Lim, D.W., Chyun, S.A., Suh, M.P.: Hydrogen storage in a potassium-ion-bound metal-organic framework incorporating crown ether struts as specific cation binding sites. Angew. Chem. Int. Ed. **53**, 7819–7822 (2014)
26. Wen, M., Mori, K., Kamegawa, T., Yamashita, H.: Amine-functionalized MIL-101 (Cr) with imbedded platinum nanoparticles as a durable photocatalyst for hydrogen production from water. Chem. Commun. **50**, 11645–11648 (2014)
27. Toyao, T., Saito, M., Dohshi, S., Mochizuki, K., Iwata, M., Higashimura, H., Horiuchi, Y., Matsuoka, M.: Development of a

Ru complex-incorporated MOF photocatalyst for hydrogen production under visible-light irradiation. Chem. Commun. **50**, 6779–6781 (2014)

28. He, J., Wang, J., Chen, Y., Zhang, J., Duan, D., Wang, Y., Yan, Z.: A dye-sensitized Pt@ UiO-66 (Zr) metal–organic framework for visible-light photocatalytic hydrogen production. Chem. Commun. **50**, 7063–7066 (2014)

29. Nohra, B., El Moll, H., Rodriguez Albelo, L.M., Mialane, P., Marrot, J., Mellot-Draznieks, C., O'Keeffe, M., Ngo Biboum, R., Lemaire, J., Keita, B., Nadjo, L., Dolbecq, A.: Polyoxometalate-based metal organic frameworks (POMOFs): structural trends, energetics, and high electrocatalytic efficiency for hydrogen evolution reaction. J. Am. Chem. Soc. **133**, 13363–13374 (2011)

30. Tian, T., Ai, L., Jiang, J.: Metal–organic framework-derived nickel phosphides as efficient electrocatalysts toward sustainable hydrogen generation from water splitting. RSC Adv. **5**, 10290–10295 (2015)

31. Yang, H.M., Song, X.L., Yang, T.L., Liang, Z.H., Fan, C.M., Hao, X.G.: Electrochemical synthesis of flower shaped morphology MOFs in an ionic liquid system and their electrocatalytic application to the hydrogen evolution reaction. RSC Adv. **4**, 15720–15726 (2014)

32. Xiang, Z., Cao, D., Shao, X., Wang, W., Zhang, J., Wu, W.: Facile preparation of high-capacity hydrogen storage metal–organic frameworks: a combination of microwave-assisted solvothermal synthesis and supercritical activation. Chem. Eng. Sci. **65**, 3140–3146 (2010)

33. Yang, S.J., Choi, J.Y., Chae, H.K., Cho, J.H., Nahm, K.S., Park, C.R.: Preparation and enhanced hydrostability and hydrogen storage capacity of CNT@ MOF-5 hybrid composite. Chem. Mater. **21**, 1893–1897 (2009)

34. Loera-Serna, S., Oliver-Tolentino, M.A., de Lourdes López-Núñez, M., Santana-Cruz, A., Guzmán-Vargas, A., Cabrera-Sierra, R., Beltrán, H.I., Flores, J.: Electrochemical behavior of [Cu3(BTC)2] metal–organic framework. The effect of the method of synthesis. J. Alloy. Compd. **540**, 113–120 (2012)

35. Searson, P.C.: Hydrogen evolution and entry in palladium at high current density. J. Acta Metal. Mater. **39**, 2519–2525 (1991)

36. Safavi, A., Kazemi, S.H., Kazemi, H.: Electrocatalytic behaviors of silver–palladium nanoalloys modified carbon ionic liquid electrode towards hydrogen evolution reaction. J. Fuel **118**, 156–162 (2014)

37. Fernandez, P., Castro, E., Real, S., Martins, M.: Electrochemical behavior of single walled carbon nanotubes—hydrogen storage and hydrogen evolution reaction. Int. J. Hydrogen Energy **34**, 8115–8126 (2009)

38. Prosini, P.P., Pozio, A., Botti, S., Ciardi, R.: Electrochemical studies of hydrogen evolution, storage and oxidation on carbon nanotube electrodes. J. Power Sources **118**, 265–269 (2013)

39. Paunović, P., Dimitrov, A.T., Popovski, O., Slavcheva, E., Grozdanov, A., Lefterova, E., Petruševski, G., Jordanov, S.H.: Effect of activation/purification of multiwalled carbon nanotubes (MWCNTs) on the activity of non-platinum based hypo-hyper d-electrocatalysts for hydrogen evolution. J. Mater. Res. Bull **44**, 1816–1821 (2009)

40. Conrad, H., Ertl, G., Latta, E.E.: Adsorption of hydrogen on palladium single crystal surfaces. Surf. Sci. **41**, 435–446 (1974)

41. Lopez, M.J., Cabria, I., Alonso, J.A.: Palladium clusters anchored on graphene vacancies and their effect on the reversible adsorption of hydrogen. J. Phys. Chem. C **118**, 5081–5090 (2014)

42. Abbaspour, A., Mirahmadi, E.: Electrocatalytic hydrogen evolution reaction on carbon paste electrode modified with Ni ferrite nanoparticles. J. Fuel **104**, 575–582 (2013)

9

Design and development of a novel magnetic camphor nanospheres core/shell nanostructure

Ali Maleki[1] ⓘ · Pedram Zand[1] · Zahra Mohseni[1]

Abstract In this work, the first synthesis, characterization and catalytic performance of Fe_3O_4/camphor core/shell nanospheres as a magnetic composite nanocatalyst were reported. The characterization was carried out by Fourier transform infrared spectroscopy, field-emission scanning electron microscopy, transmission electron microscopy, X-ray diffraction, energy-dispersive X-ray spectroscopy, thermogravimetric analysis, vibrating sample magnetometer, N_2 adsorption–desorption by Brunauer–Emmett–Teller and inductively coupled plasma analyses. The recoverable heterogeneous nanostructure catalyst was simply prepared and effectively employed in the one-pot multicomponent synthesis of β-amino carbonyl compounds in ethanol as a green solvent at room temperature. Further advantages of this new protocol are short reaction times, high yields, easy workup procedure, inexpensive and eco-friendly protocol and magnetically recoverability and several times reusability of the nanocatalyst without significant decrease in catalytic activity.

Keywords Fe_3O_4/camphor core/shell · Nanostructure · Nanocatalyst · Multicomponent reaction · β-Amino carbonyl · Mannich reaction

Introduction

Due to significance of catalysis, this area is a highly important object among various green chemistry principals [1]. The catalytic reactions may reduce energy requirements and decrease separation steps because of their increased selectivity. Moreover, catalysis is of crucial importance for the chemical industry to make key products like biologically important heterocyclic compounds and various fine chemicals [2, 3]. By utilizing catalysts, organic reactions can be more efficient and selective through decreasing of by-products and wastes [4]. The area of nanocatalysis which involves a substance or material with catalytic properties that possesses at least one nanoscale dimension in their structures is undergoing an explosive development. Nanocatalysis can help design catalysts with excellent activity, greater selectivity, and high stability [5–8]. Composite nanocatalysts, especially supported magnetic metal oxides nanoparticles (MNPs), have attracted considerable interest of researchers, because of their potential applications in chemical, biomedical and materials science. Due to intrinsic properties of Fe_3O_4 MNPs including high surface area, low toxicity and superparamagnetic behavior, they have received a great deal of attention in various academic and technological applications [9–14]. Recently, a number of such functionalized Fe_3O_4 nanoparticles, especially core/shell nanostructures have been employed in a range of organic transformations [15, 16]. They are widely used in the synthesis of biologically active compounds, such as drug delivery systems, magnetic resonance imaging (MRI), biosensors, bioseparation and magneto-thermal therapy [17]. On the other hand, some comprehensive reports have been described in the literature as micromeritics such as nanospheres, nanorods, nanofibers, nanoreefs, nanoboxes [18]. One of the most famous atom-efficient multicomponent reactions is Mannich reaction which involves carbon–carbon

✉ Ali Maleki
  maleki@iust.ac.ir

[1] Catalysts and Organic Synthesis Research Laboratory, Department of Chemistry, Iran University of Science and Technology, Tehran 16846-13114, Iran

bond formation between an enolizable ketone as a nucleophile and a Schiff base as an electrophile to provide β-amino carbonyl compounds [19]. Most of the previously described methods suffer from drawbacks, e.g., long reaction times, large amounts of catalysts, expensive reagents or catalysts, toxicity, harsh reaction conditions and low yields. As a result, design and introduction of green, biocompatible and inexpensive catalyst for the synthesis of β-amino carbonyl compounds is of prime importance.

Natural camphor is affordable and has also medicinal and catalyst properties. The cool and aromatic camphor can be used to alleviate skin itching and irritation. By applying camphor on the affected areas, the compound in camphor activates the nerve endings, and this produces a soothing sensation. It not only provides relief from itching and irritation but also reduces incidences of redness on the skin. Natural camphor includes other properties, such as soothes burns, cures acne, fungal and bacterial infections and promotes blood circulation [20]. Our research group after much research on natural camphor decided to use it for surface modification of Fe_3O_4 nanoparticles.

In continuation of our research on green nanocatalysts and organic synthesis [21–28], herein, we wish to report Fe_3O_4/camphor nanospheres as a novel, efficient, eco-friendly, superparamagnetic and heterogeneous catalyst and investigation of its application in the synthesis of β-amino carbonyl compounds **4** via the condensation reaction of aniline **1**, an aldehyde **2** and a ketone **3** (Scheme 1).

To the best of our knowledge, this is the first report of design, preparation, and characterization Fe_3O_4/camphor core/shell nanospheres and its application as a heterogeneous catalyst in the synthesis of β-amino carbonyl compounds. We have used camphor as a naturally abundant biocompatible material by combining with Fe_3O_4 nanoparticles to yield Fe_3O_4/camphor nanospheres as a recyclable, economical and inexpensive catalyst.

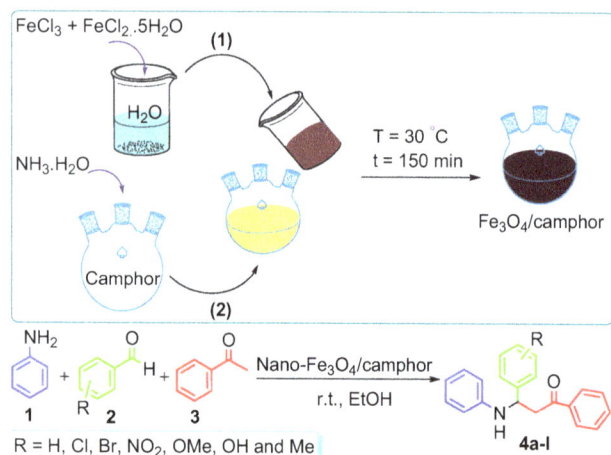

Scheme 1 Fe_3O_4/camphor-catalyzed synthesis of β-amino carbonyl products **4a-l**

Experimental

General

All the solvents, chemicals and reagents were purchased from Merck, Sigma and Aldrich. Melting points were measured on an Electrothermal 9100 apparatus and are uncorrected. Fourier transform infrared spectroscopy (FT-IR) spectra were recorded on a Shimadzu IR-470 spectrometer by the method of KBr pellet. ^{1}H and ^{13}C NMR spectra were recorded on a Bruker DRX-300 Avance spectrometer at 300.13 and 75 MHz, respectively. Field-emission scanning electron micrograph (FE-SEM) images were taken with Sigma Zeiss microscope with attached camera. Transmission electron microscopy (TEM) was performed on a Philips CM10 operated at an 80 kV electron beam accelerating voltage. Magnetic measurements of the solid samples were performed using a Lakeshore 7407 and Meghnatis Kavir Kashan Co., Iran vibrating sample magnetometer (VSMs). A Netzsch Thermoanalyzer STA 504 was used for the thermogravimetric analysis (TGA) with a heating rate of 10 °C/min under air atmosphere. Elemental analysis of the nanocatalyst was carried out by energy-dispersive X-ray spectroscopy (EDX) spectra recorded on Numerix DXP–X10P. X-ray diffraction patterns were recorded with an X' Pert Pro X-ray diffractometer operating at 40 mA, 40 kV. The specific surface area was measured via BET N_2 adsorption–desorption method using a Nansord92 instrument. Inductively coupled plasma (ICP) analyzed on a Shimadzu ICPS-7000.

Preparation of Fe_3O_4/camphor nanospheres

Fe_3O_4/camphor nanospheres employing coprecipitation in situ method has been prepared in accordance with the following conditions [26]. In a typical procedure, 2.4 g of $FeCl_3$ and 3 g of $FeCl_2·5H_2O$ were dissolved in 100 mL of deionized water. Then, 2 g of camphor and 10 mL of $NH_3.H_2O$ were mixed at 30 °C in a three-necked flask. Then, mixture of $FeCl_3$ and $FeCl_2·5H_2O$ was added in drops in ammonia containing camphor for 150 min at 30 °C. The obtained Fe_3O_4/camphor precipitate was washed repeatedly with deionized water until pH value decreased to 7 and then dried at 80 °C in an oven.

General procedure for the synthesis of β-amino carbonyl compounds 4a-l

A mixture of aromatic aldehyde (2.0 mmol) and amine (2.0 mmol) in absolute EtOH (4 mL) was added to the Fe_3O_4/camphor nanospheres (0.015 g). The mixture was stirred at room temperature for 10–15 min until the starting materials had almost disappeared (monitored by thin layer chromatograph, TLC). The ketone (3.0 mmol) was then added, and the mixture was stirred for the appropriate time until the reaction was complete as monitored by TLC. The catalyst was separated easily by an external magnet and reused as such for the next experiments. The products were isolated pure just by recrystallization from hot EtOH and no more purification was needed.

Spectral data of 3-(4-chlorophenyl)-1-phenyl-3-(N-phenylamino)propan-1-oneediates 4d

Yellow crystalline solid: mp 117 °C. FT-IR (KBr) (υ_{max}, cm^{-1}): 750, 825, 1510, 1602, 1668, 2850, 2918, 3055, 3392. ^{1}H NMR (300 MHz, DMSO-d_6) (δ, ppm): 2.09–2.36 (m, 1H, CH$_2$), 3.60–3.69 (m, 1H, CH$_2$), 5.01 (m, 1H, CH), 6.25 (m, 1H, NH), 6.65 (d, 2H, H-Ar), 6.77 (t, 1H, H-Ar), 7.13 (t, 2H, H-Ar), 7.26–7.62 (m, 7H, H-Ar), 7.87–7.91 (m, 2H, H-Ar). ^{13}C NMR (75 MHz, DMSO-d_6) (δ, ppm): 46.0,

54.3, 114.0, 118.1, 127.8, 128.1 128.7, 128.9, 129.1, 133.0, 133.4, 136.7, 141.5, 146.7, 197.8.

Results and discussion

Characterization of the prepared Fe$_3$O$_4$/camphor nanospheres

Fe$_3$O$_4$/camphor nanospheres were prepared by in situ coprecipitation method. As can be seen in Fig. 1, the FT-IR spectrum of the Fe$_3$O$_4$/camphor magnetic nanocatalyst can verify the preparation of the expected product. The bending vibration band at 584 cm^{-1} and stretching vibration band at 1627 cm^{-1} are induced by structure Fe–O vibration. A broad O–H stretch around 3419 cm^{-1} was observed in the Fe$_3$O$_4$. The C–H stretching vibrations observed at 2850 and 2921 cm^{-1} indicate that the camphor was successfully formed with Fe$_3$O$_4$ nanospheres.

To clarify the morphology of the nanocatalyst, FE-SEM images of Fe$_3$O$_4$/camphor are presented in Fig. 2. As it obvious, obtained catalyst has a uniform and spherical morphology. In addition, the claim of nanosized catalyst was proved by FE-SEM images. The average particle size distribution was about 50 nm.

Figure 3 shows the results of the EDX spectrum of the magnetite Fe$_3$O$_4$/camphor nanospheres. The presence of iron, carbon and oxygen elements clearly confirmed that Fe$_3$O$_4$/camphor nanocatalyst was synthesized.

Figure 4 shows the XRD measurements were performed with the dried powder samples of bare and mixture of the Fe$_3$O$_4$ NPs with camphor to identify the crystal phases present in the samples. The XRD pattern of a representative Fe$_3$O$_4$/camphor (curve a) along with bare Fe$_3$O$_4$ NPs

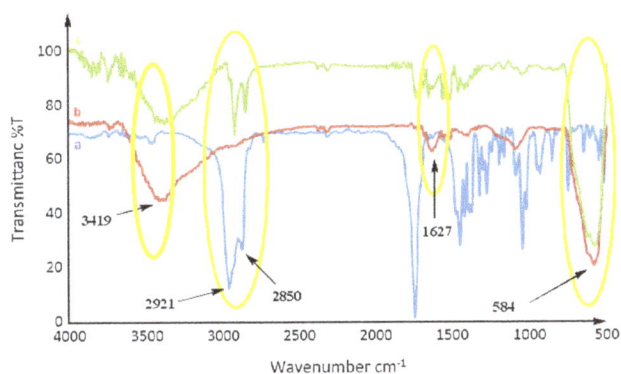

Fig. 1 The comparative FT-IR spectra of: *a* camphor, *b* Fe$_3$O$_4$ and *c* Fe$_3$O$_4$/camphor

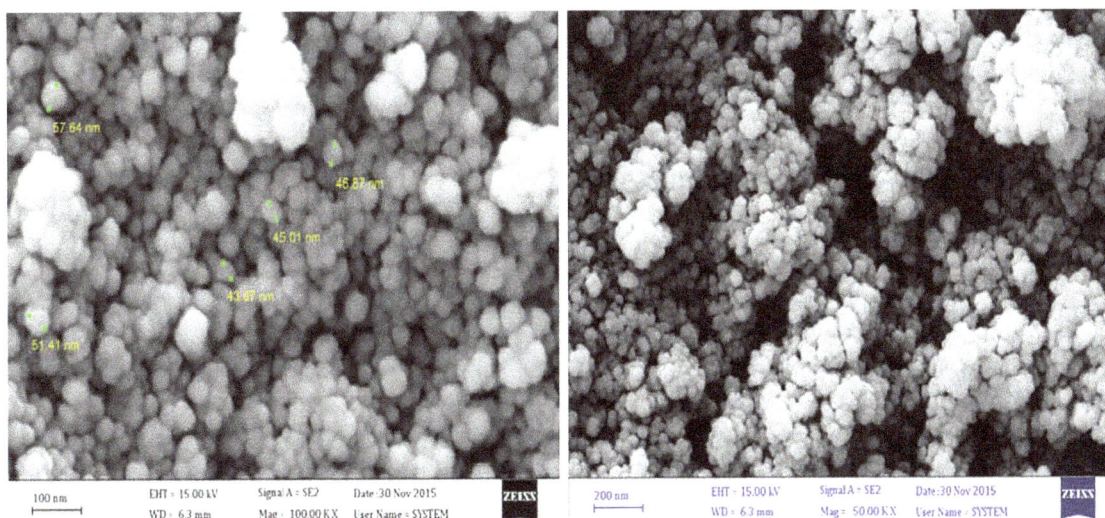

Fig. 2 FE-SEM images of Fe$_3$O$_4$/camphor nanocatalyst

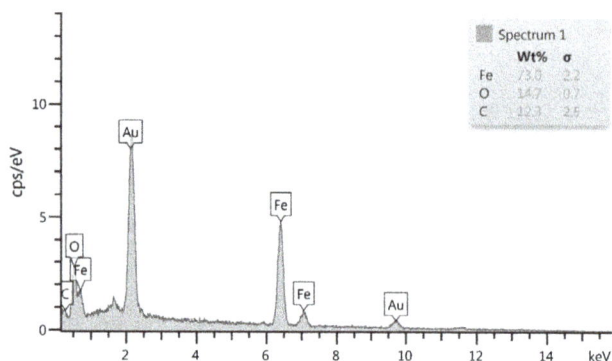

Fig. 3 EDX pattern of Fe$_3$O$_4$/camphor nanocatalyst

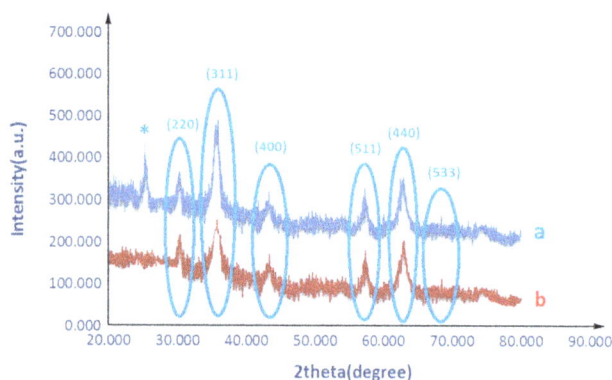

Fig. 4 XRD pattern of Fe$_3$O$_4$/camphor (a) and Fe$_3$O$_4$ (b).The symbol *asterisk* represents the camphor peak

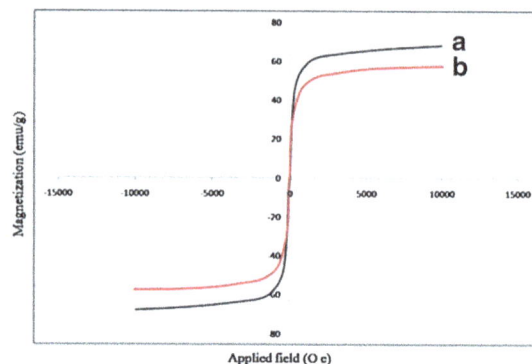

Fig. 5 VSM magnetization curves of the a Fe$_3$O$_4$ and b Fe$_3$O$_4$/camphor

(curve b) is shown. The pattern of the Fe$_3$O$_4$/camphor showed all the major peaks corresponding to Fe$_3$O$_4$. The peaks at $2\theta = 30.23, 35.63, 43.34, 57.25, 62.84$ and 74.38 can be assigned to the (2 2 0), (3 1 1), (4 0 0), (5 1 1), (4 4 0), and (5 3 3) planes, respectively. The Fe$_3$O$_4$ NPs was also clearly confirmed by corresponding reference card (JCPDS#19-0629). Additionally, one peak around $2\theta = 23°$ along with small peaks due to the camphor are observed in the case of the Fe$_3$O$_4$/camphor. This result confirmed that the mixture of Fe$_3$O$_4$ NPs with camphor was formed.

The magnetic properties of the bare Fe$_3$O$_4$ and Fe$_3$O$_4$/camphor were measured by VSM at room temperature. In Fig. 5, the hysteresis loops that are characteristic of superparamagnetic behavior can be clearly observed for all the nanoparticles. Superparamagnetism is the responsiveness to an applied magnetic field without retaining any magnetism after removal of the applied magnetic field. From the saturation magnetization value (Ms) of bare MNPs was found to be 68 emu g^{-1}. For Fe$_3$O$_4$/camphor the magnetization obtained at the same field was 49 emu g^{-1}, lower than that of bare Fe$_3$O$_4$. This is mainly attributed to the existence of camphor mixed with the nanoparticles.

TEM was used to observe the morphology of Fe$_3$O$_4$/camphor core/shell. Two different magnifications (a) and (b) TEM images of Fe$_3$O$_4$/camphor and one TEM image of bare Fe$_3$O$_4$ synthesized without the camphor (c) are shown in Fig. 6. The images (a) and (b) show that the nanostructure has nanospheres morphology. The natural camphor was suitably mixed in Fe$_3$O$_4$ nanoparticles. Therefore, natural camphor could reduce the aggregation of Fe$_3$O$_4$ nanoparticles and the particles dispersion was improved.

The thermal stability and composition ratio of camphor in the obtained Fe$_3$O$_4$/camphor nanospheres was measured through the TGA under air atmosphere with heating rate of 10 °C/min within temperature range of 0–800 °C. Figure 7 illustrates the TGA curve, depicting the variations of the residual masses of the samples with temperature. The organic material and magnetite of the sample were completely burned to generate gas products and converted into iron oxides at the elevated temperature, respectively. The composition ratio of camphor in the obtained Fe$_3$O$_4$/camphor nanospheres was estimated from the residual mass percentage. The magnetic content was about 86% and weight ratio of iron to camphor was 1:0.08. As shown in Fig. 7, the first weight loss stage (below 100 °C) can be ascribed to the evaporation of water molecules in the composition. The next weight loss stage was started from about 400 to complete at 600 °C that was because of the decomposition of natural camphor. No further obvious decomposition exhibits a high thermal stability in comparison with similar structures.

Furthermore, inductively coupled plasma (ICP) analysis of the catalysts was provided. For this purpose, the four solutions with concentrations of 20, 15, 10 and 5 ppm of iron salts as standards were prepared. Then, a solution containing the nanocatalyst was prepared. The analysis revealed that the metal contents of the catalyst was close (±0.4%) to the target metal content of 75 wt% Fe. ICP

Fig. 6 TEM images of Fe$_3$O$_4$/camphor core/shell nanospheres with different magnifications: **a** 10 nm and **b** 25 nm and **c** TEM image of bare Fe$_3$O$_4$ nanoparticles

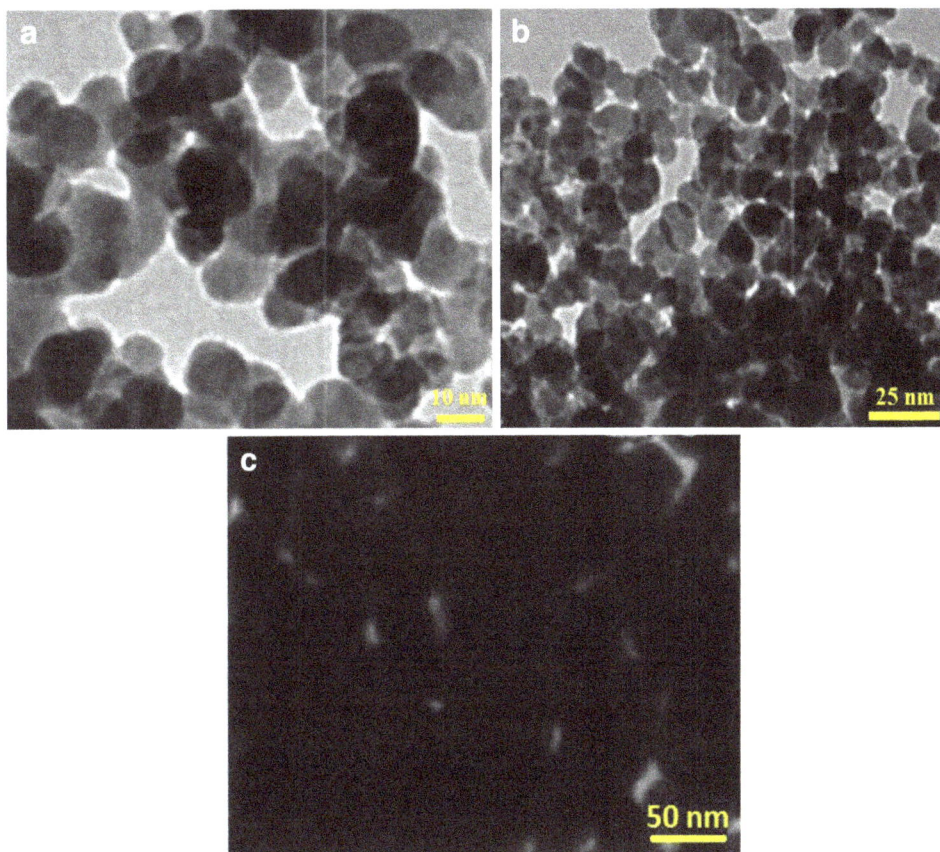

Fig. 7 TGA of Fe$_3$O$_4$/camphor nanospheres

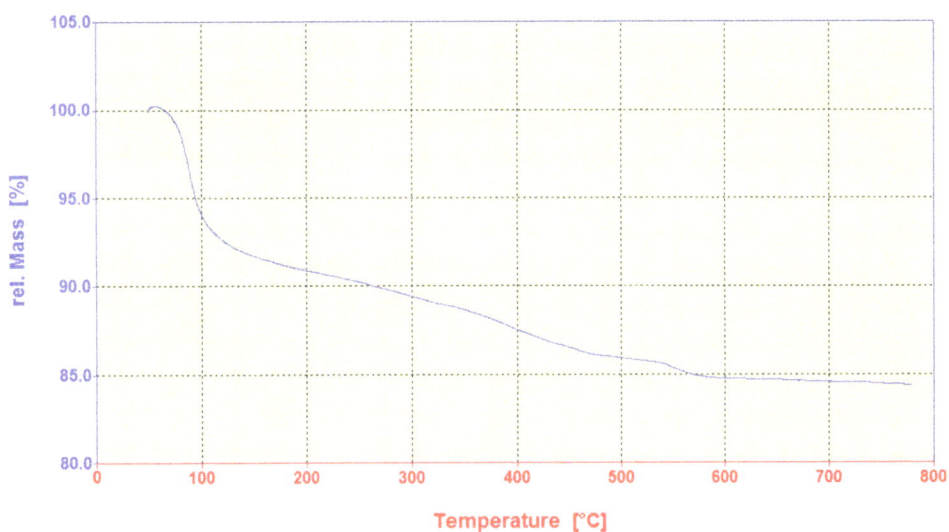

results are almost matched with the results obtained from EDX pattern analysis (73 wt% Fe).

The specific surface area of bare Fe$_3$O$_4$ nanoparticles and Fe$_3$O$_4$/camphor were measured using Brunauer–Emmett–Teller (BET) analysis with N$_2$ adsorption–desorption. BET surface areas of bare Fe$_3$O$_4$ nanoparticles and Fe$_3$O$_4$/camphor were 95.25 and 53.04 m^2 g^{-1} Respectively. Reduced specific surface area of Fe$_3$O$_4$/camphor nanospheres compared with bare Fe$_3$O$_4$ nanoparticles can confirm that the mixture of camphor with Fe$_3$O$_4$ nanoparticles was made.

Table 1 Optimization of the model reaction conditions

Entry	Fe_3O_4/camphor nanospheres (g)	Solvent	Time (min)	Yield (%)
1	0.005	EtOH	150	68
2	0.015	EtOH	35	93
3	0.020	EtOH	120	72
4	0.025	EtOH	80	75
5	0.015	MeOH	100	50
6	0.015	H_2O	80	55
7	0.015	MeCN	110	45
8	0.015	DMF	110	53
9	0.015	*n*-Hexane	120	32
10	0.015	Ether	120	35
11	0.015	Solvent-free	100	20

Table 2 Comparison of some catalyst effects with present nanocatalyst on the model reaction

Entry	Catalyst	Temperature (°C)	Time	Yield (%)	Ref.
1	Nano-Fe_3O_4/camphor	r.t.	35 min	93	This work
2	Nano-Fe_3O_4	r.t.	45 min	86	This work
3	Camphor	r.t.	1 h	75	This work
4	–	r.t.	48 h	15	This work
5	$BiCl_3$	r.t.	4 h	69	[29]
6	$HClO_4$–SiO_2	r.t	5 h	68	[30]
7	$CaCl_2$	80	2 h	46	[31]
8	Sulfated MCM-41	Reflux	7 h	83	[32]
9	ZnO-nanoparticles	60	10 h	85	[19]

Table 3 Fe_3O_4/camphor nanosphere-catalyzed synthesis of β-amino carbonyl compounds **4a-l**

Entry	R	Product	Time (min)	TON	TOF (h^{-1})	Yield[a] (%)	Mp (°C) Found	Mp (°C) Reported
1	4-OMe	**4a**	60	19.2	0.32	75	148	147–149 [33]
2	4-Br	**4b**	35	23.07	0.659	90	128	127–129 [34]
3	4-Me	**4c**	60	17.94	0.299	70	131	130–132 [35]
4	4-Cl	**4d**	35	23.84	0.681	93	117	116–119 [36]
5	H	**4e**	60	20.51	0.341	80	170	169–170 [33]
6	3-OH	**4f**	60	19.23	0.320	75	126	125–127 [37]
7	3-NO_2	**4g**	40	21.79	0.544	85	141	140–142 [38]
8	3-Br	**4h**	40	23.07	0.576	90	95	95–96 [39]
9	3-OMe	**4i**	60	17.94	0.299	70	107	106–107 [40]
10	2-NO_2	**4j**	50	20.51	0.410	80	160	158–162 [31]
11	2-Cl	**4k**	50	21.79	0.435	85	52	52–53 [40]
12	2-OMe	**4l**	60	16.66	0.277	65	122	122–124 [41]

[a] Isolated yield

Catalytic application of Fe_3O_4/camphor in the synthesis of β-amino carbonyl compounds

Initially, to optimize the reaction conditions for the synthesis of β-amino carbonyl compounds, various parameters on the reaction of aniline (2 mmol), 4-chlorobenzaldehyde (2 mmol) and acetophenone (3 mmol) as a pilot test were studied. First, the effect of catalyst amount on the reaction yield was studied. It was found that using 0.015 g of the Fe_3O_4/camphor nanocatalyst is sufficient to complete the reaction and gives **4d** after 35 min in 93% yield in 4 mL of EtOH as a green solvent at room temperature.

Scheme 2 The proposed mechanism for the formation of β-amino carbonyl compounds **4a-l**

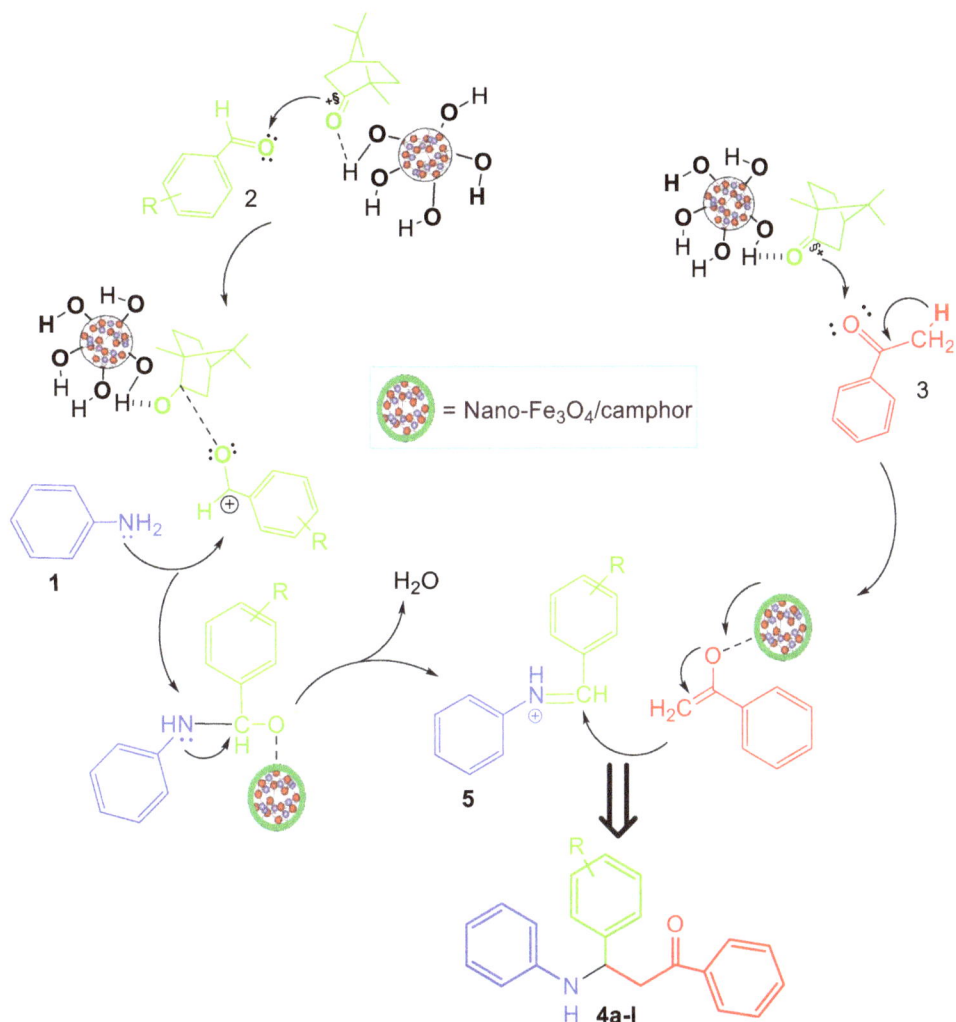

To compare the efficiency of ethanol, several solvents with different polarities were tested using the model reaction in the presence of Fe_3O_4/camphor (Table 1). As it is obvious from the results, it was found that EtOH was the superior solvent for this study than MeOH, H_2O, MeCN, DMF and also under solvent-free conditions in the presence of 0.015 g of Fe_3O_4/camphor catalyst at room temperature.

Finally, a comparison was made between the present work and others which were reported earlier for the synthesis of **4d**. The results summarized in Table 2 clearly demonstrate the superiority of the present work over all the earlier reported ones in energy consumption, high-yield products and catalyst reusability.

The scope and limitations of the application of Fe_3O_4/camphor in the synthesis of β-amino carbonyl compounds were investigated under optimized conditions. The summarized results in Table 3 show that all products, including electron-withdrawing-, electron-releasing- and halogen-substituted starting materials, were obtained in good-to-excellent yields after appropriate reaction times. The

corresponding reaction times, yields, turnover number (TON) and turnover frequency (TOF) values, and also found and literature melting points of the products are listed in Table 3.

The suggested possible mechanism for the formation of β-amino carbonyl compounds **4a-l** is shown in Scheme 2. At first, the carbonyl group of aldehyde **2** was activated by Fe_3O_4/camphor nanospheres and become electrophilic group. Then, an amine **1** attacks the activated aldehyde **2** to form iminium **5**. In addition, the enol form of ketone **3** was formed via tautomerisation in the presence of the nanocatalyst. Finally, the enol was added to **5** to yield the desired product **4** [42].

Catalyst recyclability examination for Fe_3O_4/camphor

To study the catalyst reusability, the model reaction was carried out using 0.015 g of Fe_3O_4/camphor. After completion of the reaction, nanospheres were separated using

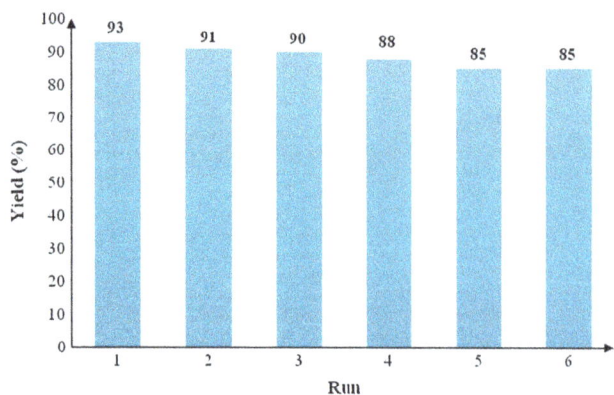

Fig. 8 Recycling of Fe_3O_4/camphor nanocatalyst for the synthesis of **4d**

an external magnet and washed with ethanol and water, dried and reused in another set of reaction. It was observed that the catalyst can be reused for six times without any significant reduction in product yield (Fig. 8). In addition, no extra care should be taken to store or handle the catalyst in recovery process; because, it does not have any air or moisture-sensitive nature.

Conclusions

In summary, eco-friendly, green and cost-effective Fe_3O_4/camphor core/shell nanospheres as a reusable superparamagnetic nanocatalyst was described through a facile preparation process. Then, its morphology, structure and properties were characterized by FT-IR, TEM, TGA, EDX, FE-SEM, VSM, XRD, BET and ICP measurements. Then, its catalytic performance was studied in the synthesis of β-amino carbonyl compounds as Mannich products in high-to-excellent yields at room temperature. The nanocatalyst was easily recovered using an external magnet and efficiently reused several times without significant loss of activity. Furthermore, to the best of our literature survey, the present work is the first report on design, preparation and catalytic investigation of magnetized camphor.

Acknowledgements The authors gratefully acknowledge the partial support from the Research Council of the Iran University of Science and Technology.

References

1. Anastas, P.T., Kirchhoff, M.M., Williamson, T.C.: Catalysis as a foundational pillar of green chemistry. Appl. Catal. A **221**, 3–13 (2001)
2. Anastas, P.T., Kirchhoff, M.M.: Origins, current status, and future challenges of green chemistry. Acc. Chem. Res. **35**, 686–694 (2002)
3. Vekariya, R.H., Patel, K.D., Patel, H.D.: Melamine trisulfonic acid (MTSA): an efficient and recyclable heterogeneous catalyst in green organic synthesis. RSC Adv. **5**, 90819–90837 (2015)
4. Zeynizadeh, B., Karimkoshteh, M.: Magnetic Fe_3O_4 nanoparticles as recovery catalyst for preparation of oximes under solvent-free condition. J. Nanostruct. Chem. **3**, 57–63 (2013)
5. Polshettiwar, V., Luque, R., Fihri, A., Zhu, H., Bouhrara, M., Basset, J.M.: Magnetically recoverable nanocatalysts. Chem. Rev. **111**, 3036–3075 (2011)
6. Shylesh, S., Schünemann, V., Thiel, W.R.: Magnetically separable nanocatalysts: bridges between homogeneous and heterogeneous catalysis. Angew. Chem. Int. Ed. **49**, 3428–3459 (2010)
7. Lu, A., Salabas, E., Schüth, F.: Magnetic nanoparticles: synthesis, protection, functionalization, and application. Angew. Chem. Int. Ed. **46**, 1222–1244 (2007)
8. Maleki, A., Kamalzare, M., Aghaei, M.: Efficient one-pot four-component synthesis of 1,4-dihydropyridines promoted by magnetite/chitosan as a magnetically recyclable heterogeneous nanocatalyst. J. Nanostruct. Chem. **5**, 95–105 (2015)
9. Montazeri, H., Amani, A., Shahverdi, H.R., Haratifar, E., Shahverdi, A.R.: Separation of the defect-free Fe_3O_4-Au core/shell fraction from magnetite-gold composite nanoparticles by an acid wash treatment. J. Nanostruct. Chem. **3**, 25–31 (2013)
10. Khoee, S., Kavand, A.: A new procedure for preparation of polyethylene glycol-grafted magnetic iron oxide nanoparticles. J. Nanostruct. Chem. **3**, 111 (2014)
11. Maleki, A., Rahimi, R., Maleki, S.: Preparation and characterization of magnetic chlorochromate hybrid nanomaterials with triphenylphosphine surface-modified iron oxide nanoparticles. J. Nanostruct. Chem. **4**, 153–160 (2014)
12. Li, X., Cao, W.C., Liu, Y.G.: The property variation of magnetic mesoporous carbon modified by aminated hollow magnetic nanospheres: synthesis, characteristic and sorption. ACS Sustain. Chem. Eng. **5**, 179–188 (2017)
13. Li, X., Wang, S.F.: Adsorption of Cu(II), Pb(II) and Cd(II) ions from acidic aqueous solutions by diethylenetriaminepentaacetic acid modified magnetic graphene oxide. J. Chem. Eng. Data **62**, 407–416 (2017)
14. Zhu, Y., Xue, J., Xu, T., He, G., Chen, H.: Enhanced photocatalytic activity of magnetic core–shell $Fe_3O_4@Bi_2O_3$–RGO heterojunctions for quinolone antibiotics degradation under visible light. J. Mater. Sci. **28**, 1–10 (2017)
15. Petranovska, A.L., Abramov, N.V., Turanska, S.P., Gorbyk, P.P., Kaminskiy, A.N., Kusyak, V.V.: Adsorption of cis-dichlorodiamineplatinum by nanostructures based on single-domain magnetite. J. Nanostruct. Chem. **5**, 257–258 (2015)
16. Abramov, N.V., Turanska, S.P., Kusyak, A.P., Petranovska, A.L., Gorbyk, P.P.: Synthesis and properties of magnetite/hydroxyapatite/doxorubicin nanocomposites and magnetic liquids based on them. J. Nanostruct. Chem. **6**, 223–233 (2016)
17. Zhu, Y., Fang, Y., Kaskel, S.: Folate-conjugated $Fe_3O_4@SiO_2$ hollow mesoporous spheres for targeted anticancer drug delivery. J. Phys. Chem. C **114**, 16382–16388 (2010)
18. Agam, M., Guo, Q.: Electron beam modification of polymer nanospheres. J. Nanosci. Nanotechnol. **7**, 3615–3619 (2007)
19. Kundu, K., Nayak, S.K.: (±)-Camphor-10-sulfonic acid catalyzed direct one-pot three-component Mannich type reaction of alkyl (hetero) aryl ketones under solvent-free conditions: application to the synthesis of aminochromans. RSC Adv. **2**, 480–486 (2011)
20. Cardoso, M.S., Correia, I., Galvao, A.M., Marques, F., Carvalho, M.F.: Synthesis of Ag(I) camphor sulphonylimine complexes and assessment of their cytotoxic properties against cisplatin-resistant

A2780cisR and A2780 cell lines. J. Inorg. Biochem. **166**, 55–63 (2017)

21. Maleki, A., Paydar, R.: Graphene oxide–chitosan bionanocomposite: a highly efficient nanocatalyst for the one-pot three-component synthesis of trisubstituted imidazoles under solvent-free conditions. RSC Adv. **5**, 33177–33184 (2015)

22. Maleki, A., Aghaei, M., Ghamari, N.: Efficient synthesis of 2, 3-dihydroquinazolin-4(1H)-ones in the presence of ferrite/chitosan as a green and reusable nanocatalyst. Chem. Lett. **44**, 259–261 (2015)

23. Maleki, A., Aghaei, M., Ghamari, N.: Facile synthesis of tetrahydrobenzoxanthenones via a one-pot three-component reaction using an eco-friendly and magnetized biopolymer chitosan-based heterogeneous nanocatalyst Appl. Organometal. Chem. **30**, 939–942 (2016)

24. Maleki, A., Akhlaghi, E., Paydar, R.: Design, synthesis, characterization and catalytic performance of a new cellulose-based magnetic nanocomposite in the one-pot three-component synthesis of α-aminonitriles. Appl. Organometal. Chem. **30**, 382–386 (2016)

25. Maleki, A., Kamalzare, M.: Green synthesis of 1,4-benzodiazepines over La_2O_3 and $La(OH)_3$ catalysts: possibility of Langmuir-Hinshelwood adsorption. Catal. Commun. **53**, 67–71 (2014)

26. Maleki, A., Zand, P., Mohseni, Z.: Fe_3O_4@PEG-SO_3H rod-like morphology along with the spherical nanoparticles: novel green nanocomposite design, preparation, characterization and catalytic application. RSC Adv. **6**, 110928–110934 (2016)

27. Maleki, A.: One-pot three-component synthesis of pyrido[2′,1′:2,3]imidazo[4,5-c]isoquinolines using Fe_3O_4@SiO_2–OSO_3H as an efficient heterogeneous nanocatalyst. RSC Adv. **4**, 64169–64173 (2014)

28. Maleki, A., Rabbani, M., Shahrokh, Sh: Preparation and characterization of a silica-based magnetic nanocomposite and its application as a recoverable catalyst for the one-pot multicomponent synthesis of quinazolinone derivatives. Appl. Organometal. Chem. **29**, 809–814 (2015)

29. Li, H., Zeng, H.Y., Shao, H.W.: Bismuth(III) chloride-catalyzed one-pot Mannich reaction: three-component synthesis of β-amino carbonyl compounds. Tetrahedron Lett. **50**, 6858–6860 (2009)

30. Bigdeli, M.A., Nemati, F., Mahdavinia, G.H.: $HClO_4$-SiO_2 catalyzed stereoselective synthesis of β-amino ketones via a direct Mannich-type reaction. Tetrahedron Lett. **48**, 6801–6804 (2007)

31. Kulkarni, P., Totawar, B., Zubaidha, P.K.: An efficient synthesis of β-aminoketone compounds through three-component Mannich reaction catalyzed by calcium chloride. Monatsh. Chem. **143**, 625–629 (2012)

32. Vadivel, P., Maheswari, C.S., Lalitha, A.: Synthesis of β-amino carbonyl compounds via Mannich reaction using sulfated MCM-41. Int. J. Innov. Tech. Expl. Eng. **2**, 267–270 (2013)

33. Wang, X.C., Zhang, L.J., Zhang, Zh, Quan, ZhJ: PEG-OSO_3H as an efficient and recyclable catalyst for the synthesis of β-amino carbonyl compounds via the Mannich reaction in PEG–H_2O. Chin. Chem. Lett. **23**, 423–426 (2012)

34. Shirini, F., Akbari-Dadamahaleh, S., Mohammad-Khah, A.: Rice-husk-supported $FeCl_3$ nano-particles: introduction of a mild, efficient and reusable catalyst for some of the multi-component reactions. C. R. Chim. **16**, 945–955 (2013)

35. Bahrami, K., Khodaei, M.M., Mohammadi, M., Babajani, N.: Ethane-1,2-diaminium hydrogen sulfate: recyclable organocatalyst for one-pot synthesis of β-amino ketones by a three-component Mannich reaction. J. Chem. Res. **38**, 223–225 (2014)

36. Maghsoodlou, M.T., Khorshidi, N., Mousavi, M.R., Hazeri, N., Habibi-Khorassani, S.M.: Starch solution as an efficient and environment-friendly catalyst for one-pot synthesis of β-aminoketones and 2,3-dihydroquinazolin-4(1H)-ones in EtOH. Res. Chem. Intermed. **41**, 7497–7508 (2015)

37. Luo, H.T., Kang, Y.R., Nie, H.Y., Yang, L.M.: Sulfamic acid as a cost-effective and recyclable catalyst for β-amino carbonyl compounds synthesis. J. Chin. Chem. Soc. **56**, 186–195 (2009)

38. Davoodnia, A., Tavakoli-Nishaburi, A., Tavakoli-Hoseini, N.: Carbon-based solid acid catalyzed one-pot Mannich reaction: a facile synthesis of β-amino carbonyl compounds. Bull. Korean Chem. Soc. **32**, 635–638 (2011)

39. Khan, A.T., Parvin, T., Choudhury, L.H.: Bromodimethylsulfonium bromide catalyzed three-component Mannich-type reactions. Eur. J. Org. Chem. **5**, 834–839 (2008)

40. Wu, Y., Chen, Ch., Jia, G., Zhu, X., Sun, H., Zhang, G., Zhang, W., Gao, Z.: Salicylato titanocene complexes as cooperative organometallic Lewis acid and Brønsted acid catalysts for three-component Mannich reactions. Chem. Eur. J. **20**, 8530–8535 (2014)

41. Sharghi, H., Jokar, M.: Highly stereoselective facile synthesis of β-amino carbonyl compounds via a Mannich-type reaction catalyzed by γ-Al_2O_3/$MeSO_3$H (alumina/methanesulfonic acid: AMA) as a recyclable, efficient, and versatile heterogeneous catalyst. Can. J. Chem. **88**, 14–26 (2010)

42. Mannich, C., Krösche, W.: About a condensation product of formaldehyde, ammonia and antipyrin. Arch. Pharm. **250**, 647–667 (1912)

10

Antibacterial activity of ZnO films prepared by anodizing

Shalaleh Gilani[1] · Mohammad Ghorbanpour[2] · Aiyoub Parchehbaf Jadid[3]

Abstract In the present work, anodizing of zinc foil was investigated in NaOH and oxalic acid electrolytes under the influence of different concentrations of the electrolyte, while temperature and voltage were kept constant. Anodized zinc plates were characterized using scanning electron microscope (SEM), UV–Vis diffuse reflectance spectroscopy (DRS UV–Vis), and X-ray diffraction (XRD) analysis. Characterization of anodized Zn plates using SEM showed that their morphology was significantly influenced by the type and concentration of anodizing electrolyte. XRD analysis indicated that the ZnO thin films were of hexagonal wurtzite structures. From contact angle measurements, it has been observed that the contact angle of anodized film is higher than that of pure zinc foil. Antibacterial results suggest that the parent zinc foil did not show the antibiotic activity, but the anodized zinc oxide is effective both toward Gram-positive bacteria and Gram-negative bacteria.

Keywords Anodizing · ZnO film · Antibacterial activity · Electrolyte

✉ Mohammad Ghorbanpour
ghorbanpour@uma.ac.ir

[1] Department of Food Science, Islamic Azad University, Sarab Branch, Sarab, Iran

[2] Chemical Engineering Department, University of Mohaghegh Ardabili, Ardabil, Iran

[3] Department of Chemistry, Islamic Azad University, Ardabil Branch, Ardabil, Iran

Introduction

Zinc oxide is a low-cost, non-toxic material with a wide bandgap of 3.37 eV, natural n-type electrical conductivity, and having a wurtzite structure. Because of its optical, electrical, and piezoelectric properties, this semiconductor is used as photocatalyst. Other applications of zinc oxide are light-emitting diodes, lasers, field-emission devices, and chemical sensors [1]. In several surveys over the past two decades, zinc oxide has been shown to possess activity against a broad spectrum of Gram-positive and Gram-negative bacteria [2–5]. Sawai et al. attributed the antimicrobial action of zinc oxide powder slurry to the liberation of hydrogen peroxide; they suggested that hydrogen peroxide crosses the microbial cell membrane, resulting in growth inhibition or destruction [2]. In subsequent work, Sawai et al. showed efficacy by zinc oxide against *S. aureus*, which was attributed to the strong affinity between zinc oxide and *S. aureus* cells [3]. Zhang et al. showed that zinc oxide exhibits bacteriostatic activity against *E. coli*. Scanning electron microscopy data suggested that zinc oxide–bacteria direct interactions may have damaging and breakdown of bacterial cell membranes [4]. Skoog et al. suggest that zinc oxide-coated nanoporous alumina membranes have activity against some microorganisms, such as *B. subtilis*, *E. coli*, *S. aureus*, and *S. epidermidis* in agar diffusion assays. On the other hand, the zinc oxide-coated membranes did not show activity against *P. aeruginosa*, *E. faecalis*, and *C. albicans* [5].

To date, varying methods were reported to fabricate ZnO nanostructures, for example, vapor-phase transport, metalorganic chemical-vapor deposition, laser ablation, thermal decomposition, a template-directed method, and chemical synthesis have been engaged to produce ZnO hexagram whiskers, quantum dots, nanorods, nanowires, etc. [1–9]. These ZnO nanostructures are generally fabricated under complex process control, high reaction temperatures, long reaction times, expensive chemicals, and a specific method for specific nanostructures [3–5]. Among these methods, electrochemical synthesis is favored by many due to its simplicity, low-temperature operation process, viability of commercial production, flexibility, and relatively low cost. Anodizing is a well-known route to synthesize low-dimensional self-organized structures. As it was discussed previously, the antibacterial effect of ZnO nanoparticles was extensively studied. On the other hand, many works have been done on the anodizing of zinc foil. But up to our knowledge, there are no reports on the antibacterial properties of anodized ZnO structures in the literature. In the present work, anodizing of the zinc foil was investigated in NaOH and oxalic acid electrolytes under the influence of different concentrations of the electrolyte, while the temperature and voltage were kept constant.

Experimental

Sample preparation

The electrolytes for zinc anodizing were 0.1, 0.3, and 0.5 M NaOH and oxalic acid aqueous solutions. A steel sheet (Dirgodazazar Co., Tabriz, Iran) was used as the cathode. A zinc foil (0.1-mm thick and purity >99.9 %, Dirgodazazar Co., Tabriz, Iran) sonicated in ethanol for 5 min with a surface area of 0.78 cm^2 and was used as the working electrode. The anodizing process was conducted at a constant voltage of 10 V for 60 min at room temperature. Immediately after anodizing, the zinc foil was washed with distilled water and dried in a warm airflow.

Characterization

The morphology of samples was observed with a scanning electron microscope (LEO 1430VP, Germany). X-ray diffraction (XRD) was carried out using Philips PW 1050 diffractometer (The Netherlands). UV–Vis diffuse reflectance spectroscopy (DRS UV–Vis) was taken in the wavelength range of 200–800 nm using a spectrophotometer (Scinco S4100, S. Korea).

Static contact angles were measured with a handmade contact angle meter at room temperature. Water droplets of approximately 5 μL were dropped gently onto the surface of a sample. Three points of each sample were tested, and the average value of the three left and right contact angles was calculated as the determined static contact angle [10, 11].

Antibacterial activity

The bacterial strains used for the antibacterial activity were Gram-negative *E. coli* (PTCC 1270) and Gram-positive *S. aureus* (PTCC 1112) obtained from the Iranian Research Organization for Science and Technology (Tehran, Iran). The antibacterial activity of the samples was tested by the agar diffusion test. Samples were exposed to bacteria in solid media (nutrient agar), and the inhibition zone around each sample was quantified and put down as the antibacterial effect. Agar plates were inoculated with 100-μL suspensions of bacteria. Then, anodized samples were placed on agar plates and covered at 37 °C for 24 h. The inhibition zone for bacterial growth was observed visually [12].

Results and discussion

Development mechanism of ZnO film, the growth mechanism, and formation of ZnO layer on the surface of Zn sheet take place under the influence of constant applied voltage. When the acidic solution (oxalic acid) was employed as the electrolyte, as described in most studies [7, 8], the possible mechanism for the process can be expressed equally:

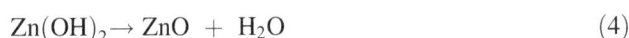

$$H_2O \rightarrow H^+ + OH^- \tag{1}$$

$$Zn \rightarrow Zn^{2+} + 2e^- \tag{2}$$

$$Zn^{2+} + 2 OH^- \rightarrow Zn(OH)_2 \tag{3}$$

$$Zn(OH)_2 \rightarrow ZnO + H_2O \tag{4}$$

The zinc sheet (anode) was converted to Zn^{2+} ions by releasing two electrons which moved toward the cathode. The water molecules were ionized into H^+ and OH^- ions. H_2 were liberated at the cathode; hence, pH of the solution was changed from the acidic to basic one. Finally, remaining OH^- ions move toward the anode resulting in the formation of $Zn(OH)_2$. However, in the basic medium (NaOH electrolyte), the possible mechanism for the process can be expressed as:

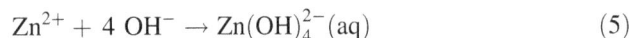

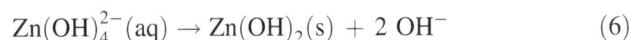

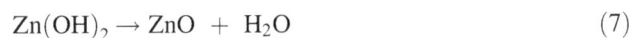

$$Zn^{2+} + 4 OH^- \rightarrow Zn(OH)_4^{2-}(aq) \tag{5}$$

$$Zn(OH)_4^{2-}(aq) \rightarrow Zn(OH)_2(s) + 2 OH^- \tag{6}$$

$$Zn(OH)_2 \rightarrow ZnO + H_2O \tag{7}$$

The initial phase of the reaction is the active dissolution of Zn, which is ascribed to the formation of $Zn(OH)_4^{2-}$. When the concentration of $Zn(OH)_4^{2-}$ exceeds the solubility product of $Zn(OH)_2$, precipitation of a compact layer of $Zn(OH)_2$ will occur on the anode surface. Finally, ZnO will form.

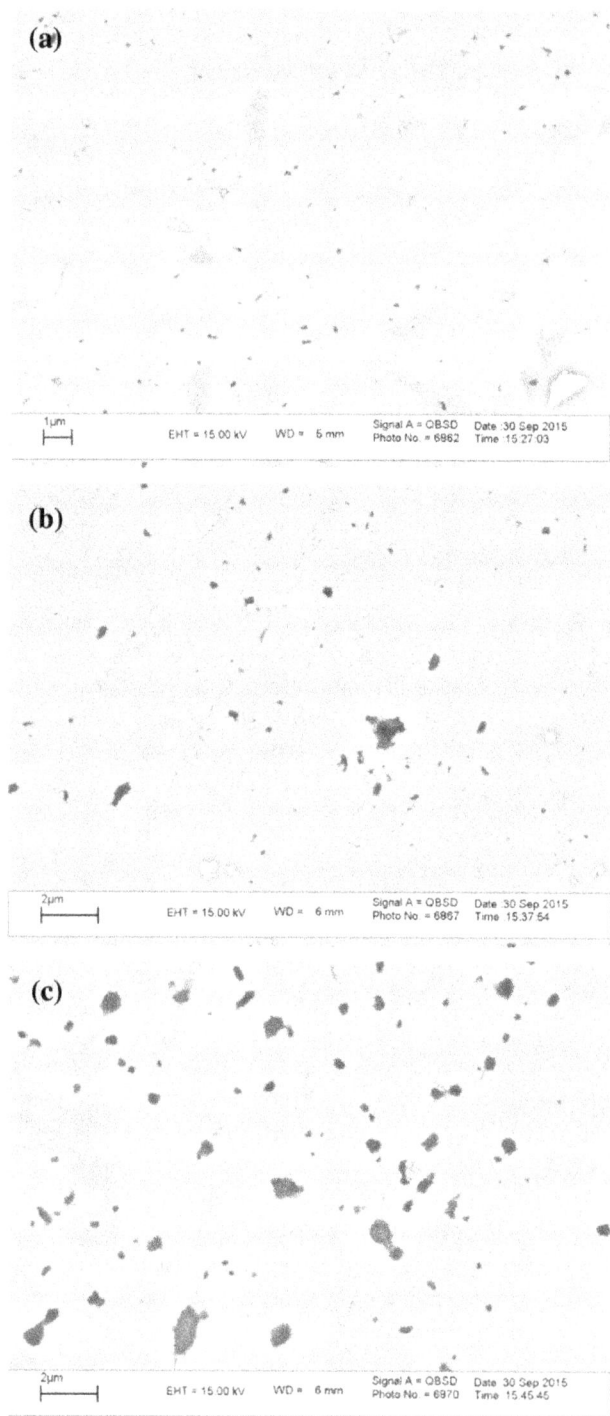

Fig. 1 SEMs of anodized zinc foil using 0.1 (**a**), 0.3 (**b**), and 0.5 M (**c**) NaOH as electrolyte at a constant voltage of 10 V for 1 h

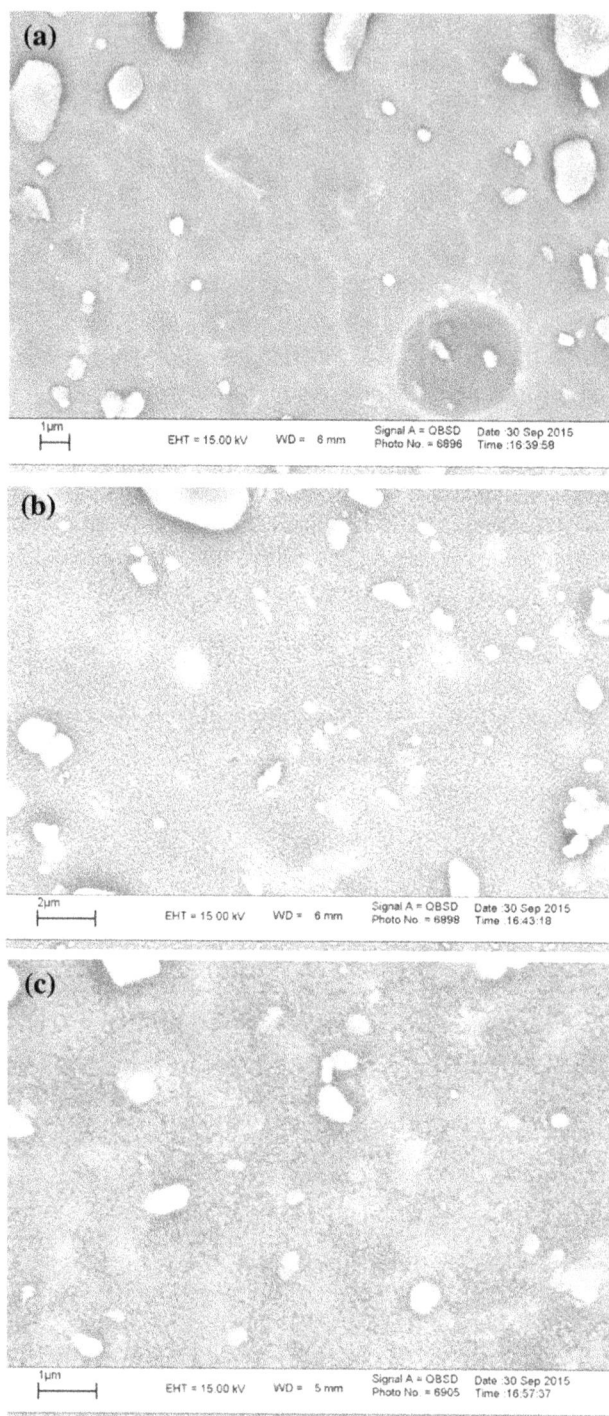

Fig. 2 SEMs of anodized zinc foil using 0.1 (**a**), 0.3 (**b**), and 0.5 M (**c**) oxalic acid as electrolyte at a constant voltage of 10 V for 1 h

Surface morphology of the anodized zinc plates

Figures 1 and 2 show the SEM images of Zn plates anodized in NaOH and oxalic acid electrolytes at 25 °C and 10 V for 1 h, respectively. In general, it seems that the electrochemical condition is an important ingredient for the organization of nanostructures. When using different electrolytes for the anodizing of the zinc foil and while keeping other parameters constant, different ZnO nanostructure arrays were observed. As presented in Fig. 1a, in an electrolyte of NaOH, a ZnO nano and micro hole array was formed, whereas in oxalic acid (Fig. 2a), a ZnO nanoparticle array was observed. This study further confirms that different types of electrolytes produce different morphologies of ZnO.

Figure 1 shows the SEM images of the Zn plates anodized in NaOH electrolyte. The obtained SEM images showed that the morphologies of the anodized Zn plates were significantly influenced by electrolyte concentration. The surfaces were relatively flat and with parallel scratches which were believed to be originated from the mechanical grinding and polishing process. As can see in Fig. 1, some holes distributed over the entire surface. These holes were randomly spread over the surface of the anodized Zn plate. The size of holes increased as the concentration of electrolyte increased. This suggests that the concentration of electrolyte plays an important role in the determination of hole size.

Figure 2 displays the SEM images of the anodic ZnO nanoparticles formed in 0.1, 0.3, and 0.5 M oxalic acid solution under anodic voltages of 10 V. It can be seen that for Zn plates anodized in oxalic acid, no holes was formed. However, nanoporous structures were discovered along the surface of the Zn plate anodized in mentioned concentration of oxalic acid. These nanostructures were distributed over the surface of the anodized Zn plate. At a lower oxalic acid concentration (0.1 M), the formation of nanoparticles was incomplete. By increasing the oxalic acid concentration to 0.5 M (Fig. 2c), a nanoparticle array was completely formed. Different from anodic ZnO structures, some larger ZnO particles can be observed to distribute on the substrate surface of the sample. Moreover, these molecules and nanostructures cannot be seen along the sample surface before anodizing. They should be produced in the anodic process.

It seems that using sodium hydroxide causes serious corrosion due to its high corrosion characteristics. Sodium hydroxide is a strong base and has the ability of creation of holes on the zinc surface. Hence, oxalic acid is a weak organic acid, and it can affect the zinc surface poorly.

The results were in good consistent with the observation reported by Shetty and Nanda, in which the Zn plates were anodized in deionized water [13]. They reported that morphologies of the ZnO thin film were significantly dependent on the anodizing voltage. However, no nanostructure was reported on their findings. Instead, wall-like construction was formed at 1 V, which changed gradually to well-defined porous structure, as the voltage increased to 9 V [6].

XRD patterns

To identify the composition of the nanoporous structures, the anodized Zn plates were characterized using the XRD. The XRD patterns of the anodized zinc foil using oxalic

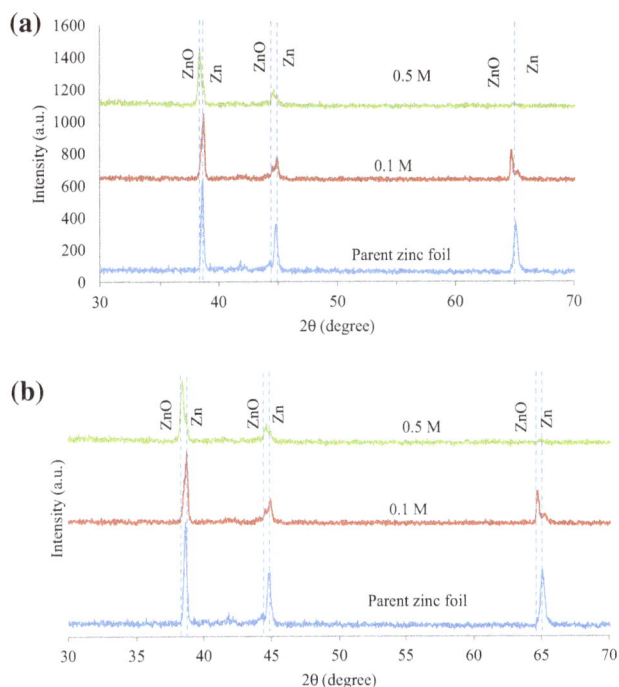

Fig. 3 XRD patterns of the anodized zinc foil **a** oxalic acid and **b** NaOH as electrolyte at a constant voltage of 10 V for 1 h

acid and NaOH as electrolytes are shown in Fig. 3. It can be observed that peaks attributed to Zn and ZnO were present in these XRD patterns. The presence of Zn peaks were due to the Zn plates which were used as substrates in this study. The peaks at scattering angles (2θ) of 38.61, 44.82, and 64.89 correspond to the reflection from 101, 102, and 103 crystal planes, respectively. The XRD pattern is identical to the hexagonal phase with wurtzite structure [5, 6, 14].

Optical properties

Optical properties of the anodized Zn plates were characterized based on UV–Vis absorption spectra shown in Fig. 4. The calculated bandgap of ZnO film is around 3.2 eV. Concentration changes have not a big effect on the bandgap, but the more concentration of the electrolyte; the more porous of produced nanostructure and, consequently, its surface is increased, and thus, the absorption of light is increased by the sample.

Contact angle measurement

Table 1 shows the change in contact angle of pure zinc foil and the anodized film. It is evident that the anodizing induces a remarkable change in the surface properties of the films. Water contact angle testing results confirmed that the surface of pure zinc foil was hydrophilic with an angle

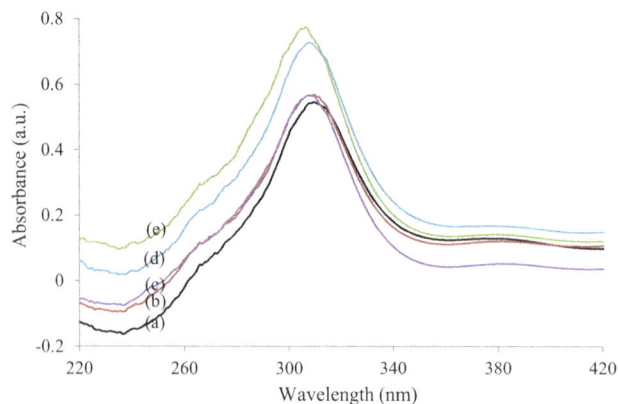

Fig. 4 Optical absorption spectra of the zinc foil (**a**) and anodized zinc foil using 0.1 (**b**) and 0.5 M (**c**) oxalic acid and using 0.1 (**d**) and 0.5 M (**e**) NaOH as electrolyte at a constant voltage of 10 V for 1 h

Table 1 Water contact angle of the anodized zinc foil

Electrolyte	Concentration of electrolyte (M)	Contact angle (°)
Parent foil	0	89.3 ± 1.3
NaOH	0.1	78.3 ± 2.5
	0.3	80.3 ± 1.5
	0.5	83.4 ± 2.6
Oxalic acid	0.1	77.8 ± 3.5
	0.3	81.7 ± 2.0
	0.5	84.7 ± 3.8

of 87.3 ± 2.3°. It has been noticed that the contact angle of the anodized film is less than that of pure zinc foil. This fact confirms that anodized films are more hydrophilic for both electrolytes.

With referring to Table 1, one can see that the contact angle for each electrolyte increases with increasing anodizing time. From the values of the contact angle and the SEM images, we can observe that the value of the contact angle is directly correlated with the surface structure of the film. The reason of the increment of contact angel has been explicated in terms of the quantity of air spaces observed in the nanoscale in an earlier work [11, 15]. The trapped air pressure balances the gravity of the water droplet, and the surface tension of the water tries to keep the shape of the droplet spherical. Hence, the film with high intensity of air displays higher hydrophobic characteristics. The biggest value of the contact angle in the case of different anodized samples can be assigned to the high intensity of air entrapped in the film surface.

Antimicrobial activity

The disinfectant properties of samples were investigated with Gram-negative *E. coli* and Gram-positive *S. aureus* at 37 °C for 24 h. Parent zinc foil did not show the antibiotic activity. The zones of growth inhibition were observed for

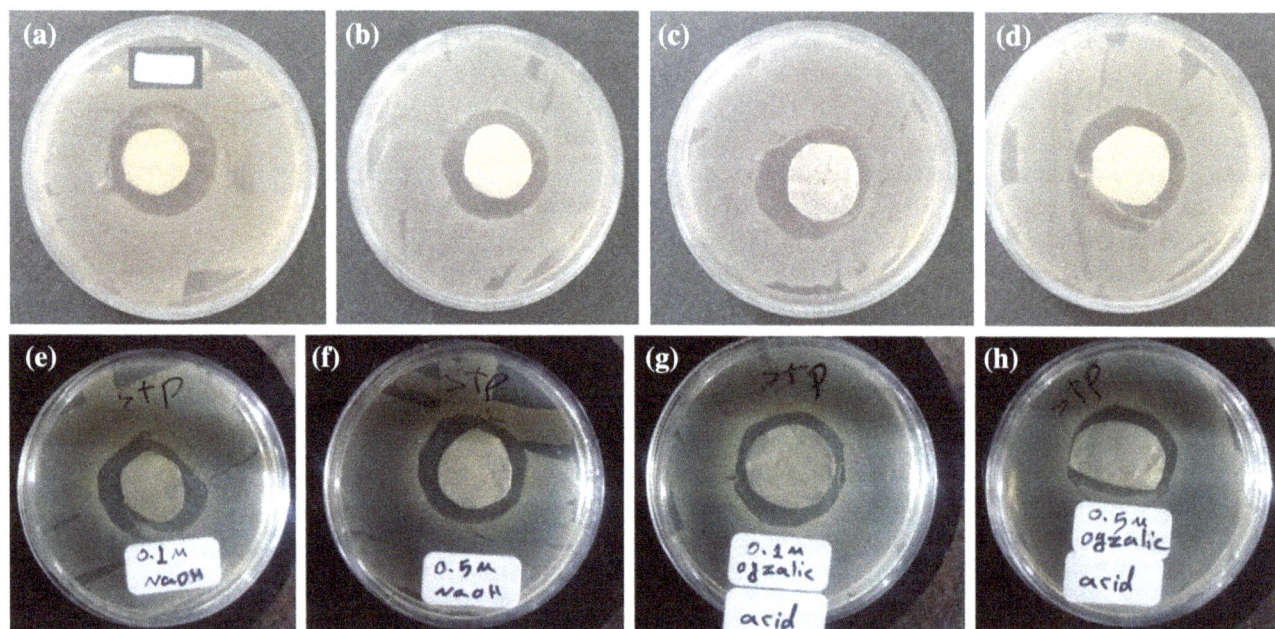

Fig. 5 Growth inhibition of the anodized zinc foil using 0.1 (**a**) and 0.5 M (**b**) NaOH and 0.1 (**c**) and 0.5 M (**d**) oxalic acid as electrolyte against *E. coli*, and 0.1 (**e**) and 0.5 M (**f**) NaOH and 0.1 (**g**) and 0.5 M (**h**) oxalic acid as electrolyte against *S. aureus*

Table 2 Growth inhibition of the anodized zinc foil

Electrolyte	Concentration of electrolyte (M)	Growth inhibition	
		E. coli (mm)	S. aureus (mm)
Parent zinc foil	0	0	0
NaOH	0.1	6.0 ± 0.2	6.4 ± 0.9
	0.3	6.1 ± 0.3	6.5 ± 0.4
	0.5	5.7 ± 0.2	6.0 ± 0.6
Oxalic acid	0.1	6.0 ± 0.2	5.8 ± 0.2
	0.3	5.9 ± 0.2	5.4 ± 0.3
	0.5	6.3 ± 0.5	5.3 ± 0.9

the anodized zinc foils in all the cases (Fig. 5). These results are summarized in Table 2. The zones of growth inhibition for *E. coli* and *S. aureus* were 5.7–6.3 and 5.3–6.5 mm, respectively. Thus, these results suggest that the anodized zinc oxide is not only effective toward Gram-positive bacteria, but also Gram-negative bacteria. On the other hand, the antibacterial activity is relatively uniform for all samples (Table 2).

Three distinct mechanisms of action have been put forward in the literature for the zinc oxide antimicrobial activity: (i) the production of reactive oxygen species (ROS) because of the semiconductor properties of ZnO, (ii) the destabilization of microbial membranes upon direct contact of ZnO particles to the cell walls, and (iii) the intrinsic antimicrobial properties of Zn^{2+} ions released by ZnO in aqueous medium [9]. Since ZnO particles were unable to disperse out of the anodized samples, antimicrobial species have necessarily been arisen from the released Zn^{2+} to the agar medium and direct contact of bacteria with anodized surface.

Conclusion

In the present work, ZnO nanostructures were successfully fabricated by anodizing a Zn sheet in sodium hydroxide (NaOH) and oxalic acid electrolytes under the influence of different concentrations of the electrolyte, while the temperature and voltage were kept constant. Characterization of anodized Zn plates using SEM showed that their morphology was significantly influenced by the type and concentration of anodizing electrolyte. The structural characterization showed that the anodic ZnO had hexagonal wurtzite structure. From contact angle measurements, it has been observed that the contact angle of anodized film is higher than that of pure zinc foil. Antibacterial results suggest that the parent zinc foil did not show the antibiotic activity, but the anodized zinc oxide is effective both toward Gram-positive bacteria and Gram-negative bacteria.

References

1. Secu, C.E., Sima, M.: Photoluminescence and thermoluminescence of ZnO nano-needle arrays and films. Opt. Mater. **31**, 876–880 (2009)
2. Sawai, J., Shoji, S., Igarashi, H., Hashimoto, A., Kokugan, T., Shimizu, M., Kojima, H.: Hydrogen peroxide as an antibacterial factor in zinc oxide powder slurry. J. Ferment. Bioeng. **86**, 521–522 (1998)
3. Sawai, J.: Quantitative evaluation of antibacterial activities of metallic oxide powders (ZnO, MgO and CaO) by conductimetric assay. J. Microbiol. Methods **54**, 177–182 (2003)
4. Zhang, L.L., Jiang, Y.H., Ding, Y.L., Povey, M., York, D.: Investigation into the antibacterial behaviour of suspensions of ZnO nanoparticles (ZnO nanofluids). J. Nanopart. Res. **9**, 479–489 (2007)
5. Skoog, S.A., Bayati, M.R., Petrochenko, P.E., Stafslien, S., Daniels, J., Cilz, N., Comstock, D.J., Elam, J.W., Narayan, R.J.: Antibacterial activity of zinc oxide-coated nanoporous alumina. Mater. Sci. Eng. B **177**, 992–998 (2012)
6. Voon, C.H., Derman, M.N., Hashim, U., Lim, B.Y., Sam, S.T., Foo, K.L., Ten, S.T.: Synthesis of nanoporous zinc oxide by anodizing of zinc in distilled water. Appl. Mech. Mater. **754–755**, 1126–1130 (2015)
7. Huang, G.S., Wu, X.L., Cheng, Y.C., Shen, J.C., Huang, A.P., Chu, P.K.: Fabrication and characterization of anodic ZnO nanoparticles. Appl. Phys. A **86**, 463–467 (2007)
8. Goh, H.S., Adnan, R., Farrukh, M.A.: ZnO nanoflake arrays prepared via anodizing and their performance in the photodegradation of methyl orange. Turk. J. Chem. **35**, 375–391 (2011)
9. Pasqueta, J., Chevalierb, Y., Pelletierb, J., Couvala, E., Bouviera, D., Bolzingerb, M.A.: The contribution of zinc ions to the antimicrobial activity of zinc oxide. Colloid Surf. A **457**, 263–274 (2014)
10. Ghorbanpour, M.: Optimization of sensitivity and stability of Au/Ag bilayer thin films used in surface plasmon resonance chips. J. Nanostruct. **3**, 309–313 (2013)
11. Ghorbanpour, M., Falamaki, C.: A novel method for the fabrication of ATPES silanized SPR sensor chips: exclusion of Cr or Ti intermediate layers and optimization of optical/adherence properties. Appl. Surf. Sci. **301**, 544–550 (2014)
12. Pouraboulghasem, H., Ghorbanpour, M., Shayegh, R., Lotfiman, S.: Synthesis, characterization and antimicrobial activity of alkaline ion-exchanged ZnO/bentonite nanocomposites. J. Cent. S. Univ. **23**, 787–792 (2016)
13. Shetty, A., Nanda, K.K.: Synthesis of zinc oxide porous structures by anodization with water as an electrolyte. Appl. Phys. A **109**, 151–157 (2012)

14. Zhao, J., Wang, X., Liu, J., Meng, Y., Xu, X., Tang, C.: Controllable growth of zinc oxide nanosheets and sunflower structures by anodization method. Mater. Chem. Phys. **126**, 555–559 (2011)
15. He, S., Zheng, M., Yao, L., Yuan, X., Li, M., Ma, L., Shen, W.: Preparation and properties of ZnO nanostructures by electrochemical anodization method. Appl. Surf. Sci. **256**, 2557–2562 (2010)

11

Synthesis of Bi_2WO_6 nanoplates using oleic acid as a green capping agent and its application for thiols oxidation

Rahmatollah Rahimi[1] · Shabnam Pordel[1] · Mahboubeh Rabbani[1]

Abstract Bi_2WO_6 nanoplates were synthesized by a simple one-step hydrothermal method using oleic acid (OA) as a green and cheap capping agent. The X-ray diffraction (XRD) analysis, scanning electron microscopy (SEM), and energy-dispersive X-ray spectroscopy (EDS) were used to characterize the products. The interaction between the precursor product and oleic acid was studied by Fourier Transform Infrared (FT-IR). In addition, the catalytic activity of prepared Bi_2WO_6 for the oxidation of thiols to disulfides as an important reaction in both biological and chemical processes was investigated. It was found to be an efficient catalyst for the selective oxidation of thiols to the corresponding disulfides, without over-oxidation, at room temperature.

Keywords Bi_2WO_6 · Oleic acid · Thiol oxidation · Green capping agent

Introduction

The oxidation of thiols to disulfides without over-oxidation is a pivotal reaction in both biological and chemical processes. A number of methods for this conversion, such as potassium dichromate [1], I_2/HI [2], sodium perborate [3], and nitric oxide [4], have been used. Most of these procedures suffer from drawbacks, such as generation of undesirable waste materials, low selectivity, over-oxidation, and low yield. Hence, developing economical, green, and mild approaches have attracted a great attention [5]. In accordance with these aims, previous studies suggested that metal oxides would be capable of oxidizing thiols under mild conditions [6].

Bi_2WO_6, the most important member in the Aurivillius family, possessing a layered structure with the perovskite-like slab of $(WO_4)^{2-}$ and $(Bi_2O_2)^{2+}$ has been widely studied because of its potential applications in dielectric, ion conductive, solar-energy-transfer, luminescent, and photocatalyst materials [7]. So far, different methods have been used to synthesize Bi_2WO_6, such as biomimetic [8], hydrothermal [9], electrospinning [10], ultrasonic spray pyrolysis [11], sol–gel [12], and microwave-assisted synthesis [13]. Among them, the hydrothermal method is an advantageous technique, since it is an environmental-friendly, low-cost method to provide highly pure products with controlled morphology [14]. The various 3D morphologies of Bi_2WO_6, including flower-like structure [15], clew- like microspheres [16], multilayered disk-like architecture [17], and octahedron-like structure [18] have been synthesized by the hydrothermal method. A 2D nanostructure, such as nanoplates and nanoparticles, is an important category of nanostructured materials, which have attracted much attention in recent years [19].

In this study, we report the synthesis of Bi_2WO_6 nanoplates by a simple hydrothermal method using oleic acid as a green and inexpensive capping agent. Furthermore, for the first time, we have studied the catalytic activity of Bi_2WO_6 nanoplates to oxidation of thiols by H_2O_2 as oxidant in non-aqueous media under neutral conditions at room temperature.

✉ Mahboubeh Rabbani
m_rabani@iust.ac.ir

[1] Department of Chemistry, Iran University of Science and Technology, 16846-13114 Tehran, Iran

Experimental

Hydrothermal synthesis of Bi_2WO_6

All chemicals were of analytical grade and used without further purification. Bi_2WO_6 nanoplates were synthesized by the hydrothermal method as follows: 0.46 g of $Bi(NO_3)_3 \cdot 5H_2O$ (0.95 mmol) was dissolved into a 30-mL $NaNO_3$ solution (2 M) using a magnetic stirrer at 75 °C for 10 min. Then, 12-mL $Na_2WO_4 \cdot 2H_2O$ (0.05 M) was added dropwise into the above solution. After 2 h, 2-mL OA was added into the above solution and stirred for 30 min. The resulted suspension was transferred into a 50-mL Teflon-lined autoclave and maintained at 175 °C for 15 h. After the autoclave cooled at room temperature, the collected yellowish white precipitate was washed with deionized water and acetone for several times and dried in oven at 100 °C for 6 h. The final product was calcined at 550 °C for 4 h.

Characterization

The morphology of Bi_2WO_6 nanoplates were observed by a VEGA/TESCAN microscope with an accelerating voltage of 30.00 kV. The FT-IR analyses were carried out on a Shimadzu FT-IR-8400S spectrophotometer using KBr pellets. Furthermore, the structure of particles were analyzed by a JEOL diffractometer with monochromatic Cu Kα radiation ($\lambda = 1.5418$ Å). An Oxford energy-dispersive X-ray spectroscopy (EDS) was used for the determination of Bi and W elements in the product.

Oxidation reactions

First, 1 mmol of thiol was dissolved in 10-mL dichloromethane, then 5-mg Bi_2WO_6 catalyst was added, and the mixture was stirred at room temperature. After the completion of the reaction (detected by TLC), the reaction mixture was filtered and the solid material was washed with 20 mL of dichloromethane. The filtrate was evaporated and the resulting material was either recrystallized or subjected to a silica-gel plate. The filtrate was then subjected to GC and GC-Mass analysis.

Results and discussion

Characterization of Bi_2WO_6 nanoplates

The composition and phase of Bi_2WO_6 nanoplates were characterized by the XRD analysis, as shown in Fig. 1a. All the diffraction peaks are indexed to the orthorhombic phase of Bi_2WO_6 and match well with the reported data (JCPDS card No. 73-1126, $a = 5.457$, $b = 5.436$, and $c = 16.42$ Å). In addition, the EDS result (Fig. 1b) showed merely the presence of Bi, W, and O elements, which is indicative of its high purity. The weight percentages of 69.19 and 30.81 were obtained for Bi and W, respectively (molar ratio of Bi:W is about 2:1).

The SEM image of Bi_2WO_6 reveals the production of Bi_2WO_6 nanoplates with the average thickness of about 90 nm and length and width of about 100–350 nm (Fig. 2).

For the investigation of the interactions between Bi_2WO_6 and oleic acid, the FT-IR spectra of pure oleic acid,

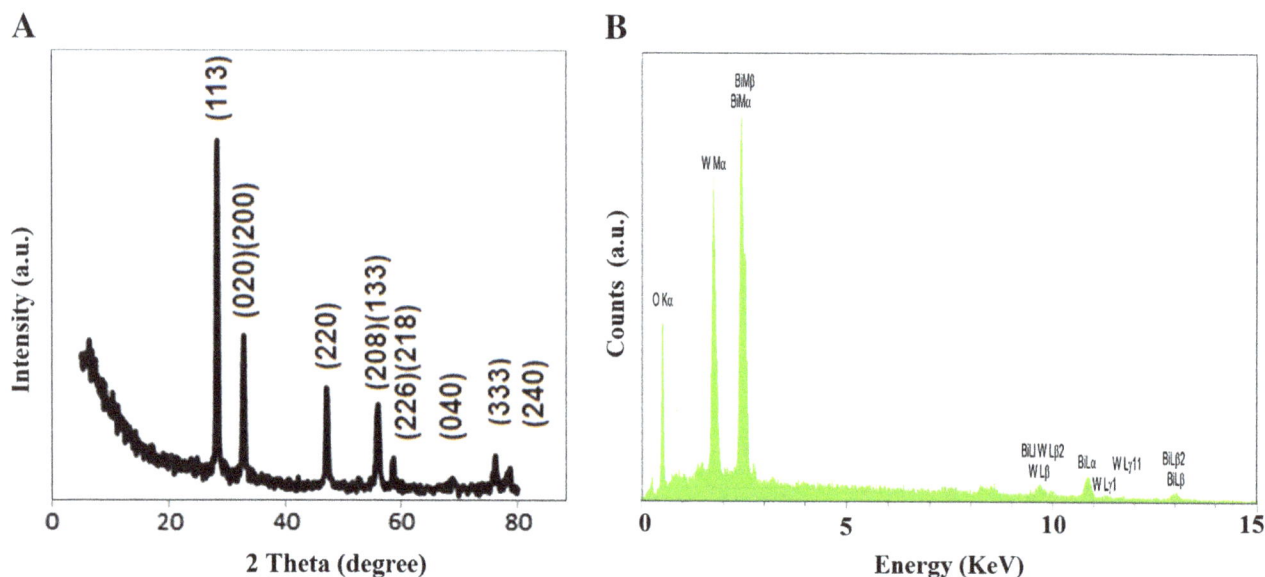

Fig. 1 **a** XRD patterns and **b** EDS analysis of Bi_2WO_6 nanoplates

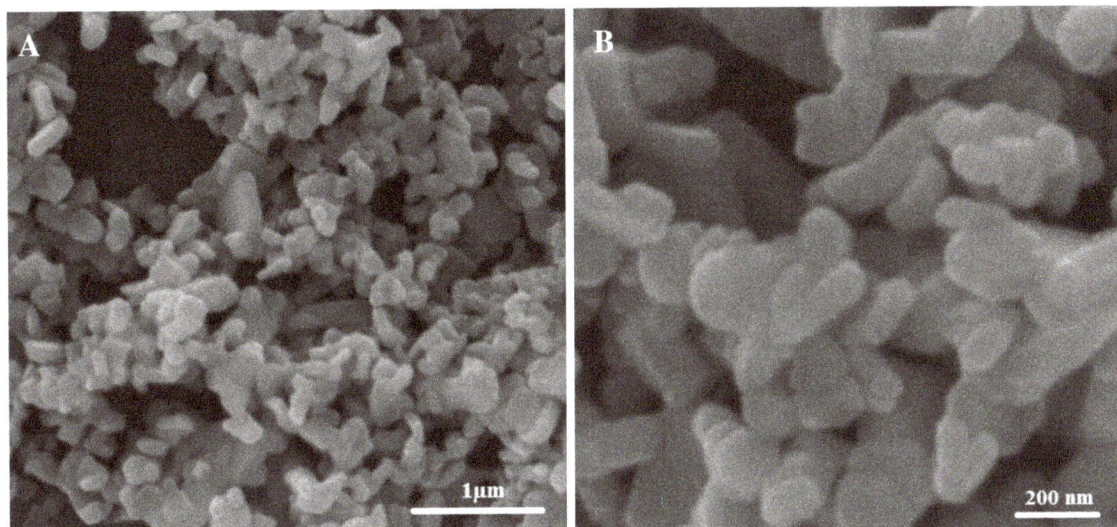

Fig. 2 Morphology evolution of Bi_2WO_6 nanoplates

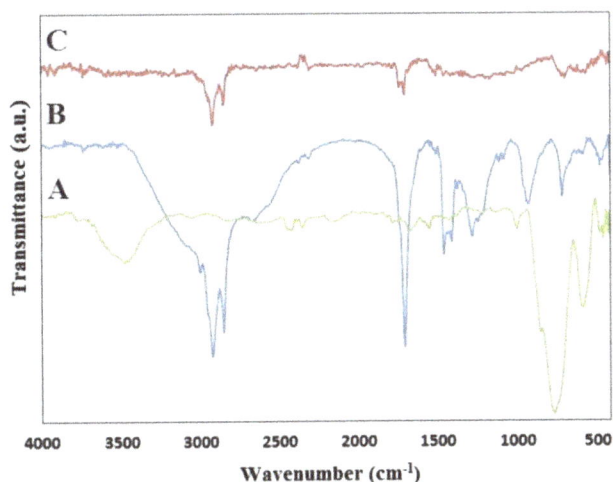

Fig. 3 FT-IR spectra of a OA and b OA adsorbed on Bi_2WO_6 surface, and c Bi_2WO_6 nanoplates after calcination

Table 1 Effect of solvents on oxidation of thiophenol

Entry	Solvent	Conv. (%)[a]	Sel. (%)[a]
1	Water	5	>99
2	Ethanol	60	>99
3	Dichloromethane	97	>99

Reaction conditions: thiols (1 mmol), Bi_2WO_6 (5 mg), solvent (4 mL), H_2O_2 (50 μL), 60 min, room temperature

[a] Conversion and selectivity were determined by GC

Table 2 Effect of catalyst loading on oxidation of thiophenol

Entry	Amount of Catalyst (mg)	Conv. (%)[a]	Sel. (%)[a]
1	Without catalyst	8	>99
2	2.5	61	>99
3	5	97	>99

Reaction conditions: thiols (1 mmol), dichloromethane (4 mL), H_2O_2 (50 μL), 60 min, room temperature

[a] Conversion and selectivity were determined by GC

Bi_2WO_6 capped by oleic acid and Bi_2WO_6 after calcination are compared in Fig. 3. The FT-IR spectrum of calcined Bi_2WO_6 (Fig. 3a) displays strong sharp bands at 734 and 579 cm^{-1}, which belong to the stretching vibrations of W–O and Bi–O bonds, respectively [20]. Figure 3b shows the characteristic peaks of the pure OA comprising the oleyl group. The peaks at about 2925 and 2854 cm^{-1} were assigned to the asymmetric and symmetric CH_2 stretching modes, respectively, and the C = O stretch of the carboxylic acid dimers was observed at about 1710 cm^{-1} [21]. It is known that in the FT-IR spectrum of the organic acids adsorbed on the metal oxides (Fig. 3c), the peaks at 1680–1710 cm^{-1} are attributed to the C = O stretching modes of the single hydrogen-bonded carboxylic acid

groups, while in the cyclic hydrogen-bonded dimeric form, the frequency of carbonyl group shifts to higher wavenumbers by ~ 35 cm^{-1} [22, 23]. Therefore, the peak at 1745 cm^{-1} can be corresponded to the adsorbed OA in the form of cyclic hydrogen-bonded dimeric structure. Although the Bi_2WO_6 treated with OA was washed several times with acetone, the peak at 1710 cm^{-1} was remained which indicates that a fraction of the OA was physically adsorbed on the Bi_2WO_6 particles through the van der Waals forces (the possible formation mechanism of Bi_2WO_6 nanoplates is discussed in SI).

Table 3 Oxidation of various thiols catalyzed by Bi_2WO_6

Entry	Thiol	Product	Conversion (%)	Selectivity (%)
1			97[a]	100[a]
2			69[b]	100[b]
3			74[a]	100[a]
4			83[a]	100[a]

Reaction conditions: thiols (1 mmol), Bi_2WO_6 (5 mg), dichlromethane (4 mL), H_2O_2 (50 μL), 60 min, room temperature
[a]Conversion and selectivity were determined by GC
[b]Conversion and selectivity were determined by chromatography

Catalytic study

The catalytic performance of Bi_2WO_6 was studied for the oxidation of thiols using H_2O_2 as oxidant. To optimize reaction conditions for the oxidation of sulfides, thiophenol was taken as a model compound and different reaction conditions including catalyst loading and solvent were studied. As shown in Table 1, the reaction was performed in various solvents. The conversion of thiophenol is 5, 60, and 97 % when water, ethanol, and dichloromethane are used as solvent, respectively (Entries 1–3). The effect of catalyst loading on the conversion of thiophenol is also investigated. As shown in Table 2, the conversion of thiophenol without catalyst is about 8 % (Entry 1); 'while, in the presence of catalyst, the conversion was increased from 61 to 97 % when the amounts of catalyst were varied from 2.5 to 5 mg (Entries 2, 3). These results suggested that the progress of thiophenol oxidation is strongly dependent upon the catalyst loading.

Under optimal conditions, aromatic thiols with different substituents have been converted to their corresponding disulfides in the presence of Bi_2WO_6 (Table 3). As can be seen, aromatic thiols with various substituents in ortho and para positions resulted in good yields of the corresponding disulfides without any other by-products (no peaks of by-products for thiophenol oxidation were observed in GC-Mass chromatogram, supporting information, Fig S1).

Conclusion

In summary, we have synthesized Bi_2WO_6 nanoplates by a simple one-spot hydrothermal method using oleic acid as a capping agent. The FT-IR studies revealed that oleic acid was chemically adsorbed by cyclic hydrogen-bonded carbonyl groups on Bi_2WO_6 surface, and also, there is physically adsorbed oleic acid via van der Waals forces. Furthermore, the catalytic activity of Bi_2WO_6 for the synthesis of disulfides via oxidative coupling of thiols was studied. For the use of a nontoxic and inexpensive catalyst, the mild conditions and relatively short reaction time and excellent yields are some of the notable advantages of this method.

References

1. Patel, S., Mishra, B.K.: Cetyltrimethylammonium dichromate: a mild oxidant for coupling amines and thiols. Tetrahedron Lett. **44**, 1371–1372 (2004)
2. Wu, X., Rieke, R.D., Zhu, L.: Preparation of disulfides by the oxidation of thiols using bromine. Synth. Commun. **26**, 191–196 (1996)
3. McKillop, A., Koyuncu, D., Krief, A., Dumont, W., Renier, P., Trabelsi, M.: Efficicient, high yield, oxidation of thiols and selenols to disulphides and diselenides. Tetrahedron Lett. **31**, 5007–5010 (1990)
4. Pryor, W.A., Church, D.F., Govindan, C.K., Crank, G.: Oxidation of thiols by nitric oxide and nitrogen dioxide: synthetic utility and toxicological implications. J. Org. Chem. **47**, 156–159 (1982)
5. Soleiman-Beigi, M., Taherinia, Z.: Simple and efficient oxidative transformation of thiols to disulfides using Cu (NO3) 2·3H2O in H2O/AcOEt. Monatsh. Chem. **145**, 1151–1154 (2014)
6. Wallace, T.J.: Reactions of thiols with metals. I. Low-temperature oxidation by metal oxides 1. J. Org. Chem. **31**, 1217–1221 (1996)
7. Huang, H., Chen, H., Xia, Y., Tao, X., Gan, Y., Weng, X.: Controllable synthesis and visible-light-responsive photocatalytic activity of Bi_2WO_6 fluffy microsphere with hierarchical architecture. J. Colloid Interf. Sci. **370**, 132–138 (2012)
8. Yao, F., Yang, Q., Yin, C., Zhu, S., Zhang, D., Moon, W.-J.: Biomimetic Bi_2WO_6 with hierarchical structures from butterfly wings for visible light absorption. Mater. Lett. **77**, 21–24 (2012)
9. Shang, M., Wang, W., Sun, S., Zhou, L., Zhang, L.: Bi_2WO_6 nanocrystals with high photocatalytic activities under visible light. J. Phys. Chem. **112**, 10407–10411 (2008)
10. Shang, M., Wang, W., Ren, J., Sun, S., Wang, L., Zhang, L.: A practical visible-light-driven Bi_2WO_6 nanofibrous mat prepared by electrospinning. J. Mater. Chem. **19**, 6213–6218 (2009)
11. Mann, A.K., Skrabalak, S.E.: Synthesis of single-crystalline nanoplates by spray pyrolysis: a metathesis route to Bi_2WO_6. Chem. Mater. **23**, 1017–1022 (2011)
12. Liu, Y., Lv, H., Hu, J., Li, Z.: Synthesis and characterization of Bi_2WO_6 nanoplates using egg white as a biotemplate through sol-gel method. Mater. Lett. **139**, 401–404 (2015)
13. Cao, X.F., Zhang, L., Chen, X.T., Xue, Z.L.: Microwave-assisted solution-phase preparation of flower-like Bi_2WO_6 and its visible-light-driven photocatalytic properties. Cryst. Eng. Comm. **13**, 306–311 (2011)
14. Yoshimura, M., Byrappa, K.: Hydrothermal processing of materials: past, present and future. J. Mater. Sci. **43**, 2085–2103 (2008)
15. Dumrongrojthanath, P., Thongtem, T., Phuruangrat, A., Thongtem, S.: Hydrothermal synthesis of Bi_2WO_6 hierarchical flowers with their photonic and photocatalytic properties. Superlattice Microst. **54**, 71–77 (2013)
16. He, D., Wang, L., Li, H., Yan, T., Wang, D., Xie, T.: Self-assembled 3D hierarchical clew-like Bi_2WO_6 microspheres: synthesis, photo-induced charges transfer properties, and photocatalytic activities. Cryst. Eng. Comm. **13**, 4053–4059 (2011)
17. Tang, P., Chen, H., Cao, F.: One-step preparation of bismuth tungstate nanodisks with visible-light photocatalytic activity. Mater. Lett. **68**, 171–173 (2012)
18. Li, Y., Liu, J., Huang, X.: Synthesis and visible-light photocatalytic property of Bi_2WO_6 hierarchical octahedron-like structures. Nanoscale Res. Lett. **3**, 365–371 (2008)
19. Zhang, C., Zhu, Y.: Synthesis of square Bi_2WO_6 nanoplates as high-activity visible-light-driven photocatalysts. Chem. Mater. **17**, 3537–3545 (2005)
20. Geng, Y., Zhang, P., Kuang, S.: Fabrication and enhanced visible-light photocatalytic activities of $BiVO_4/Bi_2WO_6$ composites. RSC Adv. **4**, 46054–46059 (2014)
21. Pawsey, S., Yach, K., Reven, L.: Self-assembly of carboxyalkylphosphonic acids on metal oxide powders. Langmuir **18**, 5205–5212 (2002)
22. Shukla, N., Liu, C., Jones, P.M., Weller, D.: FTIR study of surfactant bonding to FePt nanoparticles. J. Magn. Magn. Mater. **266**, 178–184 (2003)
23. Blyholder, G., Adhikar, C., Proctor, A.: Structure and orientation of oleic acid adsorbed onto silica gel. Colloids Surf. A **105**, 151–158 (1995)

12

A facial, scalable, and green synthesis of superparamagnetic palladium–carbon catalyst and its use in disproportionation of gum rosin

Ramin Mostafalu[1] · Akbar Heydari[1] · Marzban Arefi[1] · Maryam Kazemi[1] ·
Abbas Banaei[2] · Fatemeh Ghorbani[2]

Abstract We reported here the preparation, characterization, and catalytic ability of a Fe_3O_4–supported palladium–carbon (Fe_3O_4–Pd–C) as an efficient catalyst in disproportionation of gum rosin. The magnetic nanocatalyst was prepared by a simple method and was characterized by FTIR, XRD, TEM, N_2 adsorption–desorption, VSM, and atomic absorption analysis. The prepared catalyst displayed excellent activity in disproportionation of gum rosin. The magnetic Fe_3O_4–Pd–C catalyst was successfully recycled three times with keeping its catalytic performance. The simplicity of the nanocatalyst production method and simple separation and recyclability, on the other hand, make possible the industrial production and application of the catalyst.

Keywords Magnetic palladium–carbon · Palladium–carbon · Magnetic active carbon · Disproportionated rosin (DPR) · Gum rosin · Activated carbon

Introduction

Active carbon due to high surface area, an open pore structure, excellent thermal and chemical stability, and low price is widely used as catalyst support and adsorbent, and finds extensive uses in the chemical and pharmaceutical-manufacturing industries [1–3]. However, carbon-supported catalysts are difficult to separate from reaction mixture and need special filtration system (Fig. 1), and then using them in large scales has been limited [4].

An attractive alternative to filtration or centrifugation is magnetic separation [5–8]. Magnetic separation method due to simplicity, high efficiency, and low cost has been widely used. In this regard, many researches are concentrating on modified active carbon [9–12]. For example, recently, Schuth et al. have reported in situ preparation of magnetic-activated carbon by formation of Fe_3O_4 nanoparticles in the pores of carbon [9]. More recently, preparation of activated maize cob coated with magnetic nanoparticles have been reported by Morad et al. for methylene blue (MB) adsorption [10]. Kakavandi et al. used magnetic-activated carbon for removal of aniline [11]. Mohan et al. used magnetic-activated carbon for tri-nitro-phenol removal from aqueous solution [12]. However, in the most of these reports, magnetic active carbon is used as an absorbent, and research on modified carbon using magnetic nanoparticles as a support for application in catalytic reaction has not been reported before.

In our recently efforts, we have found that the activated carbon-supported palladium nanoparticles are an efficient nanocatalyst for the disproportionation of rosin [13]. Here in, the catalytic activity of magnetic palladium–carbon for disproportionation of rosin is reported.

Experimental section

Chemicals

All reagents and chemicals were purchased from the Daejung, Merck and Aldrich companies, and gum rosin was obtained as a gift from Padideh Shimi Jam Co.

✉ Akbar Heydari
heydar_a@modares.ac.ir

[1] Chemistry Department, Tarbiat Modarres University, P.O. Box 14155-4838, Tehran, Iran

[2] Padideh Shimi Jam Co., Karaj, Iran

Fig. 1 Filtration system for remove palladium–carbon catalyst from reaction mixture

Preparation of the activated carbon-supported palladium nanoparticles (Pd–NP–AC)

Twenty grams of activated carbon was dispersed in 50 ml distilled water at 80 °C. One gram of palladium metal was dissolved in aqua regia (4 ml) and added to the reaction mixture. After 2 h, the resulting black solid was filtered and washed by the use of distilled hot water for three times. The sample was dispersed in 60 ml distilled water, and using 2 M sodium hydroxide solution the pH was increased to 9. Next, 60 ml formaldehyde solution (37%) was added to the reaction. After 2 h at 80 °C, the resulting solid was filtered and washed using hot water and dried at 105 °C.

Preparation Fe_3O_4–palladium–active carbon (Fe_3O_4–Pd–C)

$FeCl_3 \cdot 6H_2O$ (8.7 mmol) and $FeCl_2 \cdot 4H_2O$ (4.3 mmol) were dissolved in distilled water at N_2 atmosphere. Subsequently, at 90 °C, 15 ml ammonia (25%) and 1 g of palladium–active carbon were added to the reaction mixture. After 30 min, the formed Fe_3O_4–Pd–C was collected with a magnet, washed three times with distilled hot water and dried at 105 °C. Fe_3O_4 nanoparticles were individually prepared according to the above procedure without adding palladium–carbon.

General procedure for gum rosin disproportionation by Fe_3O_4–Pd–C

In a three-neck flask fitted with a stirrer, condenser, and thermometer, 100 g of gum rosin was heated under N_2 atmosphere. Once temperature of reaction was reached to 280 °C, and a sample was withdrawn. Subsequently, Fe_3O_4–Pd–C was added to reaction flask and more samples were withdrawn every 1 h. A gas chromatography analysis was performed for a quantitative analysis of the samples [13].

Recyclability of Fe_3O_4–Pd–C catalyst

To recycle the Fe_3O_4–Pd–C catalyst, the catalyst was collected by a magnet, washed with iso-propanol, and dried at 80 °C.

Result and discussion

The Fe_3O_4–Pd–C catalyst was prepared by a simple co-precipitation of iron precursors (Fe^{2+} and Fe^{3+}) and palladium–carbon (Fig. 2).

Figure 3 shows the FTIR spectra of the magnetic Fe_3O_4–Pd–C in the 400–4000 cm^{-1} wave number range. The IR adsorption band in 567.78 cm^{-1} is attributed to the Fe–O bonds. The broad band between 3000–3700 cm^{-1} region is attributed to the hydroxyl groups. The peaks at 1094–1571 cm^{-1} are ascribed to the presence of active carbon [12, 14]. Therefore, it can be concluded that Fe_3O_4 nanoparticles were supported successfully on the palladium–carbon.

The X-ray diffraction peaks (XRD) of the Fe_3O_4 (Fig. 4) and Fe_3O_4–Pd–C are presented in Fig. 5. Six strong diffraction peaks at $2\theta = 30.3, 35.7, 43.4, 53.9, 57.8$, and 63.1, which are related to the (220), (311), (400), (422), (511), and (440) phases of Fe_3O_4 (JCPDS Card no. 88-0315, $a = 8.375$ Å), were found in the XRD patterns of Fe_3O_4 and Fe_3O_4–Pd–C. In the spectra of the Fe_3O_4–Pd–C, narrow peaks at $2\theta = 40.1°$ and $46.6°$ and $67.9°$ were related to the presence of Pd [15], and the mass loadings of Pd were determined to be 2.46% w/w by atomic absorption analysis. The broadening of the $2\theta = 10–30°$ in the spectra of the Fe_3O_4–Pd–C is related to activated carbon.

The crystal size of the Fe_3O_4 nanoparticles was calculated using Scherrer's equation [16]. The determined particle size for Fe_3O_4 nanoparticles in the Fe_3O_4 and Fe_3O_4–Pd–C came out to be 28 nm and 7.5 nm, respectively. This reveals that during the synthesis of the Fe_3O_4 nanoparticles, Pd–C prevents aggregation of these nanoparticles and because of that Fe_3O_4 nanoparticles loaded on activated carbon had the smallest sizes.

The lattice strain and dislocation density were 1.2×10^{-3} and 3.9×10^{-2} m^{-2} for Fe_3O_4 and 1.7×10^{-2} and 1.5×10^{-2} m^{-2} for Fe_3O_4–Pd–C, respectively, and were calculated from Eq. (1) and (2) where 'D' and 'β' are the crystallite size and the full width at half maximum (radian), respectively [16]. Small strain value indicates good crystallinity.

$$\delta = \frac{1}{D^2} \tag{1}$$

$$\varepsilon = \frac{\beta_{hkl}}{4\tan\theta} \tag{2}$$

Fig. 2 Schematic image to preparation Fe_3O_4–Pd–C catalyst

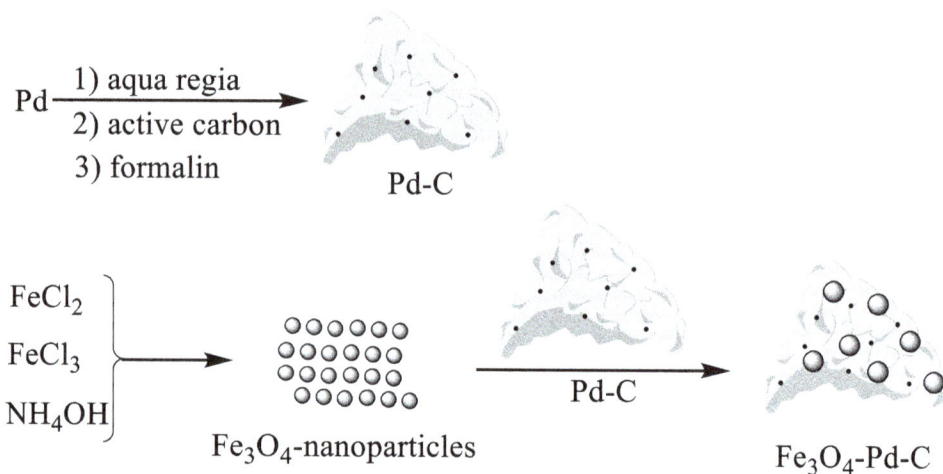

Pd $\xrightarrow{\begin{array}{l}\text{1) aqua regia}\\\text{2) active carbon}\\\text{3) formalin}\end{array}}$ Pd-C

$\left.\begin{array}{l}FeCl_2\\FeCl_3\\NH_4OH\end{array}\right\} \longrightarrow$ Fe_3O_4-nanoparticles $\xrightarrow{\text{Pd-C}}$ Fe_3O_4-Pd-C

Fig. 3 FTIR spectrum of Fe_3O_4–Pd–C

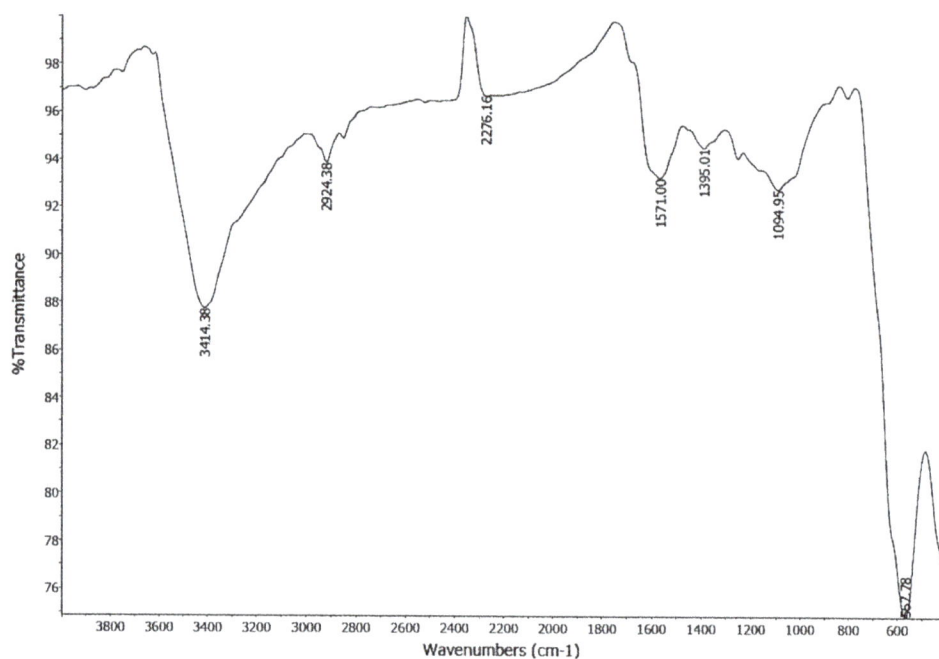

Fig. 4 XRD patterns of Fe_3O_4

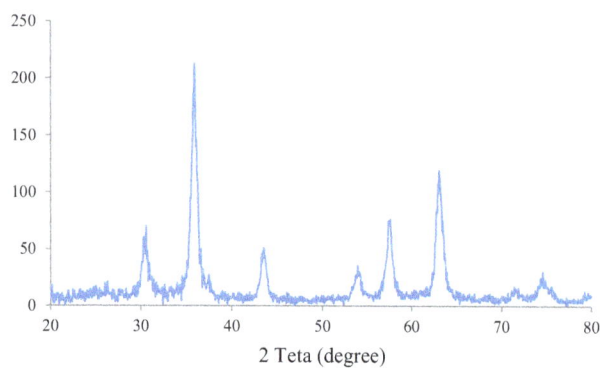

Fig. 5 XRD patterns of Fe_3O_4–Pd–C

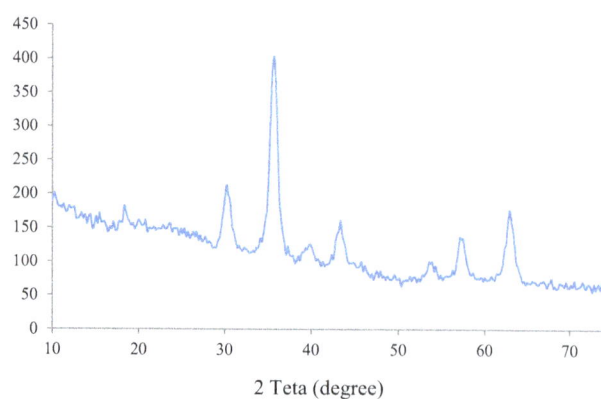

Fig. 6 **a** TEM of Fe_3O_4 nanoparticles (scale 50 nm) and **b** Palladium–carbon (scale 200 nm)

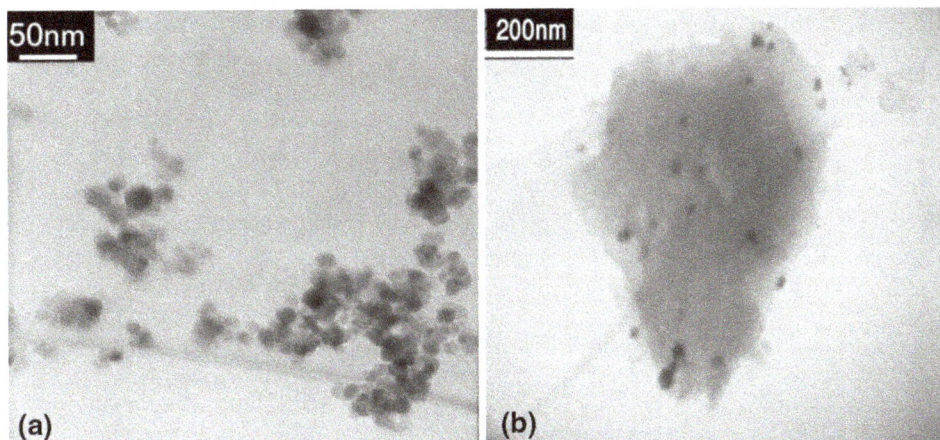

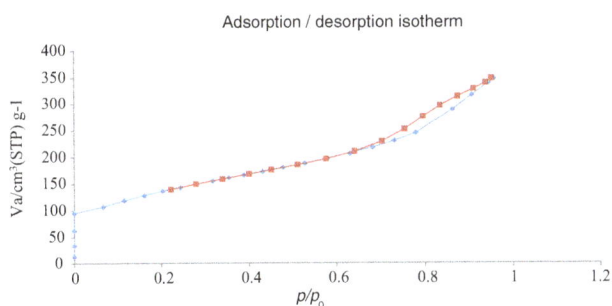

Fig. 7 Adsorption/desorption isotherm of Fe_3O_4–Pd–C catalyst

Figure 6 shows the transmission electron microscopy (TEM) images of the palladium–carbon (Pd–C) and Fe_3O_4 nanoparticles. In the TEM of Pd–C, palladium with high atomic number blocked partial electrons from TEM electron beam, and this made the palladium nanoparticles dark under TEM [5]. TEM images of Pd–C reveal that palladium nanoparticles are well dispersed on activated carbon and the size of Pd particles is about 10–45 nm. The TEM image of Fe_3O_4 clearly shows that synthesized nanoparticles are distributed uniformly and have an average size of 10–30 nm.

The N_2 adsorption–desorption isotherm of the Fe_3O_4–Pd–C show type IV pattern with hysteresis (H3) loops at relative pressure of 0.6–1.0 (Fig. 7). Brunauer–Emmett–Teller (BET) analysis indicated that surface area of the Fe_3O_4–Pd–C is 487.7 m^2/g, while total pore volume is 0.0407 cm^3/g. The pore size diameter was calculated be 2.582 nm. The high surface area (487.7 m^2/g) may be beneficial to enhancing the catalytic activity of Fe_3O_4–Pd–C catalyst.

The magnetically controllable aggregation behavior of Fe_3O_4–Pd–C was investigated by vibrating sample magnetometer (VSM) studies. Figures 8 and 9 show the VSM curves of the Fe_3O_4 and Fe_3O_4–Pd–C, respectively,

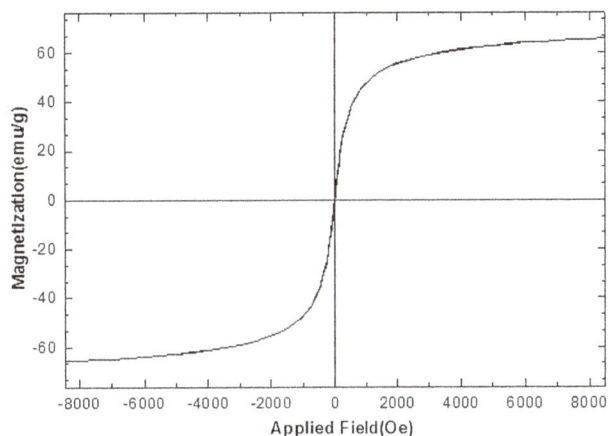

Fig. 8 VSM curve of the Fe_3O_4 nanoparticles

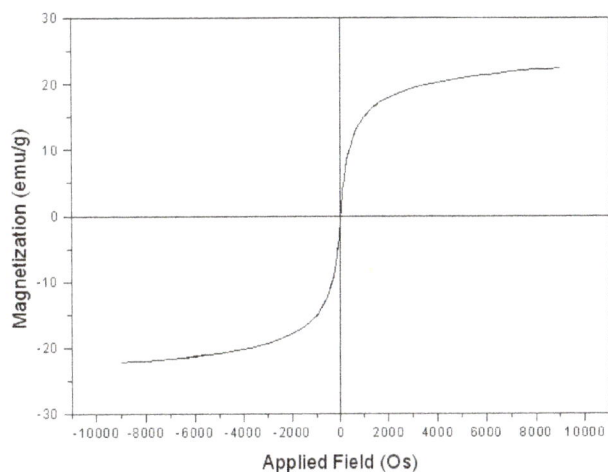

Fig. 9 VSM curve of the Fe_3O_4–Pd–C catalyst

measured by the Meghnatis Daghigh Kavir Company (Iran). Fe_3O_4 and Fe_3O_4–Pd–C show superparamagnetic behavior, and the saturation magnetization was determined to be 65.5 and 25.0 emu g^{-1}, respectively.

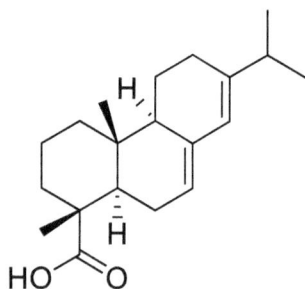

Fig. 10 Chemical structure of abietic acid

Table 1 Reusability test of catal

DAA % First run	DAA % Second run	DAA % Third run
74	69.5	68

Gum rosin is one of the important renewable forestry products and approximately 90% of the gum rosin is the abietic-type resin acid with a conjugated double bond (Fig. 10) [17]. By the use of a catalytic disproportionation reaction, the highly reactive conjugated double bond in abietic-type resin acid can be modified. Due to its advantages, i.e., high oxidation resistance, high softening point, low brittleness, and very good color stability, disproportionated rosin (DPR) has proved to be superior quietly to rosin in many applications.

Here in, the catalytic activity of Fe_3O_4–Pd–C in the synthesis of DPR from gum rosin was studied, and the reaction was checked by gas chromatography (GC) analysis [13]. A control experiment showed that Fe_3O_4 and Fe_3O_4–active carbon could not catalyze this disproportionation reaction, and the presence of palladium is essential. When the reaction was carried out with Fe_3O_4–Pd–C (0.1% w/w), dehydroabietic acid was obtained in 65% yield after 6 h. If the reaction temperature is decreased from 280 to 220 °C, dehydroabietic acid will obtain in 19.6% yield. After evaluation of the catalytic activity of Fe_3O_4–Pd–C, optimization shows that the best result was obtained by using of 0.25% w/w of Fe_3O_4–Pd–C with an optimal reaction temperature of 280 °C. In this condition, dehydroabietic acid was obtained in 74% yield after 1 h.

To recycle the Fe_3O_4–Pd–C catalyst, the magnetic catalyst was collected by a magnet, washed with iso-propanol for three times, and dried. At least, the Fe_3O_4–Pd–C catalyst was stable and reusable for three reaction runs (Table 1).

Conclusions

In conclusion, the catalytic disproportionation of gum rosin over a magnetic palladium–carbon (Fe_3O_4–Pd–C) was investigated. The catalyst obtained via the pathway described in this manuscript is essentially superparamagnetic, has high porosities and high surface areas, and displayed suitable activity in disproportionation of gum rosin. The Fe_3O_4–Pd–C catalyst was stable and reusable for at least three reaction runs. Such findings are important because the simplicity of the nanocatalyst production method and simple separation and recyclability on the other hand make possible the industrial production and application of the catalyst.

Acknowledgements We are thanks the INSF (Iran National Science Foundation), Tarbiat Modares University and Padideh Shimi Jam Co. for supporting of this work.

References

1. Marsh, H., Reinoso, F.R.: Activated carbon. Elsevier, Amsterdam (2006)
2. Edward, L.K., Ko, D.C., McKay, G.: Production of active carbons from waste tyres—a review. Carbon **42**, 2789–2805 (2004)
3. Vijayakumar, P., Senthil Pandian, M., Pandikumar, A., Ramasamy, P.: A facile one-step synthesis and fabrication of hexagonal palladium-carbon nanocubes (H–Pd/C NCs) and their application as an efficient counter electrode for dye-sensitized solar cell (DSSC). Ceram. Int. **43**, 8466–8474 (2017)
4. Holade, Y., Ege, N., Karine, S., Teko, S., Napporn, W., Kokoh, K.B.: Recent advances in carbon supported metal nanoparticles preparation for oxygen reduction reaction in low temperature fuel cells. Catalysts **5**, 310–348 (2015)
5. Lu, A.H., Salabas, E.L., Ferdi, S.: Magnetic nanoparticles: synthesis, protection, functionalization, and application. Angew. Chem. Int. Ed. **46**, 1222–1244 (2007)
6. Karimi, B., Mansouri, F., Mohammad Mirzaei, H.: Recent applications of magnetically recoverable nanocatalysts in C–C and C–X coupling reactions. ChemCatChem. **7**, 1736–1789 (2015)
7. Faraji, M., Yamini, Y., Rezaee, M.: Magnetic nanoparticles: synthesis, stabilization, functionalization, characterization, and applications. J. Iran. Chem. Soc. **7**, 1–37 (2010)
8. Arefi, M., Saberi, D., Karimi, M., Heydari, A.: Superparamagnetic Fe(OH)$_3$@Fe$_3$O$_4$ nanoparticles: an efficient and recoverable catalyst for tandem oxidative amidation of alcohols with amine hydrochloride salts. ACS Comb. Sci. **17**, 341–347 (2015)
9. Schwickardi, M., Olejnik, S., Salabas, E. L., Schmidt, W., Schuth, F.: Scalable synthesis of activated carbon with superparamagnetic properties. Chem. Commun. 3987–3989 (2006)
10. Tan, K.A., Morad, N., Teng, T.T., Norli, I., Panneerselvam, O.: Removal of cationic dye by magnetic nanoparticle (Fe$_3$O$_4$) impregnated onto activated maize cob powder and kinetic study of dye waste adsorption. APCBEE Procedia **1**, 83–89 (2012)
11. Kakavandi, B., Jonidi, A., Rezaei, R., Nasseri, S., Ameri, A., Esrafily, A.: Synthesis and properties of Fe$_3$O$_4$-activated carbon magnetic nanoparticles for removal of aniline from aqueous solution: equilibrium, kinetic and thermodynamic studies. Iran. J. Environ. Health. Sci. Eng. **10**, 19 (2013)
12. Mohan, D., Sarswat, A., Singh, V.K., Alexandre Franco, M., Pittman, C.U.: Development of magnetic activated carbon from almond shells for trinitrophenol removal from water. Chem. Eng. J. **172**, 1111–1125 (2011)
13. Mostafalu, R., Hydari, A., Banaei, A., Ghorbani, F., Arefi, M.: The use of palladium nanoparticles supported on active carbon for synthesis of Disproportionate rosin (DPR) as a useful emul-

sifier in petrochemical industries. J. Nanostruct. Chem. **7**, 61–66 (2016)

14. Ríos-Hurtado, J.C., Múzquiz-Ramos, E.M., Zugasti-Cruz, A., Cortés-Hernández, D.A.: Mechano ynthesis as a simple method to obtain a magnetic composite (activated carbon/Fe₃O₄) for hyperthermia treatment. J. Biomater. Nanobiotechnol. **7**, 19–28 (2016)

15. Zamani, F., Hosseini, S.M.: Palladium nanoparticles supported on Fe₃O₄/amino acid nanocomposite: highly active magnetic catalyst for solvent-free aerobic oxidation of alcohols. Catal. Commun. **43**, 164–168 (2014)

16. Vijayakumar, P., SenthilPandian, M., Su Pei, L., Pandikumar, A., Ming, H. N., Mukhopadhyay, S., Ramasamy, P.; Investigations of tungsten carbide nanostructures treated with different temperatures as counter electrodes for dye sensitized solar cells (DSSC) applications, Mater. Sci. Mater. Electron. **26**, 7977–7986 (2015)

17. Wang, L., Chen, X., Sun, W., Liang, J., Xu, X., Tong, Z.: Kinetic model for the catalytic disproportionation of pine oleoresin over Pd/C catalyst. Ind. Crop. Prod. **49**, 1–9 (2013)

13

Chemisorption of BH₃ and BF₃ on aluminum nitride nanocluster: quantum-chemical investigations

Ali Shokuhi Rad[1] ⓘ

Abstract In this study, two functionals (B3LYP and ωB97XD) were used for density functional theory (DFT) calculation of two major boron compounds (BH₃ and BF₃) adsorption on fullerene-like Al₁₂N₁₂ nanocluster. High values of adsorption energy, −268.6 (−244.7) for BF₃ and −224.5 (−196.4) kJ/mol for BH₃ were found using ωB97XD (B3LYP) functional, indicating strong chemisorption which is the result of Lewis acid–base interaction of adsorbent and adsorbates. The high negative values of ΔG (Gibbs free energy) and ΔH (enthalpy) confirm spontaneous exothermic adsorption process. Further studies were done by taking into account the charge analysis, FMO (frontier molecular orbitals), MEP (molecular electrostatic potential), density of states (DOS), and reactivity of resulted systems.

Keywords Fullerene-like cluster · Al₁₂N₁₂ Nanocage · Chemisorption · Boron trifluoride · Borane

Introduction

Borane (trihydridoboron) and boron trifluoride are two inorganic compounds with chemical formulas of BH₃ and BF₃, respectively. They are drab gases that are famous as substantial reagents in reaction pathway [1, 2]. In addition, they are known as significant Lewis acids (because of the electron deficiency of boron atom in their compounds) and general construction blocks for further boron complexes.

Studying the interface conception of different B–N bonds is vital for energy storage viewpoint. Group III-nitrogen compounds are well-recognized as storage energy owing to their lightweight and inherent high releasing energy [3].

Although BH₃ itself has not normally used as a reactant; however, it is a potential intermediate in adsorption of diborane, B₂H₆, that is broadly used as a source of boron [4]. Similarly, boron halides, including BF₃ and BCl₃, have been used as substitute probe molecules owing to the 2p¹ electronic structure of boron that provides a powerful Lewis acid when shared with a halogen atom [5, 6]. The boron atom of BF₃ has an unfilled pᶻ-like orbital in perpendicular to the molecular plane and has a propensity to receive electron pairs, and this property causes adsorption of boron compounds. For example in our recent study [7], we investigated the adsorption of some boron compounds on the surface of nitrogen-doped graphene. We found formation of new bond between nucleophilic atom (N) and electrophilic atom (B), whereas there was weak physisorption in the case of pristine graphene. There are various investigations on BF₃ and BH₃ in literature in this regard. For example, Xu et al. [8] used DFT to study the hydroboration of the Ge(1 0 0)−2 × 1 surface with BH₃. Based on their result, the Ge(1 0 0) surface displays rather different surface reactivity to dissociative adsorption of BH₃ in comparison with the C(1 0 0) and Si(1 0 0) surfaces. In another work, dissociative adsorption of BH₃ on the Si(100) surface has been searched with nonlocal DFT by Konecny and Doren [4]. They revealed that a Si–B bond is constructed through a nucleophilic attack on boron, leaving BH₂ and H fragments bonded to the surface.

Abee and Cox [5] searched on BF₃ as a probe molecule to interrogate the basicity of Cr₂O₃ (101̄2) surfaces.

After a successful development of CNT and fullerene C60, various spherical fullerene-like configurations

---

✉ Ali Shokuhi Rad
  a.shokuhi@gmail.com; a.shokuhi@qaemiau.ac.ir

[1] Department of Chemical Engineering, Qaemshahr Branch, Islamic Azad University, Qaemshahr, Iran

composed of inorganic non-carbon materials have been stimulated a great deal of interest [9–20]. Strout et al. revealed that the fullerene-like nano-cages $X_{12}Y_{12}$ are the most stable structures among all $(XY)_n$ ($X = B$, Al, Ga,... and $Y = N$, P, As, ...) semiconductors [9]. $Al_{12}N_{12}$, $Al_{12}P_{12}$, $B_{12}N_{12}$, and $B_{12}P_{12}$ are of great significance because of their high steadiness, large energy band gap, and outstanding chemical and physical properties. Among them, $B_{12}N_{12}$ nanoclusters were synthesized by Oku et al. in 2004 [10]. Different applications of nanostructure materials have been studied in recent literature [11–24]. Because of important roles of boron compounds in different areas of study, recently we rummaged on the chemisorption property of boron trichloride (BCl_3) on the surface of $Al_{12}N_{12}$ nanocluster [6]. Following successful potential of mentioned nanocluster as an ideal adsorbent, we persuaded to search on the adsorption of other boron compounds on this kind of nanocluster. As we found, there is no investigation in literature on the BH_3 and BF_3 adsorption on the $Al_{12}N_{12}$ nanocluster. In this paper, I have employed DFT method to investigate the adsorption of above-mentioned molecules on the surface of $Al_{12}N_{12}$ nanocluster. The attained results were analyzed by considering adsorption energies, the natural bond orbital (NBO) charge transfer, frontier molecular orbitals, density of states, and global indices of activities.

Computational details

The geometries of $Al_{12}N_{12}$ nanocluster and corresponding boron complexes were fully relaxed at the B3LYP (and $\omega B97XD$)/6–31G** level of theory as implemented in Gaussian 09 suite of program [25]. The suitability of B3LYP functional for different nanostructures has been proved [26–29]. To consider the effect of dispersion on adsorption energies, all relaxed structures were subjected to new optimization using meta-hybrid functional (wB97XD) [30]. For each calculation, the global charge of system was neutral. Frequency calculations have been performed to ensure the stability of the interaction. Adsorption energy (E_{ad}) of BX_3 ($X = F$, H) on $Al_{12}N_{12}$ nanocluster is defined as

$$E_b = E_{Al12N12-BX3} - (E_{Al12N12} + E_{BX3}), \qquad (1)$$

where $E_{Al12N12-BX3}$ is the total energy of the adsorbed BX_3 molecule on the surface of nanocluster ($Al_{12}N_{12-}$), and $E_{Al12N12}$ and E_{BX3} are the total energies of the free $Al_{12}N_{12}$ and a free BX_3 molecule, respectively.

Natural bond orbital (NBO) analysis [31] is used to follow charge distribution and charge transfer between BX_3 and $Al_{12}N_{12}$. Frontier molecular description has been studied to follow the variation in the structure of

nanocluster when BX_3 is adsorbed. Parr et al. [32] stated the electrophilicity concept. Chemical potential (μ) is defined based on the following eq. [33]:

$$\mu = -(E_{HOMO} + E_{LUMO})/2, \qquad (2)$$

where E_{HOMO} is the energy of HOMO and E_{LUMO} is the energy of LUMO. In addition, hardness (η) can be calculated using the Koopmans' theorem [33] as:

$$\eta = (E_{HOMO} + E_{LUMO})/2 \qquad (3)$$

Softness (S) [33] and electrophilicity (ω) [33] are defined as the following eqs. correspondingly.

$$S = 1/(2\eta) \qquad (4)$$

$$\omega = \mu^2/2\eta. \qquad (5)$$

Results and discussion

The electronic structure of $Al_{12}N_{12}$ nanocluster is pretty discussed in different investigations [11–14], so, a comprehensive explanation is not provided here. As shown in Fig. 1, this kind of nanocluster is performed of some tetragon (four membered) and hexagon (six membered) rings. Actually, there are two kinds of Al...N bond through the cluster, one is shared between two hexagon rings (b@66) and another is shared between a hexagon ring and a tetragon ring (b@64). The bond distances are calculated to be 1.79 and 1.86 Å for b@66 and b@64, respectively. The NBO charge allocation in $Al_{12}N_{12}$ nanocluster is also depicted in Fig. 1 (right). It is obvious from Fig. 1 that the charge is uniformly distributed through the nanocluster. Each Al atom has a positive charge of +1.84 e, whereas each N atom has a negative charge of −1.84 e through the structure. As a result, positive and negative sites of cluster correspond to nucleophilic and electrophilic attacks of adsorbates, respectively.

To study the interaction of BF_3 and BH_3 onto $Al_{12}N_{12}$ nanocluster, firstly I placed the BX_3 ($X = F$, H) molecule at different orientations above the cluster, 1-the B atom of BX_3 above the N of cluster, 2-the X atom of BX_3 above the N of cluster so that X–B bond has perpendicular orientation than the surface. All these initial configurations were used to fully optimization using two before-mentioned functionals. Despite there are different initial configurations, I found that there is only one relaxed structure (output) for BF_3 as well as BH_3 upon optimization. Figure 2 depicts side and tope views of adsorbed BF_3 and BH_3 onto nanocluster. As shown in Fig. 2, adsorption of BF_3 and BH_3 correspond to the formation of new bond between N of cluster and B of adsorbate. The B...N bond length is 1.45 and 1.59 Å for $Al_{12}N_{12}$–BF_3 and $Al_{12}N_{12}$–BH_3 complexes,

Fig. 1 Relaxed structure of
Al₁₂N₁₂ nanocluster along with
the NBO charge distribution of
nanocluster

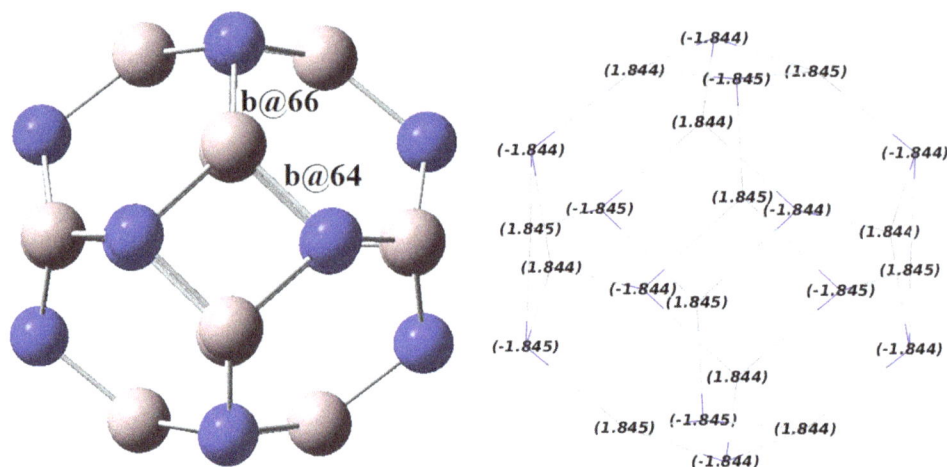

Fig. 2 Side and top views of
adsorbed BF₃ and BH₃ on
Al₁₂N₁₂ nanocluster

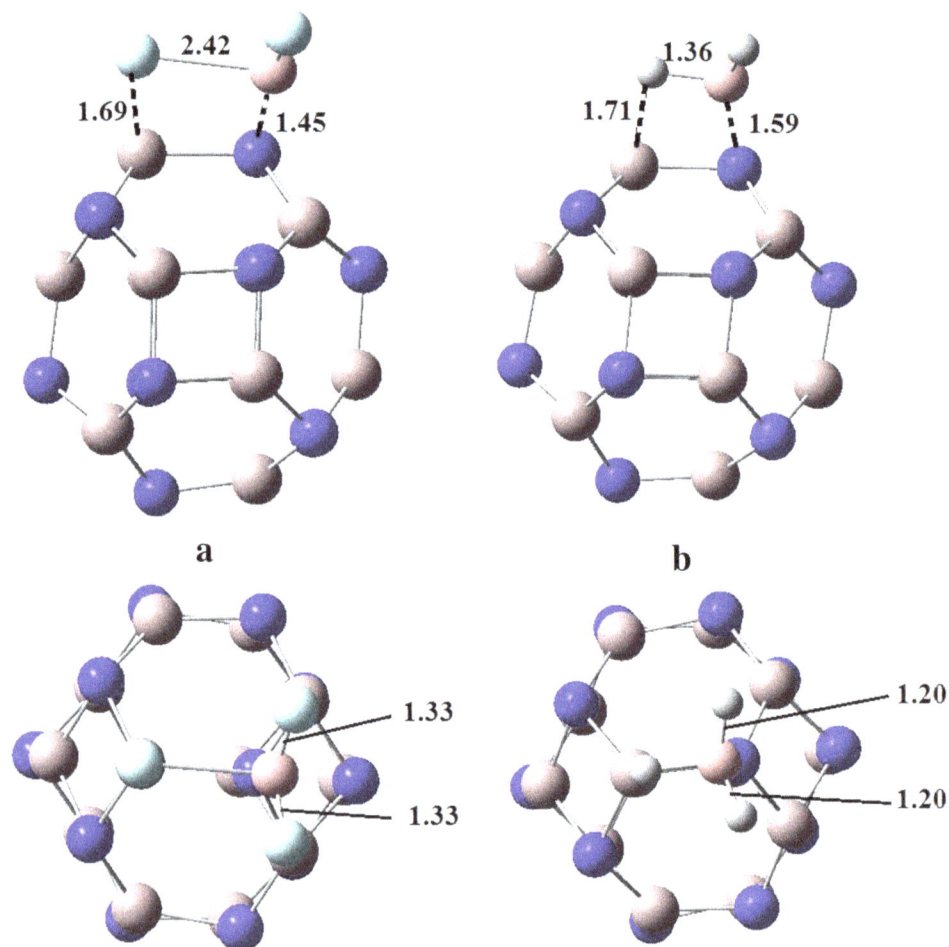

respectively. The lower distance in the earlier complex
more likely relates to the more positively charged boron in
BF₃ owing to strong electron affinity of F atoms. As a
result, more positively charged boron atom in BF₃ corresponds to stronger Lewis acid–base interaction. This bond

formation results the transferring in hybridizing of B atom
from sp^2 to sp^3. Moreover, we can see another bond formation between the F atom of BF₃ and also the H atom of
BH₃ with the Al atom of nanocluster (both F and H are
negatively charged in their related complex). These new

Table 1 The values of adsorption energy (E_{ad}, kJ/mole) using two functionals (wB97XD and B3LYP), nearest equilibrium distance of analyte- surface (d_e, Å), mean of bond distances b@66, and b@64 (Å) at the area of interaction, NBO net charge transfer (Q_{NBO}, e), change in the enthalpy and Gibbs free energy (kJ/mol)

System	E_{ad} (wB97XD)	E_{ad} (B3LYP)	d_e	b@66	b@64	Q_{NBO}	ΔH	ΔG
$Al_{12}N_{12}$	–	–	–	1.79	1.86	–	–	–
$Al_{12}N_{12}$–BF_3	−268.6	−244.7	1.45	2.03	1.92	−0.247	−240.3	−195.1
$Al_{12}N_{12}$–BH_3	−224.5	−196.4	1.59	1.92	1.87	−0.241	−187.3	−140.6

bonds formation is because of transferring the density of electron from F and H atoms (which are rich in electron) to unoccupied orbital of Al atom of nanocluster. It is obvious that the bond length of Al...F and Al...H are calculated to be 1.69 and 1.71 Å, respectively, which are comparable with B...N bond in their complexes.

The experimental bond length of B–H and B–F in free BH_3 and BF_3 molecules is reported 1.19 and 1.31 Å, respectively [5, 34], which are very close to the values of 1.194 and 1.317 Å calculated by this study. It is shown in Fig. 2 that after adsorption of BF_3 molecule, fully dissociation in one of the B-F bonds is happened, whereas partial dissociation in one of the B–H bonds is resulted. In the case of BF_3, one B...F bond is elongated to 2.42 Å (83.3% increase) and two other bonds are partially extended to 1.33 Å. So, because of the large change in one of the bonds in the molecule, we can conclude fully dissociation of BF_3 upon adsorption on $Al_{12}N_{12}$ nanocluster. On the other hand, for BH_3, the elongation is less pronounced. It is shown in Fig. 2 that one of the bond lengths of B...H reaches to 1.36 Å (13.3% increase) and two other reach to 1.20 Å upon adsorption which attribute to the partial dissociation of molecule.

The values of adsorption energy are listed in Table 1. We can see that the adsorption of BF_3 and BH_3 on the surface of $Al_{12}N_{12}$ nanocluster corresponds to releasing energy of −244.7 and −196.4 kJ/mol (based on B3LYP functional), and −268.6 and −224.5 kJ/mol (based on ωB97XD functional), respectively. The difference between the values of adsorption energy achieved by two functionals reveals that the dispersion parameter plays an important role in interaction. All of these values of adsorption are categorized in chemisorption region. In addition, considerable difference between the results of two functionals suggests that dispersion has important role in adsorption process of this study. The results of adsorption energies are totally in agreement with the results of bond distances, the higher adsorption energy corresponds to the lower bond distance. Moreover, calculated value of adsorption energy are greatly higher than what reported for BH_3 adsorption on Si(100) surface ($\simeq 180$ kJ/mole, [4]), and BF_3 adsorption on α-Cr_2O_3 ($10\overline{1}2$) ($\simeq 133$ kJ/mole, [5]).

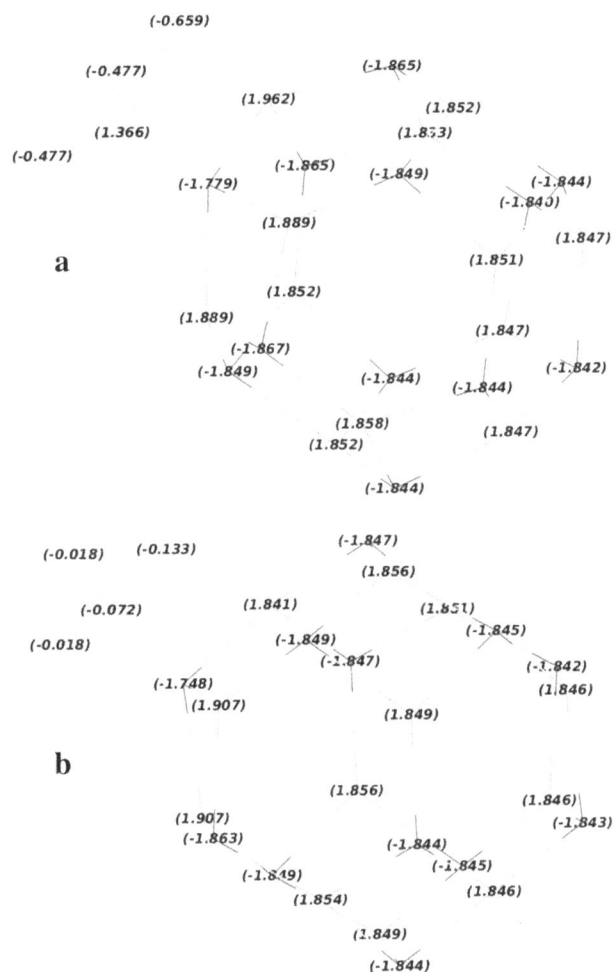

Fig. 3 The NBO charge allocations after BF_3 (**a**), and BH_3 (**b**) adsorption

Despite the high changes in the electronic structure of adsorbates, the changes in the electronic structure of nanocluster are also important. It is shown in Table 1 that the Al...N bonds in b@66 (b@64) are changed from 1.79 (1.86) Å for pristine nanocluster to 2.03 (1.92) for BF_3 and 1.92 (1.87) for BH_3 adsorbed systems. It is obvious that the effect of BF_3 adsorption on the electronic structure of nanocluster is more pronounced compared to BH_3 which is attributed to the stronger interaction in the earlier complex. By adsorption of BF_3 and BH_3, not only the bond lengths through the nanocluster are changed, but also the charge

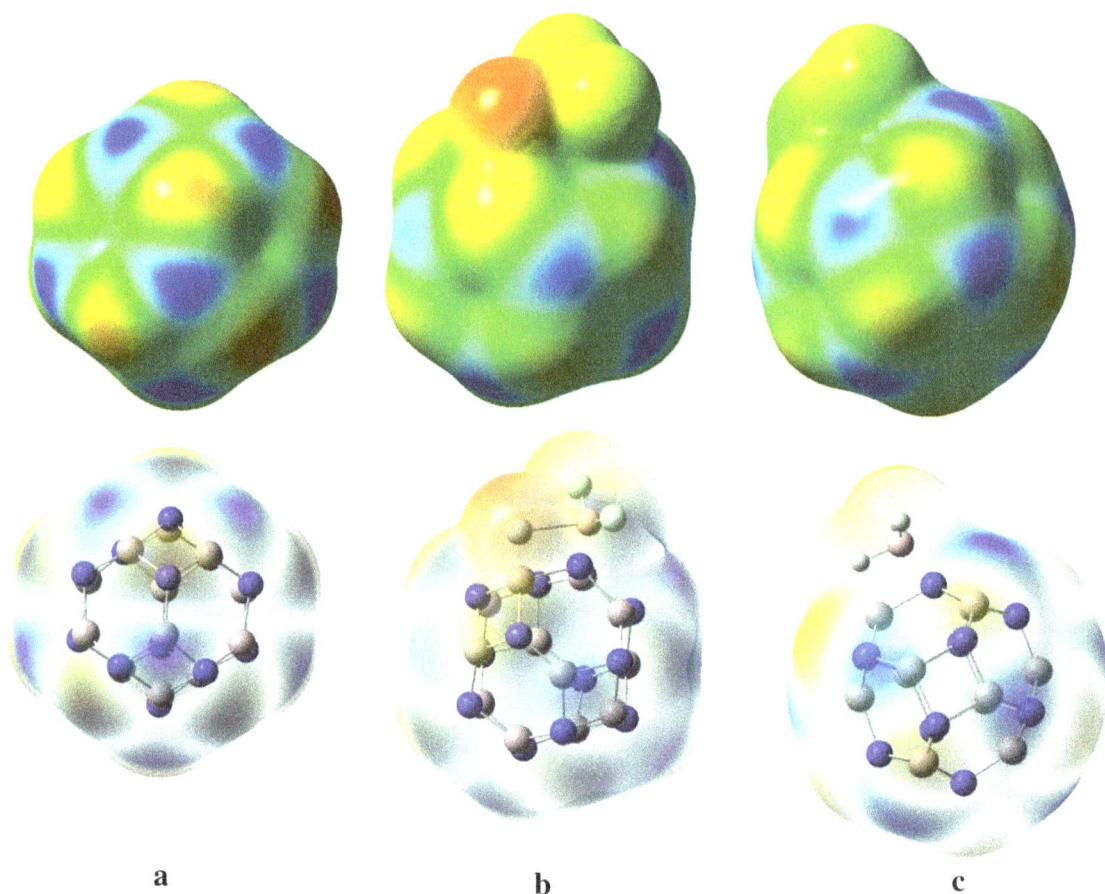

Fig. 4 The MEP of pristine $Al_{12}N_{12}$ nanocluster (**a**) along with its complexes with BF_3 (**b**) and BH_3 (**c**)

allocation on each atom significantly is effected (see Fig. 3). By comparing the NBO charge distribution of pristine nanocluster (Fig. 1, right) with that of complexes (Fig. 3), one can conclude significant change especially in the area of interaction, which is considerable in the case of BF_3 compared to BH_3 adsorption. Along with the stronger interaction of N (of nanocluster) with boron atom of BF_3 rather than BH_3, the higher change in the electronic structure of nanocluster upon adsorption of BF3 is also effected by the stronger Lewis acid–base interaction between F atom (of BF_3) and Al atom (of nanocluster) compared to that interaction between the same Al atom and H atom (of BH_3). Therefore, the difference in the NBO charge distribution of two complexes is a result of two different kinds of interaction between corresponding atoms. The values of net charge transfer are also listed in Table 1. As expected, direction of charge transfer is from nanocluster to adsorbates, confirming transfer of charge from nucleophilic atom (N) to electrophilic atom (B). On the other hand, more charge transfer for BF_3 adsorption is in accordance to its higher interaction. However, direction of net charge transfer are the resultant of two revers charge transfer; one from N (of cluster) to B, and another from F

(H) of adsorbate to Al (of cluster), but the value of the earlier is much higher as expected, resulting overall direction of charge transfer from $Al_{12}N_{12}$ nanocluster to adsorbates.

In order to examine the thermodynamic feasibility of BF_3 and BH_3 adsorption on the $Al_{12}N_{12}$ at ambient temperature and pressure ($T = 298.14$ K, and $P = 1$ atm), I have calculated free energies (ΔG) and enthalpy changes (ΔH) of each system using the results of vibrational frequency calculations,. The computed values of ΔH are -240.3 and -187.3 kJ/mol, and those of ΔG are -195.1 and -140.6 kJ/mol, respectively. However, the lower value of ΔG compared to that of ΔH is because of the entropic influence. The negative value of Gibbs free energy as well as enthalpy of each system confirms that adsorption of above-mentioned adsorbates is a spontaneous exothermic process.

To better understand the binding sites, I have calculated the molecular electrostatic potentials (MEP) of each system. For this determination, MEP of $Al_{12}N_{12}$ nanocluster and its complexes ($Al_{12}N_{12}$–BF_3 and $Al_{12}N_{12}$–BH_3) were compared and are given in Fig. 4. For pristine $Al_{12}N_{12}$ nanocluster, Aluminum has positive charge (blue color)

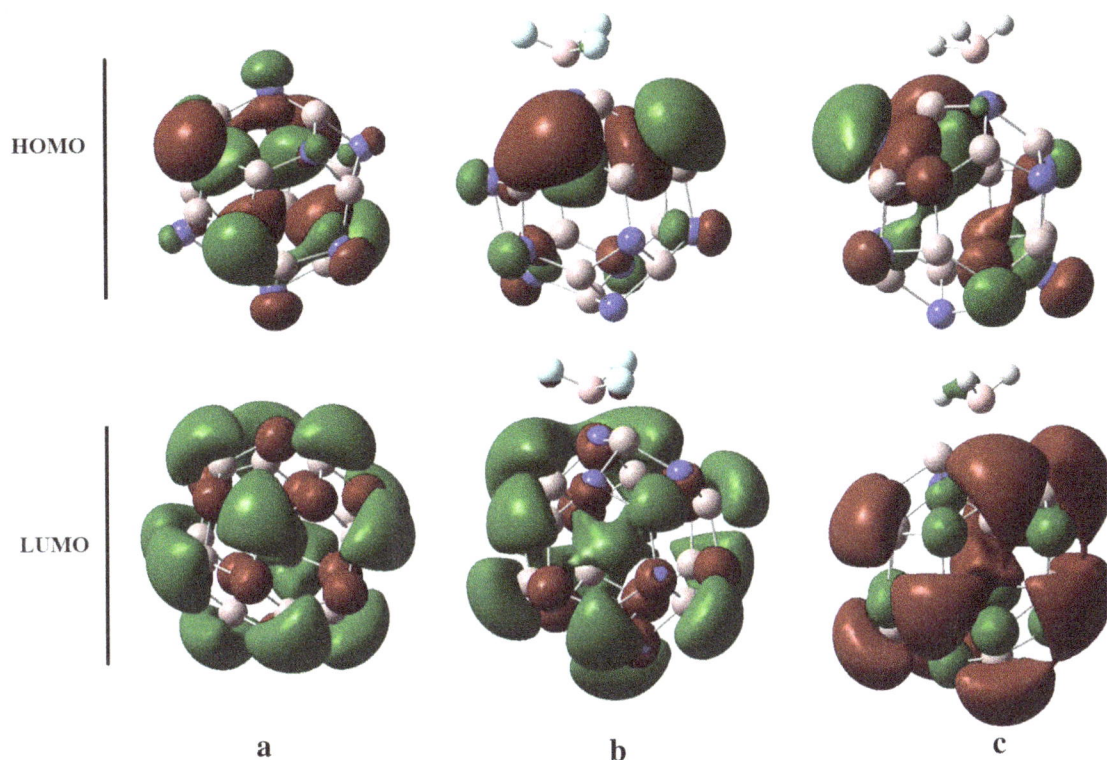

Fig. 5 HOMO and LUMO distributions of pristine $Al_{12}N_{12}$ (**a**) $Al_{12}N_{12}$–BF_3 (**b**), and $Al_{12}N_{12}$–BH_3

Table 2 The electronic properties of free $Al_{12}N_{12}$ and its complexes

System	E_{HOMO}	E_{LUMO}	E_g	η	μ	S	ω
free $Al_{12}N_{12}$	−6.472	−2.541	3.931	1.965	−4.506	0.254	5.157
$Al_{12}N_{12}$–BF_3	−6.380	−2.762	3.618	1.809	−4.571	0.276	5.767
$Al_{12}N_{12}$–BF_3	−6.474	−2.605	3.869	1.934	−4.539	0.258	5.315

$[E_g = (E_{HUMO} − E_{LUMO})]/ E_{LUMO})]eV.eV, \ [\eta = −(E_{HUMO} − E_{LUMO})/ E_{LUMO})2.2]/eV, \ [\mu = (E_{HOMO} + E_{LUMO})/2]/E_{LUMO})/2]eV.eV, S = [1/2\eta]eV^{-1}, \omega = [\mu^2/2\eta]$

while the nitrogen has negative charge. The negative charge on N in the nanocluster is not much pronounced (yellow color). After adsorption of BF_3 and BH_3 molecules, we can see important changes in the blueish and yellowish part of cluster, especially at the area of interaction. The red part of $Al_{12}N_{12}$–BF_3 complex attributes to the electronegative F atom of BF_3 while no red part can be seen in $Al_{12}N_{12}$–BH_3 complex. This is another proving for happening two strong Lewis acid–base interactions (N...B and Al... F) during the earlier complex formation.

Frontier molecular orbitals are also studied to search the effect of BF_3 and BH_3 adsorption on the electronic properties of $Al_{12}N_{12}$ nanocluster. The outlooks of HOMOs and LUMOs are shown in Fig. 5 and their values are given in Table 2. The spreading of densities also delivers important information concerning the adsorbent capability of $Al_{12}N_{12}$ nanocluster for BF_3 and BH_3 molecules. As shown in

Fig. 5, the HOMO in $Al_{12}N_{12}$ nanocluster is confined on nitrogen atoms, while the LUMO has density homogeneously circulated on the whole nanocluster. Upon BF_3 and BH_3 adsorption, the densities are redistributed confirming important alterations in the electronic structure of $Al_{12}N_{12}$ nanocluster.

The HOMO–LUMO energies of $Al_{12}N_{12}$ nanocluster are −6.472 and −2.541 eV, correspondingly. As shown in Fig. 5 and the data of Table 2, the adsorption of BF_3 and BH_3 causes some changes in the electronic structure of the $Al_{12}N_{12}$ nanocluster. While the energy gap of pristine $Al_{12}N_{12}$ nanocluster is 3.931 eV, upon BF_3 and BH_3 adsorption, the energy gaps are reduced to 3.618 and 3.869 eV, respectively.

For BF_3 adsorption, the HOMO of resulted system increases while the LUMO decreases; however, the change in the LUMO is more pronounced than HOMO, resulting in

Fig. 6 Density of states of
pristine $Al_{12}N_{12}$ (a) $Al_{12}N_{12}$–
BF_3 (b), and $Al_{12}N_{12}$–BH_3

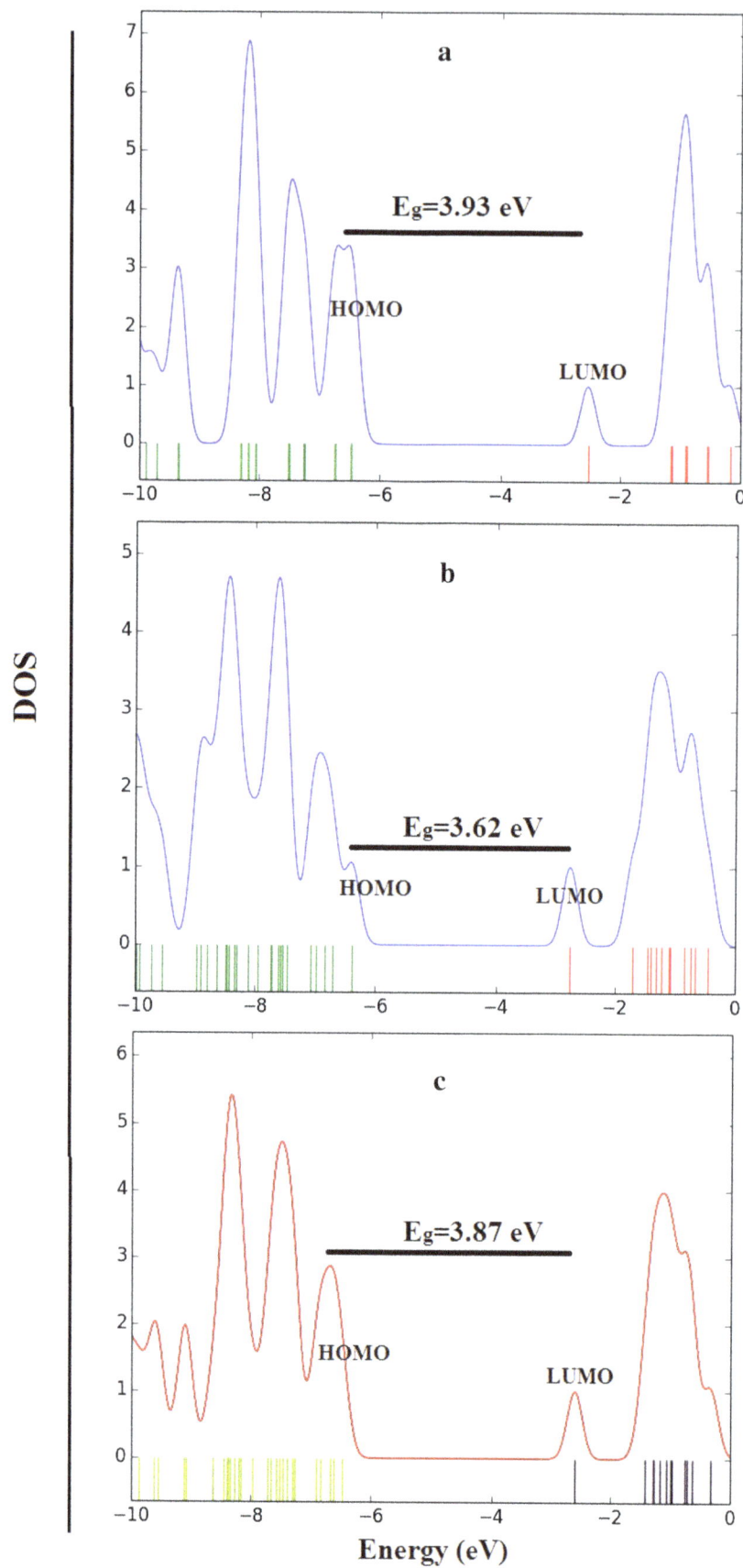

a decrease in the energy gap of system. On the other hand, upon adsorption of BH_3, the energy of HOMO approximately remains unchanged while the LUMO decreases, resulting in a little decrease in the energy gap of system.

Density of states (DOS) of pristine $Al_{12}N_{12}$ and $Al_{12}N_{12}$–BF_3 and $Al_{12}N_{12}$–BH_3 complexes is studied in order to obtain more understanding the effect of adsorption (See Fig. 6). Some new energy states appear upon adsorption of both BF_3 and BH_3 molecules. These new states appear near the frontier molecular orbitals, which confirm case of strong hybridizing. The alteration in electronic properties is very essential for the development of sensors. Alteration in HOMO–LUMO causes change in electrical conductivity, which is the prime factor in the design of electrochemical sensors. Association between conductivity and E_g can was determined as Eq. (6) [35].

$$\sigma \alpha \exp\left(-E_g/kT\right), \tag{6}$$

where it could be understood that a small lessening in E_g leads to meaningfully greater electrical conductivities.

The global indices of reactivity for pristine $Al_{12}N_{12}$ and its complexes are given in Table 2. They are pretty vital parameters owing to they exemplify the reactivity and steadiness of a system. Entertainingly, the hardness of nanocluster (1.965 eV) decreases on complexation with BF_3 (1.809 eV) and BH_3 (1.934 eV). One can see that the change in hardness is comparatively small for BH_3 adsorption ($\Delta \eta = -0.031$ eV) whereas meaningfully large alteration is observed for BF_3 adsorption ($\Delta \eta = -0.156$). Meanwhile, hardness attributes to the steadiness of a complex toward distortion in the attendance of electrical field, based on the result, the $Al_{12}N_{12}$ complexes are more susceptible to deformation under electrical field than pristine nanocluster. Softness inversely changes compared to hardness, hence, it is predictable that softness for $Al_{12}N_{12}$–BF_3 and $Al_{12}N_{12}$–BH_3 will increase, resulting increase in the reactivity of complexes.

Conclusion

BF_3 and BH_3 with a low-lying empty orbital are strong Lewis acids, so their capability to accept electrons along with their planar geometries suggests that nucleophilic attack by a nucleophilic atom is performed. The goal of this research is investigation on the adsorption properties of BF_3 and BH_3 molecules on $Al_{12}N_{12}$ nanocluster considering DFT method. I found fully dissociative adsorption for BF_3 and partial dissociative adsorption for BH_3 molecules. The computed values of ΔH are -240.3 and -187.3 kJ/mol, and those of ΔG are -195.1 and -140.6 kJ/mol, for BF_3 and BH_3 adsorption, respectively. The negative value of Gibbs free energy approves that

adsorption of above-mentioned adsorbates is a spontaneous process. Results of frontier molecular orbitals confirm some important changes in the electronic structure of the nanocluster upon adsorption.

Acknowledgements I highly acknowledge the financial support received from the Iran Nanotechnology Initiative Council, Iran.

References

1. Rasul, G., Prakash, S.G.K., Olah, G.A.: Complexes of CO2, COS, and CS2 with the super Lewis Acid BH4+ contrasted with extremely weak complexations with BH3: theoretical calculations and experimental relevance. J. Am. Chem. Soc. **121**, 7401–7404 (1999)
2. Brinck, T., Murray, J.S., Politzer, P.: A Computational analysis of the bonding in boron trifluoride and boron trichloride and their complexes with ammonia". Inorg. Chem. **32**, 2622–2625 (1993)
3. Leskiw, B.D., Castleman Jr., A.W., Ashman, C., Khanna, S.N.: Reactivity and electronic structure of aluminum clusters: the aluminum–nitrogen system. J. Chem. Phys. **114**, 1165–1169 (2001)
4. Konecny, R., Doren, D.J.: Adsorption of BH_3 on Si (100)−(2×1). J. Phys. Chem. B **101**, 10983–10985 (1997)
5. Abee, M.W., Cox, D.F.: BF3 Adsorption on α-Cr₂O₃ ($10\overline{1}2$): Probing the Lewis basicity of surface oxygen anions. J. Phys. Chem. B **105**, 8375–8380 (2001)
6. Rad, A.S., Ayub, K.: DFT study of boron trichloride adsorption on the surface of Al12N12 nanocluster. Mol. Phys. **115**, 879–884 (2017)
7. Rad, A.S., Shadravan, A., Soleymani, A.A., Motaghedi, N.: Lewis acid-base surface interaction of some boron compounds with N-doped graphene, first principles study. Curr. Appl. Phys. **15**, 1271–1277 (2015)
8. Xu, Y.J., Li, J.Q.: The dissociative adsorption of borane on the Ge (100)–2×1 surface: a density functional theory study. Appl. Surf. Sci. **252**, 5855–5860 (2006)
9. Strout, D.L.: Structure and stability of boron nitrides: isomers of $B_{12}N_{12}$. J. Phys. Chem. A **104**, 3364–3366 (2000)
10. Oku, T., Nishiwaki, A., Narita, I.: Formation and atomic structure of $B_{12}N_{12}$ nano-cage clusters studied by mass spectrometry and cluster calculation. Sci. Tech. Adv. Mater. **5**, 635–638 (2004)
11. Rad, A.S., Ayub, K.: A comparative density functional theory study of guanine chemisorption on $Al_{12}N_{12}$, $Al_{12}P_{12}$, $B_{12}N_{12}$, and $B_{12}P_{12}$ nano-cages. J. Alloys Compd. **672**, 161–169 (2016)
12. Rad, A.S., Ayub, K.: Adsorption of pyrrole on $Al_{12}N_{12}$, $Al_{12}P_{12}$, $B_{12}N_{12}$, and $B_{12}P_{12}$ fullerene-like nano-cages, a first principles study. Vacuum **131**, 135–141 (2016)
13. Rad, A.S.: Study on the surface interaction of Furan with $X_{12}Y_{12}$ (X = B, Al, and Y = N, P) semiconductors: DFT calculations. Heteroat. Chem. **27**, 316–322 (2016)
14. Ayub, K.: Are phosphide nano-cages better than nitride nano-cages? A kinetic, thermodynamic and non-linear optical properties study of alkali metal encapsulated X12Y12 nano-cages. J. Mater. Chem. C **4**, 10919–10934 (2016)
15. Wang, Q., Sun, Q., Jena, P., Kawazoe, Y.: Potential of AlN nanostructures as hydrogen storage materials. ACS Nano **3**, 621–626 (2009)
16. Rad, A.S., Ayub, K.: Ni adsorption on $Al_{12}P_{12}$ nano-cage: a DFT study. J. Alloys Compd. **678**, 317–324 (2016)
17. Rad, A.S., Ayub, K.: Detailed surface study of adsorbed nickel on $Al_{12}N_{12}$ nano-cage. Thin Solid Films **612**, 179–185 (2016)
18. Rad, A.S., Ayub, K.: Enhancement in hydrogen molecule

adsorption on $B_{12}N_{12}$ nano-cluster by decoration of nickel. Int. J. Hydrogen Energy **41**, 22182–22191 (2016)

19. Rad, A.S., Ayub, K.: Coordination of nickel atoms with $Al_{12}X_{12}$ (X = N, P) nano-cages enhances H_2 adsorption: a surface study by DFT. Vacuum **133**, 70–80 (2016)

20. Iqbal, J., Ayub, K.: Theoretical study of the non linear optical properties of alkali metal (Li, Na, K) doped aluminum nitride nanocages. RSC Advances **687**, 94228–94235 (2016)

21. Rad, A.S., Ayub, K.: Adsorption properties of acetylene and ethylene molecules onto pristine and Nickel-decorated $Al_{12}N_{12}$ nanoclusters. Mat. Chem. Phys **194**, 337–344 (2017)

22. Rad, A.S., Ayub, K.: Adsorption of thiophene on the surfaces of X12Y12 (X = Al, B, and Y = N, P) nanoclusters; A DFT study. J. Mol. Liq. **238**, 303–309 (2017)

23. Rad, A.S., Ayub, K.: O_3 and SO_2 sensing concept on extended surface of $B_{12}N_{12}$ nanocages modified by Nickel decoration: a comprehensive DFT study. Solid State Sci. **69**, 22–30 (2017)

24. Rad, A.S., Zareyee, D.: Adsorption properties of SO_2 and O_3 molecules on Pt-decorated graphene: a theoretical study. Vacuum **130**, 113–118 (2016)

25. Frisch, M. J.: Gaussian 09, Revision E.01, Wallingford CT (2009)

26. Chen, L., Xu, C., Zhang, X.-F., Zhou, T.: Raman and infrared-active modes in MgO nanotubes. Physica E **41**, 852–855 (2009)

27. Rad, A.S.: Study of dimethyl ester interaction on the surface of Ga-doped graphene: Application of density functional theory. J. Mol. Liq. **229**, 1–5 (2017)

28. Rad, A.S., Sani, E., Binaeian, E., Peyravi, M., Jahanshahi, M.: DFT study on the adsorption of diethyl, ethyl methyl, and dimethyl ethers on the surface of gallium doped graphene. Appl. Surf. Sci. **401**, 156–161 (2017)

29. Rad, A.S., Jouibary, Y.M., Foukolaei, V.P., Binaeian, E.: Study on the structure and electronic property of adsorbed guanine on aluminum doped graphene: first principles calculations. Curr. Appl. Phys. **16**, 527–533 (2016)

30. Chai, J.D., Head-Gordon, M.: Long-range corrected hybrid density functionals with damped atom–atom dispersion corrections. Phys. Chem. Chem. Phys. **10**, 6615–6620 (2008)

31. Reed, A.E., Weinstock, R.B., Weinhold, F.: Natural population analysis. J. Chem. Phys. **83**, 735–746 (1985)

32. Parr, R.G., Szentpaly, L., Liu, S.: Electrophilicity Index. J. Am. Chem. Soc. **121**, 1922–1924 (1999)

33. Koopmans, T.: Ordering of wave functions and eigenenergies to the individual electrons of an Atom. Physica **1**, 104–113 (1934)

34. Kawaguchi, K.: Fourier transform infrared spectroscopy of the BH_3 ν_3 band". The J. Chem. Phys. **96**, 3411 (1992)

35. Li, S.: Semiconductor Physical Electronics, 2nd edn. Springer, Berlin (2006)

Synthesis and properties of magnetite/hydroxyapatite/doxorubicin nanocomposites and magnetic liquids based on them

N. V. Abramov[1] · S. P. Turanska[1] · A. P. Kusyak[1,2] · A. L. Petranovska[1] · P. P. Gorbyk[1]

Abstract Core–shell magnetosensitive nanocomposites (NC) based on single-domain magnetite (Fe_3O_4, core), with a shell consisting of hydroxyapatite (HA) and cytotoxic drug doxorubicin (DOX) layers have been synthesized. The processes of DOX adsorption on Fe_3O_4/HA surface from physiologic solution have been studied. DOX release into saline was found to decrease with growing of its quantity on NC surface. It has been determined that cytotoxic influence and antiproliferative activity of Fe_3O_4/HA/DOX NC with respect to *Saccharomyces cerevisiae* cells are characteristic for interaction of these cells with a free form of doxorubicin. Magnetic liquids containing Fe_3O_4/HA/DOX NC stabilized by sodium oleate and polyethylene glycol were prepared and investigated. It is shown that using the ensemble of Fe_3O_4 carriers as a superparamagnetic probe, the Langevin's paramagnetism theory, and the values of density of nanocomposite constituents, one can evaluate the size parameters of their shell, which has been corroborated by independent measurements of specific surface area of nanostructures and kinetic stability of the corresponding magnetic liquids. The obtained results may be useful for development and optimization of novel forms of magnetocarried medical remedies of targeted delivery and adsorbents based on nanocomposites of superparamagnetic core–shell type with multilevel nanoarchitecture, as well as for determination and control of the size parameters of its components.

Keywords Magnetite · Doxorubicin adsorption · Magnetite/hydroxyapatite/doxorubicin nanocomposites · Magnetic liquids · Dry residues · Size parameters of nanostructure

Introduction

Magnetic nanoparticles are widely used for the preparation of novel multifunctional therapeutic and diagnostic agents for different fields of medicine [1–3], including oncology. Conjugation of nanoparticles with corresponding antibody provides nanoparticles with the ability to recognize and "mark" specific microbiological objects, cell populations, microorganisms, and so on. On this basis, the concept has been substantiated for chemical construction of magnetosensitive nanocomposites with multilevel hierarchical architecture, characterized with functions of "nanoclinics" [2] and biomedical nanorobots [3–5] (recognition of microbiological objects in biological environments; the aimed delivery of drugs to target cells and organs using an external magnetic field and deposition; complex chemo-, immuno-, neutron capture, real-time hyperthermic therapy and diagnostics).

To produce magnetosensitive multifunctional nanocomposites (NC), considerable interest of researchers is drawn by magnetite (Fe_3O_4)/hydroxyapatite (HA) nanostructures of core–shell type, characterized with a unique set of physical, chemical, and biological properties, the ability to create on their basis magnetic liquids (ML) containing anticancer remedies for different functional destination, including cytotoxic drug, anthracycline

✉ P. P. Gorbyk
phorbyk@ukr.net

[1] Chuiko Institute of Surface Chemistry, National Academy of Sciences of Ukraine, 17 General Naumov Str., Kiev 03164, Ukraine

[2] Ivan Franko Zhytomyr State University, 40V. Berdychevska Str., Zhytomyr 10008, Ukraine

antibiotic doxorubicin (DOX) [6–17]. In this regard, the urgent task is to study the characteristics of the processes of adsorptive immobilization of DOX on the surface of Fe_3O_4/HA nanocomposites and its release into the saline, which is used to create ML for medical purposes, while maintaining biological activity of the cytotoxic drug.

It has been shown [18] that for a magnetite-based polydisperse colloidal ML the coordination of experimental and theoretical magnetization curves is possible when assuming that Fe_3O_4 particles have complex magnetic structure, namely, a low-magnetic surface layer with thickness $h_1 \sim 0.83$ nm (the lattice constant of magnetite at 300 K is 0.824 nm). Emergence of the said layer is attributed to chemical interaction of a particle with a stabilizing surfactant [19]. In [11, 20], it was found that the calculations of magnetization curve for ML based on single-domain Fe_3O_4 in the framework of Langevin's paramagnetism theory coordinate satisfactorily with the experimental results in assumption that saturation magnetization of magnetite particles depends on their sizes, and from experimentally measured distributions of the nanoparticles in ensemble one can calculate the magnetization curve for ML based on them.

An important issue is to find the size distribution for ensemble of superparamagnetic nanoparticles with complex shell structure from experimental measurements of the magnetization curve. Its successful resolution could open the way for determination of size parameters of nanoarchitecture elements that make multicomponent shell structure of the nanocomposite, built on superparamagnetic nanosized carriers, including those in the composition of magnetic liquids.

Therefore, from the above, it can be argued that investigations of the possibility of using approaches applied in [11, 20], based on the use of Langevin's paramagnetism theory, for the description of nanocomposites with superparamagnetic cores and complex shell structures of different chemical nature and magnetic liquids based on them, are urgent.

The aim of this work is to investigate the DOX adsorption on the surface of magnetite/hydroxyapatite nanostructures, to synthesis bioactive nanocomposites magnetite/hydroxyapatite/doxorubicin of core–shell type and ML based on them, to study the magnetic properties of nanocomposites and liquids, to analyze the results using Langevin's paramagnetism theory, and to determine the size parameters of multicomponent shell structure of the nanocomposites.

Experimental

Synthesis of initial magnetite, magnetite/hydroxyapatite nanocomposites, their properties and parameters, doxorubicin adsorption calculation technique are described in

[11]. In this work, we used Fe_3O_4 and Fe_3O_4/HA samples with the specific surface area $S_{sp} \sim 110$ and 100 ± 5 % m^2/g, respectively. Presence of HA shell did not practically change the magnetic properties [21] of initial magnetite (the nanocomposite core).

Investigation of nanomaterials biocompatibility was carried out by studying their influence on cell viability of baker's yeast *Saccharomyces cerevisiae* [22, 23] with the help of Goryaev chamber using optical microscopy (biological microscope of Bresser Erudit type), and methylene blue dye by registration of concentration change for cells growing at the temperature of 22 °C in suspensions containing nanocomposites, yeast cells, minimal synthetic nutrient medium (MSM) [24], saline (PS). Numerically viability (K) was evaluated by the formula: $K = M_1/(M_1 + M_2) \times 100$ %, where M_1 is a number of living cells, M_2 is a number of dead cells. The data obtained was compared to the results of control samples studies.

Bioactivity of NC Fe_3O_4/HA modified with DOX was evaluated from their cytotoxic influence on *S. cerevisiae* cells [25, 26] and a decrease in cell proliferation rate [24]. These effects are, in particular, due to participation of doxorubicin in redox cyclic reactions and a corresponding increase in quantity of free radical molecules, to induction of oxidative stress, and to cell cycle delay in G_1- and S-phase. The concentration of cells (n, mL^{-1}) was calculated by the formula of Goryaev chamber: $n = N \times 2.5 \times 10^5$, where N is a number of cells above a large square of the chamber.

Investigation of isotherm of DOX adsorption on the surface of Fe_3O_4/HA NC was performed as follows. Samples (g) of Fe_3O_4/HA NC of 30 mg were mixed with DOX solutions ($V = 5$ mL) of different concentration. DOX adsorption was carried out in saline (PS) for 2 h in dynamic mode at room temperature and pH $= 7.0$. The quantity of the substance adsorbed on the surface of nanocomposites was determined by measuring DOX concentration in the contact solutions before and after adsorption. The concentration was measured with the help of a spectrophotometer Spectrometer Lambda 35 UV/Vis Perkin Elmer Instruments at $\lambda = 480$ nm using a calibration graph. In this work the lyophilized preparation DOXORUBICIN-TEVA (Pharmachemie BV, The Netherlands) was used.

To study the time dependences of DOX adsorption on the surface of Fe_3O_4/HA NC, as well as to study the isotherms, we used the samples (30 mg) of Fe_3O_4/HA nanocomposite, which were mixed with DOX solutions in saline ($V = 5$ mL) of varying concentration, DOX adsorption was carried out in dynamic regime using a shaker at room temperature. The quantity of the substance adsorbed on the surface of nanocomposites was determined by measuring DOX concentration in contact solutions in fixed time (from 30 min to 24 h).

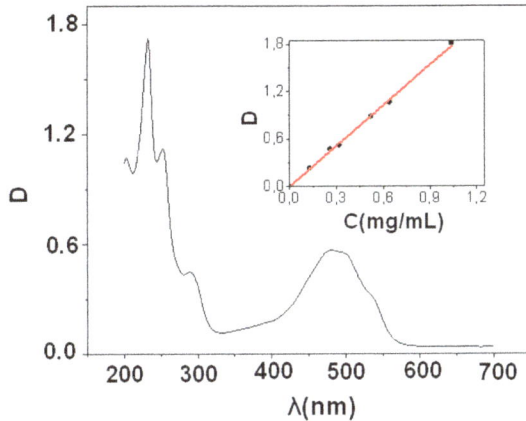

Paint (Word)

Fig. 1 Absorption spectrum of doxorubicin in PS environment. **a** Calibration graph

Calculation of the hydroxyl groups concentration on the HA surface in Fe_3O_4/HA nanocomposite was determined by thermogravimetric analysis using a derivatograph Q-1500.

Results and discussion

Adsorption studies

Doxorubicin [27] is the antitumor antibiotic agent of anthracycline type, widely used in modern oncotherapy. The mechanism of action is interaction with DNA, formation of free radicals, and direct influence on cell membranes with suppression of nucleic acids synthesis. It is characterized by an appreciable antiproliferative effect.

The DOX absorption spectrum measured in a physiological liquid environment is shown in Fig. 1.

The spectrum has several maxima: 204, 233, 254, 290, 480 cm^{-1}, a slope angle of the line of calibration graph for doxorubicin in the saline environment was optimal for the wavelength $\lambda = 480$ nm (Fig. 1a), at which we carried out the quantitative measurements of doxorubicin concentration.

Analysis of the isotherm of DOX adsorption on the surface of Fe_3O_4/HA NC (Fig. 2) shows that increase in equilibrium DOX concentration does not lead to adsorption saturation of the surface of Fe_3O_4/HA adsorbent. Concavity (S-similarity) of an initial part of isotherm relatively to the concentration axis and lack of saturation in the investigated range of equilibrium concentrations can be attributed to the polymolecular nature of adsorption and low porosity of the nanocomposite surface. In addition, S-similarity of isotherm can be caused to some extent by co-adsorption of sodium chloride because DOX adsorption was performed from saline.

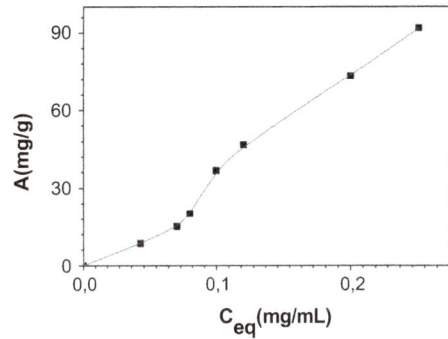

Origin 8.0 (Word)

Fig. 2 Isotherm of doxorubicin adsorption on the surface of Fe_3O_4/HA NC

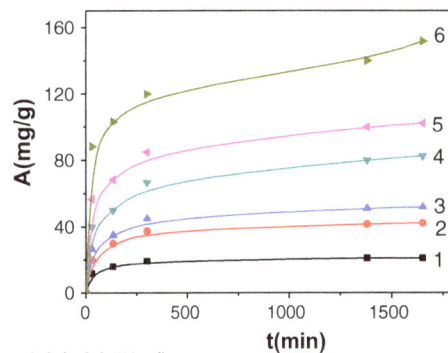

Origin 8.0 (Word)

Fig. 3 Time dependence of DOX adsorption on the surface of Fe_3O_4/HA NC in PS. Initial concentration of DOX solutions C, mg/mL: *1* 0.13, *2* 0.26, *3* 0.32, *4* 0.52, *5* 0.64, *6* 1.04

Distribution coefficient (E, mL/g) of doxorubicin between the surface of nanocomposite and solution was 366.8 mL/g at $A = 91.7$ mg/g.

Upon studies of DOX adsorption on the surface of Fe_3O_4/HA NC as function of time (Fig. 3), it was revealed that during the first 2 h 60–70 % of the substance is adsorbed, and for 24 h the adsorption was almost complete (93–97 %). And this applies to the entire range of the investigated DOX concentrations. The dependence of the extraction extent (R, %) of doxorubicin on concentration of solutions and adsorption time is given in Table 1.

The results of studies on the dependence of desorption (A_D, mg/g) on time and the percentage of desorbed substance (A_D, %) are given in Fig. 4 and Table 2. Experimental dependencies of desorption on time indicate that the doxorubicin release decreases with increasing its quantity on the surface of NC. When the quantity of adsorbed DOX is 20–50 mg/g, 80–60 % of DOX, respectively, is desorbed, whereas at large quantities of adsorbed DOX (100–150 mg/g) the release almost do not occur. This situation can be explained by the peculiarities of interaction and the emergence of sufficiently strong bonds between the

Table 1 Extraction extent of doxorubicin (R) on the surface of Fe$_3$O$_4$/HA NC depending on the concentration of solutions and adsorption time (t)

C_0, mg/mL	t, min				
	30	135	300	1380	1620
	R, %				
0.13	53.8	73.1	88.5	97.0	97.0
0.26	46.2	69.2	86.5	95.8	97.7
0.32	50.0	65.6	84.4	95.6	97.2
0.52	46.2	57.7	77.0	92.3	95.2
0.64	53.1	64.0	79.7	93.8	96.1
1.04	49.0	59.6	69.2	80.8	87.5

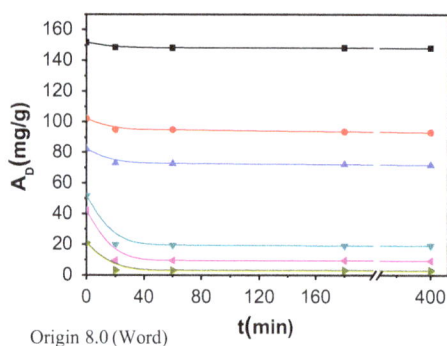

Origin 8.0 (Word)

Fig. 4 DOX desorption (A_D) from Fe$_3$O$_4$/HA/DOX NC surface in PS vs time (t) at different initial amounts of immobilized DOX

Table 2 Dependence of DOX (A_D) desorption from Fe$_3$O$_4$/HA/DOX NC surface in PS on time (t) at different initial amounts (A) of immobilized DOX

A, mg/g	t, min			
	20	60	180	400
	A_D, %			
21.0	83.8	84.3	84.3	84.8
42.3	76.4	76.6	76.8	77.0
51.9	61.5	62.1	62.4	62.6
82.5	11.6	11.5	12.1	12.7
102.5	7.0	6.1	8.3	8.7
151.7	2.2	2.3	2.3	2.4

specific functional groups on HA surface and doxorubicin molecules: hydroxyl and carbonate groups of Fe$_3$O$_4$/HA NC surface can form strong hydrogen bond with hydroxyl and amino groups of DOX; upon the desorption, in samples with lower concentration of DOX, the drug is desorbed faster through partial dissociation of hydrogen bonds [28].

The basic amount of DOX is desorbed during the first 20 min for all the investigated concentrations.

Biocompatibility and bioactivity of nanocomposites

Biocompatibility of Fe$_3$O$_4$/HA NC is a known fact [3–11]. In this paper, biocompatibility and bioactivity of the produced samples was monitored by their effect on viability of baker's yeast *S. cerevisiae* cells [11].

In the study of bioactivity of the original drug doxorubicin, we revealed experimentally that its solution in saline at the concentration of 0.5 mg/mL resulted in almost total death of yeast cells (95 %) for 3.5 days. In the technique for determination of cytotoxicity, it is accepted to use IC$_{50}$ dose at which there is a death of ~50 % of the cells [29]. Therefore, to test bioactivity, a quantity of Fe$_3$O$_4$/HA/DOX nanocomposite material (~20 mg) with immobilized doxorubicin (~50 mg/g), that was used to form a suspension, was chosen from the data of Fig. 4 and Table 2 in calculation that the concentration of released DOX in research suspensions was ~0.25 mg/mL.

Fifteen samples in total were investigated, five in each series

1. suspension of yeast cells (initial concentration $n_0 \approx$ 2.5 × 10^7 mL^{-1}) in PS with MSM (Fig. 5a);
2. suspension of yeast cells (initial concentration $n_0 \approx$ 2.5 × 10^7 mL^{-1}) in PS with MSM, containing 20 mg of Fe$_3$O$_4$/HA NC;
3. suspension of yeast cells (initial concentration $n_0 \approx$ 3.5 × 10^7 mL^{-1}) in PS with MSM, containing 20 mg of Fe$_3$O$_4$/HA/DOX NC.

All the samples contained 1.3 mL of PS (0.9 % NaCl) and 1 mL of MSM. The samples of series 1 and 2 were used for control and comparison, and series 3—for studies on bioactivity of Fe$_3$O$_4$/HA/DOX NC.

Research data analysis shows that in the yeast suspensions ($n_0 \approx$ 2.5 × 10^7 mL^{-1}) in the PS with MSM (control series of type 1) there is a characteristic for yeast [30] cell division that leads to an increase in their concentration in 16 h twice (5 × 10^7 mL^{-1}). Further rate of their reproduction slowed (possibly because of nutrient reduction). In 3.5 days their concentration was ~10^8 mL^{-1}. The viability of yeast cells in the experiments of series 1 was not significantly changed and reached ~98–99 %.

Investigation of suspensions of type 2 control series showed rather active division by which the yeast concentration in 16 h was ~6.5 × 10^7 mL^{-1}, and in 3.5 days, as in the previous case, reached ~10^8 mL^{-1} (Fig. 5b). Cell viability, as in the previous case, at all stages of the series 2 samples study was ~98–99 %. These data indicate biocompatibility of Fe$_3$O$_4$/HA NC with respect to yeast cells under the experiment conditions.

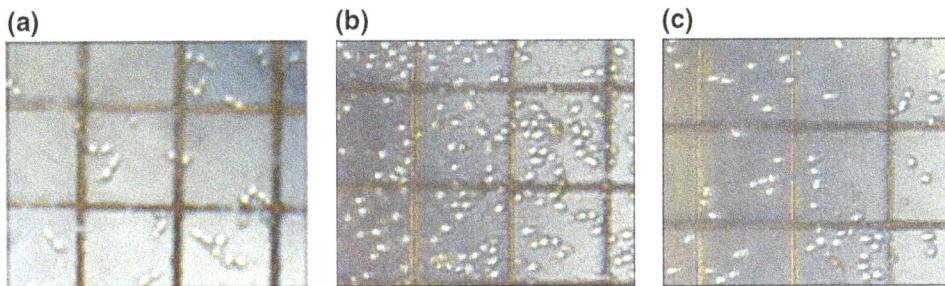

Fig. 5 Typical images of fragments of Goryaev chamber with yeast cells: in beginning of studies (**a**), after interaction of yeast cells with Fe_3O_4/HA NC (**b**), after interaction of yeast cells with Fe_3O_4/HA/ DOX NC (**c**). The interaction time is 3.5 days, $T \sim 300$ K, the concentration of cells: **a** 2.5×10^7 mL^{-1}, **b** 1×10^8 mL^{-1}, **c** 3×10^7 mL^{-1}

Upon the studies of series 3 sample suspensions, a significant inhibition of cell proliferation was revealed (Fig. 5c). So, the concentration of yeast cells in the beginning of the experiment was $\sim 3.5 \times 10^7$ mL^{-1} and was practically not changed for 16 h, only in 3.5 days their number increased up to $\sim 4 \times 10^7$ mL^{-1}. During the study, the quantity of dead cells in the samples of series 3 (Fig. 5c) was about 10 %.

Thus, analyzing the results of experiments of series 1 and 2 (Fig. 5b) and comparing them to the data of series 3 (Fig. 5c), we can conclude that Fe_3O_4/HA/DOX nanocomposites have cytotoxic influence on *S. cerevisiae* cells and reduce the rate of their proliferation.

It should be noted that the observed features of magnetosensitive NC Fe_3O_4/HA/DOX influence on yeast cells are characteristic for the interaction of these cells with the free form of doxorubicin [24–26].

Synthesis of magnetic liquids

In the next stage of research, samples of the three types of magnetic liquids based on saline (ML$_{1-3}$) were synthesized: ML$_1$ − Fe_3O_4/sodium oleate (Na ol.)/polyethylene glycol (PEG) + PS, ML$_2$ − Fe_3O_4/HA/Na ol./PEG + PS, ML$_3$ − Fe_3O_4/HA/DOX/Na ol./PEG + PS. Fe_3O_4 nanoparticles, as well as Fe_3O_4/HA NC and Fe_3O_4/HA/ DOX NC particles were stabilized by sodium oleate [31] and polyethylene glycol. It is known that PEG prevents adsorption interactions of components of a liquid with proteins [32], which is important in medical applications of magnetic liquids. For stabilization of the NP and NC surface in the ML composition, the sodium oleate samples of weight m were calculated taking into account the concentration of hydroxyl groups on the surface of magnetite and hydroxyapatite. The calculation was performed using the formula: $m = BMg$, where B is the concentration of hydroxyl groups (2.2 mmol/g on the surface of initial nanosized magnetite and 1.8 mmol/g on the surface of Fe_3O_4/HA nanocomposite, as determined from the data of

thermogravimetric analysis using a derivatograph Q–1500), M is the molecular weight of sodium oleate (304 g/mol), g is a sample weight of Fe_3O_4 or NC. Additional modifying with PEG-2000 was carried out in dynamic mode using a shaker; the amount of polymer was 10–15 % of the weight of Fe_3O_4 NP or nanocomposite sample.

Study of magnetic and structural properties of nanocomposites in composition of magnetic liquids

Further researches are based on using the ensemble of superparamagnetic carriers as a probe for determination of parameters and control of nanostructures with complex construction, in particular, in composition of the magnetic liquids [11, 20, 33].

Implementation of the said approach can be achieved using the method of magnetic granulometry [34], based on the comparison of the experimental magnetization curve to Langevin's curve at the given laws of size distribution of particles and their magnetic parameters, including saturation magnetization of the particles and the thickness of "demagnetized layer".

To analyze the magnetization curve of ML containing superparamagnetic nanoparticles, a known equation [35] was applied [11]:

$$\frac{M(H)}{\varphi_\rho M_s} = \frac{\sum\limits_{i=1}^{k} n_i(d_i - 2h_1)^3 L\left(\frac{M_s H}{k_B T} \frac{\pi}{6}(d_i - 2h_1)^3\right)}{\sum\limits_{i=1}^{k} n_i d_i^3}, \quad (1)$$

where $M(H)$ is the magnetization of ML in magnetic field of strength H; M_s is the saturation magnetization of a bulk magnetite; φ_ρ is the volumetric concentration of solid phase in ML, defined by the density of ML; d_i, n_i is the average diameter and quantity of Fe_3O_4 NP in the i interval of the diameter variation row; k is the number of intervals; h_1 is the thickness of "demagnetized layer" of magnetite;

$L(\xi) \equiv cth\xi - 1/\xi$ is the Langevin function; k_B is the Boltzmann constant; T is the temperature.

(a)   Determination of the size distribution of ensemble of nanosized magnetite particles and the thickness of their demagnetized layer in terms of the shape of magnetization curves of magnetic liquid.

The ensemble of Fe_3O_4 particles that are characterized by the sizes of 3–23 nm being in superparamagnetic state, has non-hysteresis demagnetization curve, and therefore, zero values of coercive force (H_c) and residual magnetization (M_r) [11]. These features of magnetization are mainly observed in experiment also for samples of ML based on nanocomposites Fe_3O_4/HA/DOX/Na ol./PEG + PS (ML_3) (Fig. 6a).

The specific saturation magnetization σ_s of typical ensembles of Fe_3O_4 NP synthesized for research in this paper was 62.6 ± 2.5 % Gs cm^3/g (Fig. 6b, the upper insertion). In the study of static magnetic characteristics (the measurement time was ~ 100 s), Fe_3O_4 NP or dry residues (DR) of magnetic liquids were distributed in the matrix of paraffin (to prevent interparticle interaction) under condition $m_{DR}/m_p \sim 0.1$ (m_{DR}—weight of DR, m_p—weight of paraffin). Calculation times of Neel's relaxation of magnetic moment of Fe_3O_4 NP with diameters of 3–22 nm are $(10^{-9}–10^2)$ s, respectively.

According to the experimental curve (Fig. 6a), the coercive force (H_c) of ML_3 is equal to (2 ± 0.5) Oe. The samples of Fe_3O_4 NP and DR_3 of ML_3, distributed in paraffin, are characterized by H_c 89.7 Oe and 90.0 Oe, respectively (Fig. 6b). The availability of coercive force in the investigated samples in a liquid state, probably, is due to the presence of a small number of aggregates, joint by dipole–dipole interaction, and in paraffin matrices—a small number of Fe_3O_4 NP with diameter >22 nm.

The upper insertion in Fig. 6a shows the diameter distribution of Fe_3O_4 NP, obtained in experiment by statistical processing (a program Get Data Graph Digitizer 2.24) of TEM images of initial magnetite (1), and lognormal diameter distribution (2), calculated for the same ensemble using a probability density function, similar to [11]. The average diameter value $d_0 = (\Sigma n_i d_i)/N$ and the standard deviation (s) of diameter of Fe_3O_4 NP for a choice of volume $N = 274$ was 10.77 nm ($s = 3.083$ nm), logarithm of diameter: 2.33 ($s_{lnd} = 0.298$), logarithm of volume: 6.37 ($s_{lnV} = 0.894$). For mathematical expectation of Fe_3O_4 particle diameter $M(d)$ and logarithm of diameter $M(lnd)$, the correlation is just: $M(d) = \exp[M(lnd) + (\sigma_{lnd})^2/2]$, where σ_{lnd} is the standard deviation of logarithm of a particle diameter [36]. The value of thickness of "demagnetized" surface layer h_1 of Fe_3O_4 NP, calculated according formula (1) was ~ 0.83 nm.

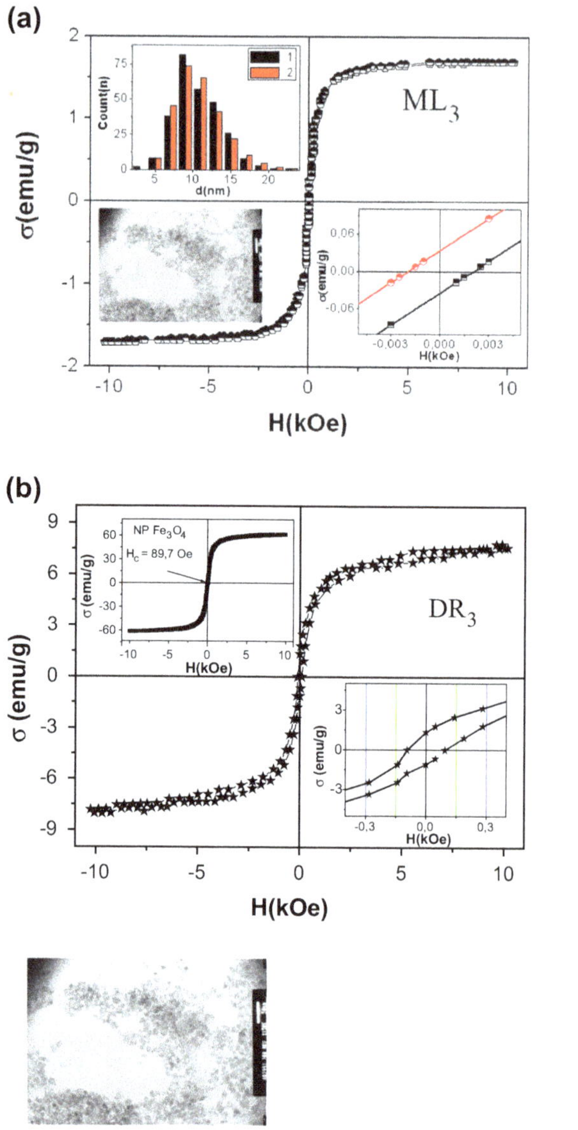

Fig. 6 Hysteresis loops: **a** ML3 (for insertions: the *top 1, 2*, histograms of experimental and lognormal (2.33, 0.298) distribution of Fe_3O_4 NP by diameters, respectively; the lower ones, TEM images (*scale* 50 nm) of magnetite nanoparticles ensemble and initial part of hysteresis loop of ML3); **b** DR3 in paraffin matrix (for insertions: the *top*, hysteresis loop of Fe_3O_4 NP; the *bottom*, initial part of the loop of DR3)

Figure 7 shows a model of Fe_3O_4/HA/DOX/Na ol./PEG NC particle with a multilayer shell in which: $d = d_s + 2h_1$ is the diameter of a spherical magnetite particle; d_s is the diameter of the area of Fe_3O_4 NP with σ_s which is the characteristic for a bulk magnetite (≈ 92 emu/g at 300 K); h_1 is the thickness of "demagnetized" layer of Fe_3O_4 particles; h_2, h_3, h_4 are the thickness of spherical layers of modifier (HA), drug (DOX) and stabilizer (Na ol./PEG), respectively.

Using the model (Fig. 7), in terms of the results of experimental measurements and ensemble parameters

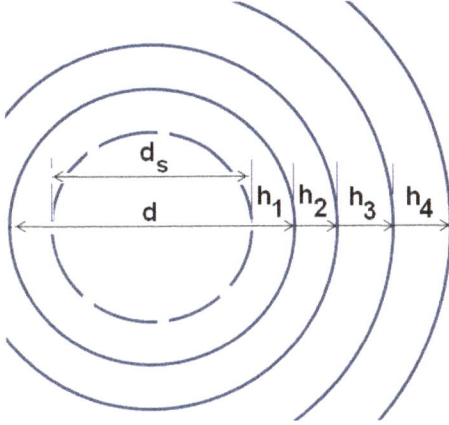

Paint (Word)

Fig. 7 Model of a particle of NC with a multilayer shell. It is marked: $d = d_s + 2h_1$ is the diameter of a spherical Fe_3O_4 NP; d_s is the diameter of the area of Fe_3O_4 NP with σ_s, that is characteristic for a bulk magnetite; h_1 is the thickness of the "demagnetized" surface layer of Fe_3O_4 NP; h_2, h_3, h_4 are the thickness of the modifying layer (HA), drug (DOX), and combined stabilizer (Naol./PEG) in NC structure, respectively

calculation for nanoparticles of magnetite and dry residues of magnetic liquid of corresponding composition, the shell layers sizes of Fe_3O_4/HA/DOX/Na ol./PEG nanostructure were determined.

In [11] it has been shown that for ensembles of nanoparticles of "core–shell" type, under condition of cores to be in a superparamagnetic state, formula (1) practically unequivocally binds the size distribution of Fe_3O_4 NP in ensemble with a shape of hysteresis curve. Taking into account that in conditions of magnetic saturation the Langevin function $L(\xi) \to 1$, and in DR a volumetric concentration of solid phase ~ 1, formula (1) can be written as

$$\frac{M_s^{NCPM}}{M_s^{NPM}} = \frac{\langle\rho_{NCPM}\rangle\sigma_s^{NCPM}\int_0^\infty V_{NCPM}f(V_{NCPM})\mathrm{d}V}{\langle\rho_{NPM}\rangle\sigma_s^{NPM}\int_0^\infty V_{NPM}f(V_{NPM})\mathrm{d}V}, \quad (2)$$

where M_s^{NCPM}, M_s^{NPM} are the saturation magnetization of ensembles of particles of NC (NCPM) and Fe_3O_4, respectively; $\langle\rho_{NCPM}\rangle$, $\langle\rho_{NPM}\rangle$ are the average density of ensembles of particles of NC and Fe_3O_4, respectively; σ_s^{NCPM}, σ_s^{NPM} are the specific saturation magnetization of ensembles of particles of NC and Fe_3O_4, respectively; $f(V_{NCPM})$, $f(V_{NPM})$ are the probability density functions for the volumes of ensembles of particles of nanocomposite and magnetite, respectively; V_{NCPM}, V_{NPM} are the volume of a particle of NC and Fe_3O_4, respectively.

(b) Determination of thickness of the layer of combined stabilizer (Na ol./PEG)

The synthesized samples of initial Fe_3O_4 and ML_1 composed of Fe_3O_4/Na ol./PEG + PS were dried at room temperature. The mass proportion of magnetite in the dry residue of ML_1 (DR_1) was experimentally defined as $\alpha_{Fe_3O_4}^{exp} = \sigma_s^{DR1}/\sigma_s^{NPM} \pm 5\,\%$, and calculated by the formula

$$\alpha_{Fe_3O_4}^{calc} = \frac{\upsilon_{Fe_3O_4}\rho_{Fe_3O_4}}{\langle\rho_{NCPM}\rangle}, \quad (3)$$

where $\upsilon_{Fe_3O_4} = \sum n_i d_i^3/\sum n_i(d_i + 2\delta)^3$ is the volume fraction of magnetite in the sample; $\delta = h_2 + h_3 + h_4$ is the thickness of the shell; $\rho_{Fe_3O_4}$ is the density of magnetite, $\langle\rho_{NCPM}\rangle$ is the average density of the ensemble of NC particles, which was found from the formula

$$\langle\rho_{NCPM}\rangle = \frac{\sum\limits_{i=1}^{k} n_i\rho_{NCPM_i}(d_i + 2\delta)^3}{\sum\limits_{i=1}^{k} n_i(d_i + 2\delta)^3} \quad (4)$$

The density of nanocomposite particles of i interval ρ_{NCPMi} in composition of DR_{1-3} was determined by formulas (5.1–5.3):

$$\rho_i^{12} = \alpha_2\rho_1 + (1 - \alpha_2)\rho_2, \text{where } \alpha_2 = \left(\frac{d_i}{d_i + 2h_2}\right)^3, \quad (5.1)$$

$$\rho_i^{123} = \alpha_3\rho_i^{12} + (1 - \alpha_3)\rho_3, \text{where } \alpha_3 = \left(\frac{d_i + 2h_2}{d_i + 2h_2 + 2h_3}\right)^3, \quad (5.2)$$

$$\rho_i^{1234} = \alpha_4\rho_i^{123} + (1 - \alpha_4)\rho_4, \text{where } \alpha_4 = \left(\frac{d_i + 2h_2 + 2h_3}{d_i + 2h_2 + 2h_3 + 2h_4}\right)^3, \quad (5.3)$$

and ρ_1, ρ_2, ρ_3, ρ_4 are the density of magnetite, HA, DOX and Na ol./PEG, respectively; ρ_i^{12}, ρ_i^{123}, ρ_i^{1234} are the density of a particle of i interval of NC Fe_3O_4/HA, Fe_3O_4/HA/DOX and Fe_3O_4/HA/DOX/Na ol./PEG, respectively. For calculations, we used the values $\rho_1 \approx 5.19$ g/cm^3 [37], $\rho_2 \approx 2.71$ g/cm^3 [28], $\rho_3 \approx 1.00$ g/cm^3, $\rho_4 \approx 1.13$ g/cm^3 [33]). We considered that the size distribution of Fe_3O_4 NP of the initial ensemble and its dry residues was identical.

Specific surface area of the ensemble of NC particles was determined by the formula

$$S_{sp}^{calk} = 6\frac{\sum\limits_{i=1}^{k} n_i(d_i + 2\delta)^2}{\sum\limits_{i=1}^{k} \rho_{NCPM_i}(d_i + 2\delta)^3}. \quad (6)$$

According to the model (Fig. 7) in a particle of DR_1 only h_4 shell is filled (h_2, $h_3 = 0$). The results of experimental measurements and calculations of parameters of ensemble of NP Fe_3O_4 and DR_1 are shown in Table 3.

Table 3 Results of experimental measurements and calculation of parameters of ensemble of NP Fe_3O_4 and DR_1

	Experimental values				Calculated values			
Sample	d_0, nm	σ_s, Gs cm^3/g (%)	α_{Fe3O4}^{exp} (%)	S_{sp}^{exp}, m^2/g (%)	h_4, nm	$<\rho_{NCPM}>$, g/cm^3 (%)	α_{Fe3O4}^{calc}	S_{sp}^{cal}, m^2/g
Fe_3O_4	10.8	62.6 ± 2.5	1.00 ± 5	107.0 ± 5	0	5.19 ± 1	1.00	107.0
DR_1	10.8	36.6 ± 2.5	0.58 ± 5	161.0 ± 5	3.4 ± 3	2.07 ± 1	0.58	161.0

The value of $<\rho_{NCPM}>$ was calculated by formulas (4)–(5.1–5.3), α_{Fe3O4}^{calc}—by formula (3). h_4 value was found, at which $\alpha_{Fe3O4}^{calc} = \alpha_{Fe3O4}^{exp}$. In terms of the obtained h_4 and formulas (5.1–5.3)–(6), S_{sp}^{calc} was determined

As shown in Table 3, thickness of the shell of combined stabilizer Na ol./PEG in composition of dry residue of magnetic liquid Fe_3O_4/Na ol./PEG + PS is (3.4 ± 0.1) nm.

(c) Determination of the thickness of HA layer

To determine the thickness of HA layer we studied the ensemble of NP Fe_3O_4 and ML_2 composed of Fe_3O_4/HA/Na ol./PEG + PS obtained on its basis, the samples were dried at room temperature, dry residue DR_2 was obtained and their parameters were investigated by the technique described above. The results are shown in Table 4.

As shown in Table 4, the found value of thickness of hydroxyapatite layer h_2 in the structure of Fe_3O_4/HA/Na ol./PEG is 3.5 ± 0.3 nm, which in our view is satisfactorily consistent with the value of ~4 nm, determined by an independent technique in studies of Fe_3O_4/HA nanocomposites by photoelectron spectroscopy method [11]. The obtained data may indicate reliability of the results of determination of shell parameters in complex nanoarchitecture of multifunctional magnetosensitive nanocomposites.

(d) Determination of the thickness of DOX layer

The ensemble of magnetite nanoparticles and ML_3 composed of Fe_3O_4/HA/DOX/Na ol./PEG + PS obtained on their basis, were dried at room temperature. The dried samples Fe_3O_4 and dry residue DR_3 were studied, as in the previous cases. The research results are given in Table 5.

According to the data of Table 5, the found value of thickness of the layer of medical drug doxorubicin h_3 in the structure of Fe_3O_4/HA/DOX/Na ol./PEG is 2.0 ± 0.3 nm.

(e) Study of sedimentation stability of magnetic liquids

The investigated ML are nanoheterogeneous systems, where diffusion fluxes of particles dominate over sedimentation ones. Over a long period of time (years) in monodisperse sols, the fluxes become equal and a state of diffusion-sedimentation equilibrium (DSE) is set, at which the distribution of particles on height of a vessel is subjected to the hypsometric law [37]:

$$\frac{v_h}{v_0} = \exp\left[-\frac{V_{cp}(\rho_{cp} - \rho_{lc})gh}{k_BT}\right], \tag{7}$$

where v_h, v_0 are the concentrations of particles at height h and at the level of bottom of a vessel, respectively; V_{cp}, ρ_{cp} are the volume and density of colloidal particles, respectively; ρ_{lc} is the density of a liquid carrier; g is the acceleration of gravity.

The height at which concentration of particles varies in e times, characterizes the thermodynamic sedimentation stability (TDS) of a colloidal system [37] (hypsometric height L). From Eq. (7) it follows that

$$L = \frac{k_BT}{V_{cp}(\rho_{cp} - \rho_{lc})g} = \frac{6k_BT}{\pi(d + 2\delta)^3(\langle\rho_{NCPM}\rangle - \rho_{lc})g}, \tag{8}$$

where $<\rho_{NCPM}>$ is the mean density of a nanocomposite particle with diameter of core d and thickness of shell δ, calculated by formulas (5.1–5.3).

In polydisperse systems, DSE is set for each fraction of particles. Time of setting DSE (t_b) in ML was calculated by the technique given in [33]: we used formula $t_b = L_0^2/<D>$, where $<D> = (1/N)\sum k_BT/[3\pi\eta(d_i + 2\delta)]$ is the average diffusion coefficient; $L_0 = k_BT/[<V_{NCPM}> (<\rho_{NCPM}> - \rho_{lc})g]$ is the average hypsometric height; η is the dynamic viscosity, determined for the concentrated ML by a laboratory viscometer (the time of leakage of the fluid through a glass capillar with diameter of 0.2 mm was ≈300 s); the value η of diluted ML, according to experimental data [18], was calculated by Einstein formula: $\eta/\eta_0 = 1 + 5\varphi/2$, where η_0 is the dynamic viscosity of the liquid carrier (η_0 of PS is approximately equal to 0.890 mPa s at the

Table 4 Results of experimental measurements and calculation of parameters of ensemble of NP Fe_3O_4 and DR_2

Sample	Experimental values				Calculated values				
	d_0, nm	σ_s, Gs cm^3/g (%)	α_{Fe3O4}^{exp} (%)	S_{sp}^{exp}, m^2/g (%)	h_2, nm	h_4, nm	$<\rho_{NCPM}>$, g/cm^3 (%)	α_{Fe3O4}^{calc}	S_{sp}^{cal}, m^2/g
Fe_3O_4	10.8	62.6 ± 2.5	1.00 ± 5	107.0 ± 5	0	0	5.19 ± 1	1.00	107.0
DR_2	10.8	13.2 ± 2.5	0.21 ± 5	114.0 ± 5	3.5 ± 3	3.4 ± 3	2.07 ± 1	0.20	114.6

Table 5 Results of experimental measurements and calculation of parameters of magnetite nanoparticles ensemble and dry ML3 residues

	Experimental values				Calculated values					
Sample	d_0, nm	σ_s, Gs cm^3/g (%)	α_{Fe3O4}^{exp} (%)	S_{sp}^{exp}, m^2/g (%)	h_2, nm (%)	h_3, nm (%)	h_4, nm (%)	$<\rho_{NCPM}>$, g/cm^3 (%)	α_{Fe3O4}^{calc} (%)	S_{sp}^{cal}, m^2/g
Fe$_3$O$_4$	10.8	62.6 ± 2.5	1.00 ± 5	107 ± 3	0	0	0	5.19 ± 1	1.00	107.0
DR$_3$	10.8	9.9 ± 2.5	0.16 ± 5	120 ± 3	3.5 ± 3	2.0 ± 3	3.4 ± 3	1.74 ± 1	0.15	120.1

temperature of 25 °C). The estimated values t_b in ML$_{1-3}$ are nine years and more. The experimental value of L for ML$_3$ is $L^{exp} = 2.4 \pm 8$ % cm.

Using the found parameters of Fe$_3$O$_4$/HA/DOX/Na ol./PEG nanostructure in composition of magnetic liquid, the dependencies of hypsometric height (Fig. 8, curve 1) and specific surface area of DR (curve 2) on thickness h_3 of DOX layer were plotted for the model magnetic liquid of ML$_3$ type, in which Fe$_3$O$_4$/HA/DOX/Na ol./PEG NC was characterized by the fixed size of the core and layers of hydroxyapatite and combined stabilizer (d_{Fe3O4} $_{NP} = 10.8$ nm, $h_2 = 3.5$ nm, $h_4 = 3.4$ nm). Calculation dependence $L(h_3)$ was built by formula (8), $S_{sp}(h_3)$—(9):

$$S_{sp} = \frac{6}{\rho_{NCPM}(d + 2\delta)} \quad (9)$$

The value $<\rho_{NCPM}>$ was obtained by the formulas (5.1–5.3).

Dependence $L(h_3)$, shown in Fig. 8, curve 1, is typical for the colloidal systems [37]. The experimental value of L for ML$_3$ ($L^{exp} = 2.4 \pm 8$ % cm) corresponds to $h_3 = 2.2 \pm 8$ % nm.

Dependence $S_{sp}(h_3)$ given in Fig. 8 (curve 2) has a maximum caused by that S_{sp} is a complex function of ρ_{NCPM} and $(d + 2\delta)$. Its position (h_{3cr}) can be found analytically by equating to zero the expression $dS_{sp}/d\delta$ (the criterion for finding thickness of the shell, which corresponds to the maximal specific surface area of the core–shell structure S_{sp}). Ordinate corresponding to the experimental value $S_{sp}^{exp} = 120 \pm 3$ % for DR$_3$ (Table 5), twice crosses the calculated dependence S_{sp} with abscissas $h_3 = 1.94$ nm and $h_3 = 5.0$ nm. The value $h_3 = 5.0$ nm contradicts the data of magnetic measurements (Table 5) and the value obtained by L^{exp} (Fig. 8, curve 1).

So, using three independent experimental methods for measurement of σ_s, L, S_{sp}, we obtained three values of the thickness of doxorubicin layer h_3 in the structure of Fe$_3$O$_4$/HA/DOX nanocomposite: 2.0 nm (Table 5), 2.2 nm (Fig. 8, curve 1), 1.94 nm (Fig. 8, curve 2), respectively. The found values h_3 are rather close, which shows their reliability.

We note that analysis of the sizes of superparamagnetic iron oxide nanoparticles coated with a layer of carboxydextran, which are applied as contrast agents in magnetic resonance imaging (a commercial product Resovist and SH U555C) using a magnetization curve and images of ensembles of particles, obtained by transmission electron microscopy, has been performed in [38].

The results of experimental investigations and calculations given in this paper, their testing by different ways and comparison suggest that using ensembles of magnetic carriers as a superparamagnetic probe and Langevin's paramagnetism theory, one can evaluate the sizes of the components of complex shell structure of nanocomposites. The obtained data may be useful in optimizing chemical composition, structure and properties of new magnetic liquids and adsorbents containing magnetosensitive nanocomposites with complex construction of shell [39, 40].

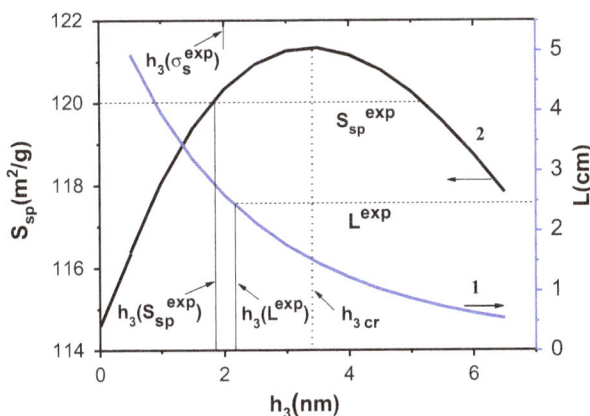

Origin 8.0 (Word)

Fig. 8 Calculation dependences: hypsometric height L (*curve 1*) and specific surface area Ssp of DR (*curve 2*) on the thickness $h3$ of DOX layer for the model magnetic liquid of ML3 type, in which Fe$_3$O$_4$/HA/DOX/Na ol./PEG NC is characterized by the fixed size of the core and layers of hydroxyapatite and combined stabilizer (dFe$_3$O$_4$ NP = 10.8 nm, $h_2 = 3.5$ nm, $h_4 = 3.4$ nm)

Conclusions

The processes of DOX adsorption on Fe$_3$O$_4$/HA NC surface from solution in physiologic liquid were studied. It was determined that an increase in equilibrium concentration of DOX within the studied concentration range does not lead to adsorption saturation of the surface of Fe_3O_4/HA adsorbent. It was revealed that release of DOX into saline decreases with growing of its quantity on NC surface.

Magnetosensitive Fe_3O_4/HA/DOX NC were synthesized. It has been determined that cytotoxic influence and antiproliferative activity of NC with respect to *S. cerevisiae* cells are typical for interaction of these cells with a free form of doxorubicin.

Magnetic liquids containing Fe_3O_4/HA/DOX NC stabilized by sodium oleate and polyethylene glycol were produced and investigated. Using the ensemble of Fe_3O_4 carriers as a superparamagnetic probe, Langevin's paramagnetism theory, the values of density of nanocomposite constituents, the size parameters of their shell were evaluated, which was corroborated by independent measurements of specific surface area of nanostructures and kinetic stability of the corresponding magnetic liquids. The obtained results may be used for the development of novel forms of magnetocarried medical remedies for targeted delivery and adsorbents based on nanocomposites of superparamagnetic core–shell type with multilevel nanoarchitecture, as well as for determination, control, and optimization of the size parameters of its components.

Acknowledgments The work was carried out with support of goal complex programs of fundamental investigations of the National Academy of Sciences of Ukraine "Fine Chemicals" (project 31/16) and "Fundamental problems in creation of novel nanomaterials and nanotechnologies" (project 38/16-N).

Compliance with ethical standards

Conflict of interest The authors declare that they have no competing interests.

References

1. Roco M.C., Williams, R.S., Alivisatos, P.: Vision for Nanotechnology R&D in the Next Decade. Kluwer Academic, Dordrecht. http://www.wtec.org/loyola/nano/IWGN.Research.Directions/IWGN_rd.pdf (2002)
2. Levy, L., Sahoo, Y., Kim, K.-S., Bergey, J.E., Prasad, P.: Synthesis and characterization of multifunctional nanoclinics for biological applications. Chem. Mater. **14**, 3715–3721. http://pubs.acs.org/doi/abs/10.1021/cm0203013 (2002)
3. Gorbyk, P.P., Dubrovin, I.V., Petranovska, A.L., Abramov, M.V., Usov, D.G., Storozhuk, L.P., Turanska, S.P., Turelyk, M.P., Chekhun, V.F., Lukyanova, N.Y., Shpak, A.P., Korduban, O.M.: Chemical construction of polyfunctional nanocomposites and nanorobots for medico-biological applications. In: Shpak, A.P., Gorbyk, P.P. (eds.) Nanomaterials and Supramolecular Structures. Physics, Chemistry, and Applications, pp. 63–78. Naukova dumka, Springer, Kiev. http://www.springer.com/us/book/9789048123087(2009)
4. Gorbyk, P.P., Chekhun, V.F.: Nanocomposites of medicobiologic destination: reality and perspectives for oncology. Funct. Mater. **19**(2), 145–156. http://functmaterials.org.ua/contents/19-2/fm192-01.pdf (2012)
5. Gorbyk, P.P., Lerman, L.B., Petranovska, A.L., Turanska, S.P.: Magnetosensitive nanocomposites with functions of medico-biological nanorobots: synthesis and properties. In: Adorno, D.P., Pokutnyi, S. (eds.) Advances in Semiconductor Research: Physics of Nanosystems, Spintronics and Technological Applications, pp. 161–198. Nova Science Publishers, N. Y. www.novapublishers.com/catalog/product_info.php?products_id=51113 (2014)
6. Huang, C., Zhou, Y., Tang, Z., Guo, X., Qian, Z., Zhou, S.: Synthesis of multifunctional Fe_3O_4 core/hydroxyapatite shell nanocomposites by biomineralization. Dalton Trans. **40**(18), 5026–5031. http://www.ncbi.nlm.nih.gov/pubmed/21455509 (2011)
7. Iwasaki, T.: Mechanochemical synthesis of magnetite/hydroxyapatite nanocomposites for hyperthermia. In: Mastai, Y. (ed.) Materials Science—Advanced Topics, pp. 175–194. InTech (2013). doi:10.5772/54344
8. Gopi, D., Ansari, M.T., Shinyjoy, E., Kavitha, L.: Synthesis and spectroscopic characterization of magnetic hydroxyapatite nanocomposite using ultrasonic irradiation. Spectrochimica Acta Part A Mol. Biomol. Spectrosc. **87**, 245–250. http://www.sciencedirect.com/science/article/pii/S1386142511010420 (2012)
9. Mir, A., Mallik, D., Bhattacharyya, S., Mahata, D., Sinha, A., Nayar, S.: Aqueous ferrofluids as templates for magnetic hydroxyapatite nanocomposites. Aqueous ferrofluids as templates for magnetic hydroxyapatite nanocomposites. J. Mater. Sci: Mater. Med. **21**, 2365–2369. http://www.researchgate.net/publication/44633635_Aqueous_ferrofluids_as_templates_for_magnetic_hydroxyapatite_nanocomposites (2010)
10. Feng, C., Chao, L., Ying-Jie, Z., Xin-Yu, Z., Bing-Qiang, L., Jin, W.: Magnetic nanocomposite of hydroxyapatite ultrathin nanosheets/Fe_3O_4 nanoparticles: microwave-assisted rapid synthesis and application in pH-responsive drug release. Biomater. Sci. **1**, 1074–1081. http://pubs.rsc.org/en/content/articlelanding/2013/bm/c3bm60086f/unauth#!divAbstract (2013)
11. Petranovska, A.L., Abramov, N.V., Turanska, S.P., Gorbyk, P.P., Kaminskiy, A.N., Kusyak, N.V.: Adsorption of *cis*-dichlorodiammineplatinum by nanostructures based on single-domain magnetite. J. Nanostruct. Chem. **5**, 275–285. http://link.springer.com/article/10.1007/s40097-015-0159-9?wt_mc=alerts.TOCjournals (2015)
12. Petranovska, A.L., Turelik, M.P., Pylypchuk, I.V., Gorbyk, P.P., Korduban, A.M., Ivasishin, O.M.: Formirovaniye biomimeticheskogo gidroksiapatita na poverkhnosti titana. Metallofizika i Noveishyye Tekhnologii. **35**(11), 1567–1584. http://mfint.imp.kiev.ua/ru/toc/v35/i11.html (2013)
13. Pylypchuk, I.V., Petranovska, A.L., Turelyk, M.P., Gorbyk, P.P.: Formation of biomimetic hydroxyapatite coating on titanium plates. Mat. Science (Med.) **20**(3), 328–332 (2014). doi:10.5755/j01.ms.20.3.4974
14. Pylypchuk, I.V., Petranovska, A.L., Gorbyk, P.P., Korduban, A.M., Markovsky, P.E., Ivasishin, O.M.: Biomimetic hydroxyapatite growth on functionalized surfaces of Ti-6Al-4 V and Ti-Zr-Nb alloys. Nanoscale Res. Lett. **10**, 338. http://www.nanoscalereslett.com/content/10/1/338/abstract (2015)
15. Davaran, S., Alimirzalu, S., Nejati-Koshki, K., Nasrabadi, H.T., Akbarzadeh, A., Khandaghi, A.A., Abbasian, M., Alimohammadi, S.: Physicochemical characteristics of Fe_3O_4 magnetic nanocomposites based on poly(N-isopropylacrylamide) for anti-cancer drug delivery. Asian Pac. J. Cancer Prev. **15**(1), 49–54. http://www.apjcpcontrol.org/paper_file/issue_abs/Volume15_No1/49-54%208.14%20Soodabeh%20Davaran.pdf (2014)
16. Anirudhan, T.S., Sandeep, S.: Synthesis, characterization, cellular uptake and cytotoxicity of a multi-functional magnetic nanocomposite for the targeted delivery and controlled release of

doxorubicin to cancer cells. J. Mater. Chem. **22**, 12888–12899. http://pubs.rsc.org/en/Content/ArticleLanding/2012/JM/C2JM31794J#!divAbstract (2012)

17. Sadighian, S., Hosseini-Monfared, H., Rostamizadeh, K., Hamidi, M.: pH-Triggered magnetic-chitosan Nanogels (MCNs) for doxorubicin delivery: physically vs. chemically cross linking approach. Adv. Pharm. Bull. **5**(1), 115–120. http://journals.tbzmed.ac.ir/APB/Manuscript/APB-5-115.pdf (2015)

18. Rozentsveyg, R.: Ferrogidrodinamika. Mir, Moskva. http://www.twirpx.com/file/179455/ (1989)

19. Araújo-Neto, R.P., Silva-Freitas, E.L., Carvalho, J.F., Pontes, T.R.F., Silva, K.L., Damasceno, I.H.M., Egito, E.S.T., Dantas, A.L., Morales, M.A., Carriço, A.S.: Monodisperse sodium oleate coated magnetite high susceptibility nanoparticles for hyperthermia applications. J. Magn. Magn. Mater. **364**, 72–79. http://www.sciencedirect.com/science/journal/03048853/364 (2014)

20. Abramov, N.V., Gorbyk, P.P.: Svoystva ansambley nanochastits magnetita i magnitnykh zhydkostey dlya primeneniy v onkoterapii. Poverkhnost'. **4**(19), 246 http://surfacezbir.com.ua/images/Arhiv/N19/3/3-8Abramov246-265.pdf (2012)

21. Borisenko, N.V., Bogatyrev, V.M., Dubrovin, I.V., Abramov, N.V., Gayevaya, M.V., Gorbyk, P.P.: Sintez i svoistva magnitochuvstvitel'nykh nanokompositov na osnove oksidov zheleza i kremniya. In: Shpak, A.P., Gorbyk, P.P. (eds.) Fiziko-khimiya Nanomaterialov i Supramolekulyarnykh Struktur. **1**, pp. 394–406. Naukova dumka, Kiev. http://www.irbis-nbuv.gov.ua/cgi-bin/irbis_nbuv/ (2007)

22. Turov, V.V., Gorbik, S.P.: Opredeleniye sil adgezii na mezhfaznoy granice kletka/voda iz dannykh ^{1}H YMR spektroskopii. Ukrainskiy Khimicheskiy Zhurnal. **69**(6), 80–85 (2003)

23. Turov, V.V., Gorbik, S.P., Chuiko, A.A.: Vliyaniye dispersnogo kremnezema na svyazannuyu vodu v zamorozhennykh kletochnykh suspenziyakh. Problemy Kriobiologii. **3**, 16–23 (2002)

24. Saenko, Y.V., Shutov, A.M., Rastorgueva, E.V.: Doxorubicin i menadion vyzyvayut zaderzhku kletochnoy proliferacii *Saccharomyces cerevisiae* s pomoshchyu razlichnykh mekhanizmov. Citologiya. **52**(5), 407–411. http://www.tsitologiya.cytspb.rssi.ru/52_5/saenko.pdf (2010)

25. Huang, R.Y., Kowalski, D., Minderman, H., Gandhi, N., Johnson, E.S.: Small ubiquitin-related modifier pathway is a major determinant of doxorubicin cytotoxicity in Saccharomyces cerevisiae. Cancer Res. **67**(2), 765–772. http://www.pubfacts.com/detail/17234788/Small-ubiquitin-related-modifier-pathway-is-a-major-determinant-of-doxorubicin-cytotoxicity-in-Sacch (2007)

26. Patel, S., Sprung, A.U., Keller, B.A., Heaton, V.J., Fisher, L.M.: Identification of yeast DNA topoisomerase II mutants resistant to the antitumor drug doxorubicin: implications for the mechanisms of doxorubicin action and cytotoxicity. Mol. Pharmacol. **52**(4), 658–666. http://www.ncbi.nlm.nih.gov/pubmed/9380029 (1997)

27. Tacar, O., Sriamornsak, P., Dass, C.R.: Doxorubicin: an update on anticancer molecular action, toxicity and novel drug delivery systems. J. Pharm. Pharmacol. **65**(2): 157–170 (2013). doi:10.1111/j.2042-7158.2012.01567.x

28. Biswanath, K., Debasree, G., Mithlesh, K.S., Partha, S.S., Vamsi, K.B., Nirmalendu, D., Debabrata, B.: Doxorubicin-intercalated nano-hydroxyapatite drug-delivery system for liver cancer: an animal model. Ceram. Int. **39**(8), 9557–9566. http://www.sciencedirect.com/science/article/pii/S0272884213005890 (2013)

29. Shpak, A.P., Chekhun, V.F., Gorbyk, P.P., Turov, V.V.: Nanomaterialy i Nanokompozity v Medizyne, Biologii, Ekologii. Naukova dumka, Kiev. http://www.irbis-nbuv.gov.ua/ (2011)

30. Babyeva, I.P., Chernov, I.Y.: Biologiya Drozhzhey. T-vo nauch. izd. KMK, Moskva. http://ashipunov.info/shipunov/school/books/babjeva2004_biologija_drozhzhej.pdf (2004)

31. Sun, S., Zeng, H., Robinson, D.B.: Monodispersed MFe2O4 (M = Fe, Co, Mn) nanoparticles. J. Am. Chem. Soc. **126**, 73–279. http://www.researchgate.net/publication/8931334_Monodisperse_MFe2O4_(M__Fe_Co_Mn)_nanoparticles (2004)

32. Mornet, S., Vasseur, S., Grasset, F., Veverka, P., Goglio, G., Demourgues, A., Portier, J., Pollert, E., Duguet, E.: Magnetic nanoparticle design for medical applications. Prog. Sol. St. Chem. **34**, 237–247. http://science.report/author/s-mornet/ (2006)

33. Abramov, N.V.: Magnitnyye zhidkosti na osnove doxorubicina dlya primeneniy v onkoterapii. Poverkhnost'. **6**, 241–258. http://surfacezbir.com.ua/images/Arhiv/N21/20.3.7.pdf (2014)

34. Banerjee, S.K., Moskowitz, B.M.: Ferrimagnetic properties of magnetite. In: Kirschvink, J.L., Jones, D.S., MacFadden, B.J. (eds.) Magnetite Biomineralization and Magnetoreception in Organisms (1st Edition). A New Biomagnetism (Topics in Geobiology). **5**, pp. 17–41. Plenum Press, New York. http://link.springer.com/chapter/10.1007/978-1-4613-0313-8_2 (1985)

35. Kim, T., Shima, M.: Reduced magnetization in magnetic oxide nanoparticles. J. Appl. Phys. **101**, 09M516. http://connection.ebscohost.com/c/articles/25114951/reduced-magnetization-magnetic-oxide-nanoparticles (2007)

36. Sahoo, P.: Probability and Mathematical Statistics. University of Louisville, Louisville. http://www.math.louisville.edu/~pksaho01/teaching/Math662TB-09S.pdf (2008)

37. Frolov, Y.G.: Kurs Kolloidnoy Khimii. Khimiya, Moskva http://t-library.org.ua/showBook.php?id=3276 (1989)

38. Chen, D.-X., Sun, N., Gu, H.-C.: Size analysis of carboxydextran coated superparamagnetic iron oxide particles used as contrast agents of magnetic resonance imaging. J. App. Phys. **106**(6), 063906–063906-9 http://dx.doi.org/10.1063/1.3211307 (2009)

39. Pokutnyi, S.I.: Theory of excitons and quasimolecules formed from spatially separated electrons and holes in quasi-zero-dimensional semiconductor nanosystems. In: Adorno, D.P., Pokutnyi, S.I. (eds.) Advances in Semiconductor Research: Physics of Nanosystems, Spintronics and Technological Applications, pp. 73–90. Nova Science Publishers, New York. http://www.amazon.com/Advances-Semiconductor-Research-Technological-Applications/dp/1633217558 (2014)

40. Gorbyk, P.P., Lerman, L.B., Petranovska, A.L., Turanska, S.P., Pylypchuk, I.V.: Magnetosensitive nanocomposites with hierarchical nanoarchitecture as biomedical nanorobots: synthesis, properties, and application. In: Grumezescu, A. (ed.) Fabrication and Self-assembly of Nanobiomaterials, Applications of Nanobiomaterials, pp. 289–334. William Andrew, Elsevier. http://store.elsevier.com/product.jsp? (2016)

15

Antibacterial silver-doped bioactive silica gel production using molten salt method

Roya Payami[1] · Mohammad Ghorbanpour[2] · Aiyoub Parchehbaf Jadid[3]

Abstract Due to broad-spectrum antimicrobial activity, silver nanoparticles have great application potential in disinfection of contaminated water. The aim of this research was the introduction of a fast and simple method titled as "molten salt method" for the production of silver-doped bioactive silica gel (SG) nanocomposite. In this method, SG was imposed into the molten salt of silver nitrate at 150 and 300 °C for various times. Interestingly, molten salt method was not utilized any reducing reagent or other chemicals unless molten silver nitrate. The synthesis and fixing of nanoparticles into the support were done in <60 min. The prepared silver/SG nanocomposite was evaluated using scanning electron microscope (SEM), energy dispersive X-ray fluorescence, leaching test and antibacterial test. SEM images showed that the contact of SG with the molten salt caused the formation of nanoparticles on the SG. On the other hand, increasing the contact time, it led to a larger and increased number of particles. The antibacterial tests demonstrated that this composite is suitable for using as antibacterial material. The test of elution with water indicated that the prepared nanocomposite is stable and the amount of the released silver in the water was negligible.

Keywords Antibacterial · Silver · Silica gel · Molten salt method

Introduction

Silver nanoparticles have recently attracted great interest due to their distinctive properties such as large surface areas, unique physical, chemical and biological properties [1, 2]. It is widely known that materials containing silver show antibacterial property [3, 4]. Now, the silver nanoparticles are emerging as a new generation of antibacterial agent, which has been used in medical applications and antibacterial water filter [5]. The use of silver nanoparticles is particularly potential to treat drinking water, which is frequently infected by antibiotic resistant bacteria. Treatment of this water usually requires using high concentrated chlorine compounds, which may cause a high risk of human cancer [6].

However, aggregation of silver nanoparticles and leaching of Ag^+ ions in aquatic system restricted the application of silver nanoparticles. Therefore, demand for the production of solid supported silver nanomaterials has been increased. Generally, the nanoparticle-doped materials have several advantages, such as high performance, low price (compared to pure silver), high chemical durability and low release silver ions for a long period of time [7, 8]. There are several methods for preparing silver-doped silica including multi-target sputtering [2, 9], sol-gel process [4] and ion exchange process [10, 11]. For example, spherical nanoparticles with a silver core and an amorphous silica shell were successfully fabricated using tetraethoxysilane as silica precursor and reducing silver nitrate with ascorbic acid. These nanoparticles had excellent antibacterial effects against *E. coli* and *S. aureus* [12]. In a similar approach,

✉ Mohammad Ghorbanpour
Ghorbanpour@uma.ac.ir

[1] Department of Food Science, Sarab Branch, Islamic Azad University, Sarab, Iran

[2] Chemical Engineering Department, University of Mohaghegh Ardabili, Ardabil, Iran

[3] Department of Chemistry, Ardabil Branch, Islamic Azad University, Ardabil, Iran

silver nanoparticles were immobilized onto the surface of magnetic silica composite to prepare magnetic disinfectant that exhibited enhanced stability and antibacterial activity [13].

However, the synthesis and fabrication of supported silver nanomaterial via these methods require extensive use of toxic and highly cost chemicals and organic solvents, which frequently raise health and safety issues [14]. This has led to increasing interest on the development of inexpensive preparation methods suitable for large economic industrial applications. This may significantly boost the industrial production of the inexpensive silver-doped silica products for various applications.

In this work, silver-doped bioactive silica gel (SG) was prepared by a new approach titled as "molten salt method". To optimize the process condition while guaranteeing the antibacterial action, different synthesis conditions (time and temperature) were selected. The obtained samples were characterized by means of scanning electron microscope (SEM) and energy dispersive X-ray fluorescence (μ-EDXRF) observation. The effect of surface modifications on the antibacterial activity of samples has been investigated.

Methods

Materials

AgNO$_3$, SG (Kieselgel 60, 0.063–0.200 mm), Mueller–Hinton broth and nutrient agar were purchased from the Merck Company (Tehran, Iran). All reagents were used without further purification. The bacterial strain used for the antibacterial activity was Gram-negative *E. coli* (PTCC 1270) and Gram-positive *S. aureus* (PTCC 1112) received from the Iranian Research Organization for Science and Technology.

Preparation of Ag/SG nanocomposite

This process was performed in the molten salt bath of AgNO$_3$ for different periods ranging from 1 to 30 min. AgNO$_3$, after weighting, were grounded to get a homogeneous mixture. The temperature of the bath was maintained at about 150 and 300 °C. A quartz beaker was partially filled with the mixture of silver nitrate (AgNO$_3$) and SG (1:1, wt/wt), and placed in a furnace which was electrically heated up to the temperature of the process. The samples were then taken out of the molten beaker, and were cooled in air. Later, they were ultrasonically cleaned with distilled water. After molten salt process, a slight yellow coloration was observed.

Characterizations

The microstructures of the samples were observed by SEM (LEO 1430VP, Germany). μ-EDXRF analysis was performed using a XMF-104 X-ray Microanalyzer (Unisantis S.A., Switzerland) equipped with a 50 W molybdenum tube and a high resolution two-stage Peltier-cooled Si-PIN detector. The samples were positioned in definite places and at a constant height of the holder base. The temperature was controlled between 32 and 34 °C throughout the experiments. The voltage and current were 30 kV and 300 μA, respectively. Each μ-EDXRF analysis was performed in 50 s to obtain sufficient counts. For homogeneity tests, three different points of approximate constant orientation with respect to the nanocomposites were analyzed [15].

Water elution test

For each composite, 0.2 g of composite was immersed in 10 mL of distilled water and vigorous agitation in a shaking water bath (30 °C, 200 RPM) for 2, 6 and 24 h. Supernatants from each test tube were collected by centrifugation at 4,000 RPM for 10 min. Silver ions released from the nanocomposites were qualitatively determined by an atomic absorption spectrometer (AA800, Perkin Elmer). Water elution test was replicated twice.

Antibacterial activity

The antibacterial activity of the composites against both *E. coli* (Gram-negative) and *S. aureus* (Gram-positive) was tested by agar diffusion test. Samples were exposed to bacteria in solid media (nutrient agar), and the inhibition zone around each sample was measured and recorded as the antibacterial effect of composites. Agar plates were inoculated with 100 μL suspensions of bacteria. The composites and parent SG were placed on agar disks and incubated at 37 °C for 24 h. The inhibition zone was measured at three different points. Antibacterial activity test was replicated twice.

Results

Characterization

After molten salt process, the color of the SG surface changed from pale yellow to bright yellow, amber and dark amber depending on the processing time and temperature, as shown in Fig. 1. For high temperature process, the coloring becomes darker. The coloration of the sample prepared at 150 °C for 30 min matches with the color of

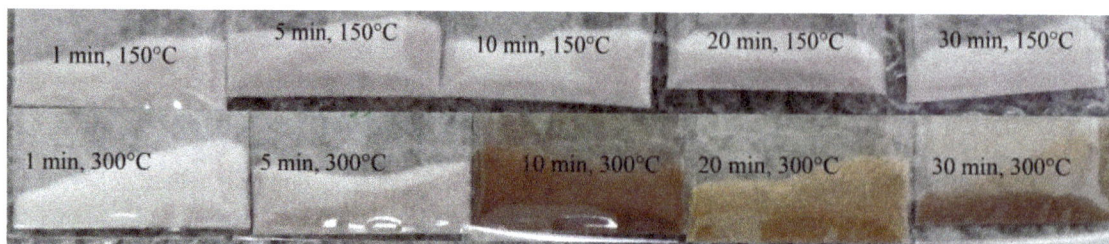

Fig. 1 The color of silver/silica gel nanocomposite at various temperatures and times

the sample processed at 300 °C for 5 min. The colors of the products obtained in this work were similar to previous reports [4, 11].

The un-embedded SG as well as the Ag^+ appears colorless. On the other hand, Ag^0 is yellow or brown/gray depending on the concentration and size of silver nanoparticles in the silica matrix [4]. The appearance and disappearance of colors is claimed be associated with the change in the state of silver (Ag^0 or Ag^+) at the various preparation time and temperatures.

Figure 2 shows the morphology of producing SG and composites. One understands that there is no remarkable change on the surface after coating with silver when compared SG surface (Fig. 2b) with produced composites (Figs. 2c, e). The only difference is the developmental changes in the micro pores on the initial structure of SG. It seems that these pores have been filled during the molten salt process with very small particles of silver. For further investigation, composites were heated at 100 °C for 3 h. Based on the available reports, heating causes the movement of silver particles on the support and consequently their adhesion. The result of this action is the formation of larger particles [16]. The comparison of composites images before and after heating resulted that the heating caused the formation of larger particles on the surface (Fig. 2c with d and f with g). Inasmuch as it has not been added any silver particle to the system during the heating process, the formation of these particles is caused due to adhesion the loaded silver on the surface of SG during the molten salt process. In other words, during the process of molten salt at temperatures and times of utilized in this research, particles that formed on the surface had not enough time to move on the SG and join together and form larger particles. The heating of composites prepares this situation. Another point is the size and the amount of formed nanoparticles. In reference to Fig. 2d and e and two samples, which prepared at 300 °C, the increasing of the time of molten salt process caused to the formation of larger nanoparticles on the surface. These results correspond to the changes of formal colors of the composites (Fig. 1).

For the better investigation of the amount of fixed silver on the SG in the composites, μ-EDXRF was used (Fig. 3).

On the basis of the past researches, the intensity of the peak of the plot for each of the elements is directly related to their concentration [15]. The peck at the energy of 2.9 eV represents silver in the composite. With due attention to this energy and Fig. 3, in 300 and 150 °C, the amount of silver in the composite has been increased with increasing the time of molten salt process. On the other hand, at constant time, the increasing of temperature causes the same result. These results verify SEM results.

Water elution test

The results of water elution of silver/SG nanocomposites are given in Fig. 4. This nanocomposite has been prepared for bactericide action. The importance of this test is that whether this nanocomposite causes water pollution or not. On the contrary, the more stability of nanocomposites, the more aged. This test exerts more intense condition to the nanocomposites than industrial use condition. It is usually utilized laminar flow in industrial systems; therefore, it exerts less tension than turbulent flow on the nanocomposites which used in this experiment. As Fig. 4 indicates, the amount of releasing silver from all of the nanocomposites is negligible. On the other hand, during the first 2 h, the most amount of silver is released, but after this period the releasing of silver is decreased so that the amount of silver after 6 h becomes constant. Thus, one can claim that the stability of prepared nanocomposite is suitable. Also the increasing of time and temperature caused the increasing of released silver. The reason is mounted silver on the support, which has not strong binding with the support. It may be eluted better from the composites after preparation and this causes elution of these particles and decreases the released silver.

Antibacterial test

Table 1 shows the antibacterial activity of the nanocomposites against *E. coli* and *S. aureus*. Generally, all silver-doped products showed antibacterial activity, while the parent SG has no inhibitory effect. It can be seen that inhibition of sample prepared in 150 °C is less than

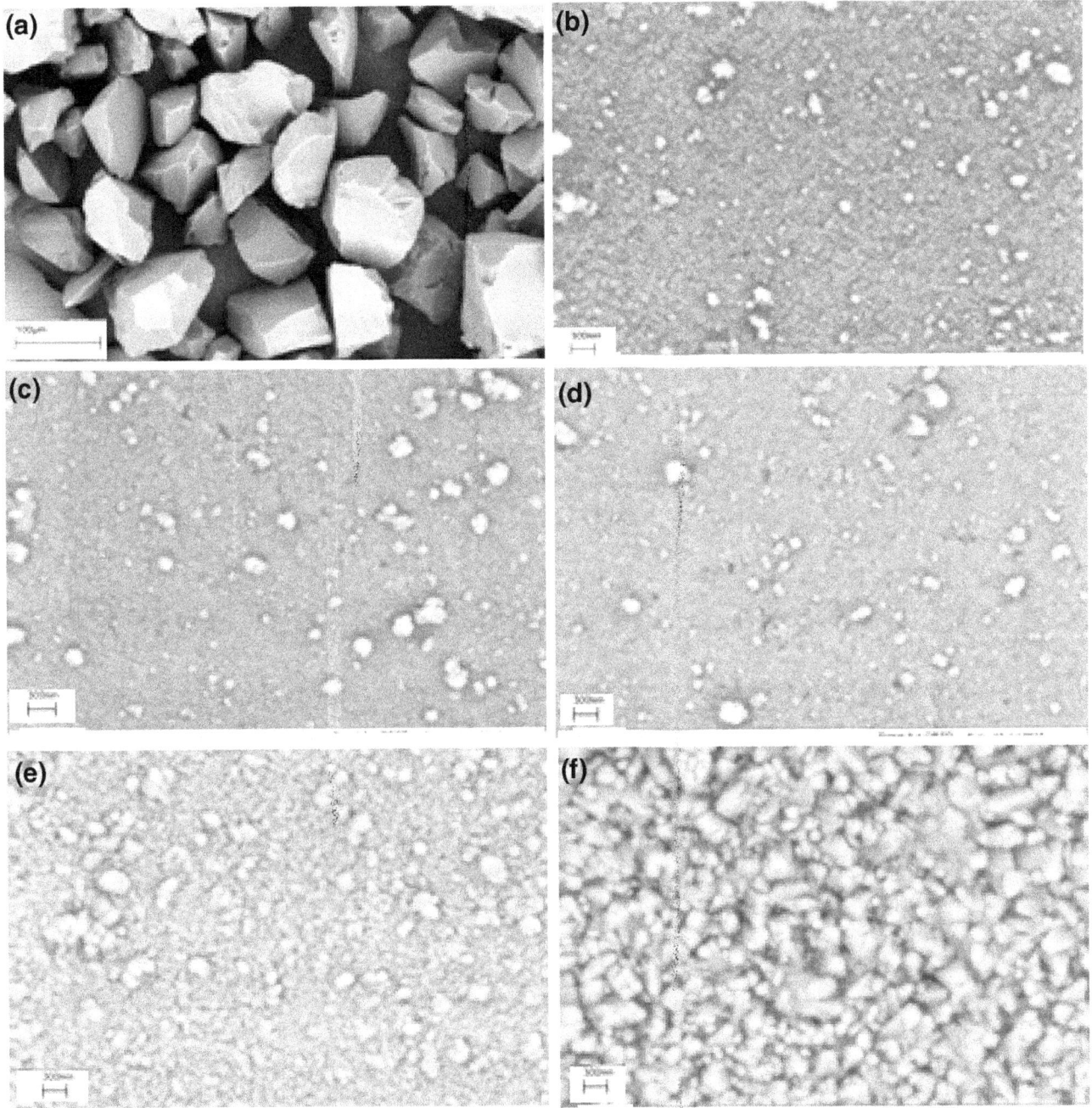

Fig. 2 The SEM micrographs of parent silica gel (**a, b**) and silver/silica gel nanocomposites prepared at 150 °C (**c**) and 300 °C (**d**) for 30 min then samples (**c, d**) annealed at 100 °C for 3 h (**e, f**), respectively

prepared samples in 300 °C (Table 1). In particular, for prepared samples in 300 °C, the zone of inhibition is almost similar for all silver-doped products.

The results of this research are comparable with the results of Hilonga et al. [4] study and our research has higher antibacterial effects relatively. They prepared the Ag/SG nanocomposite using the sol-gel method and in several steps, however, their method is time consuming and more complicated. In 2011, Quang et al. [5] prepared Ag/SG nanocomposite via chemical reduction of silver nitrate and the obtained hallow against *E. coli* was equal to 6 mm. Referring to Table 1, the prepared samples in our study at 300 °C had hollows about 6 mm against *E. coli*.

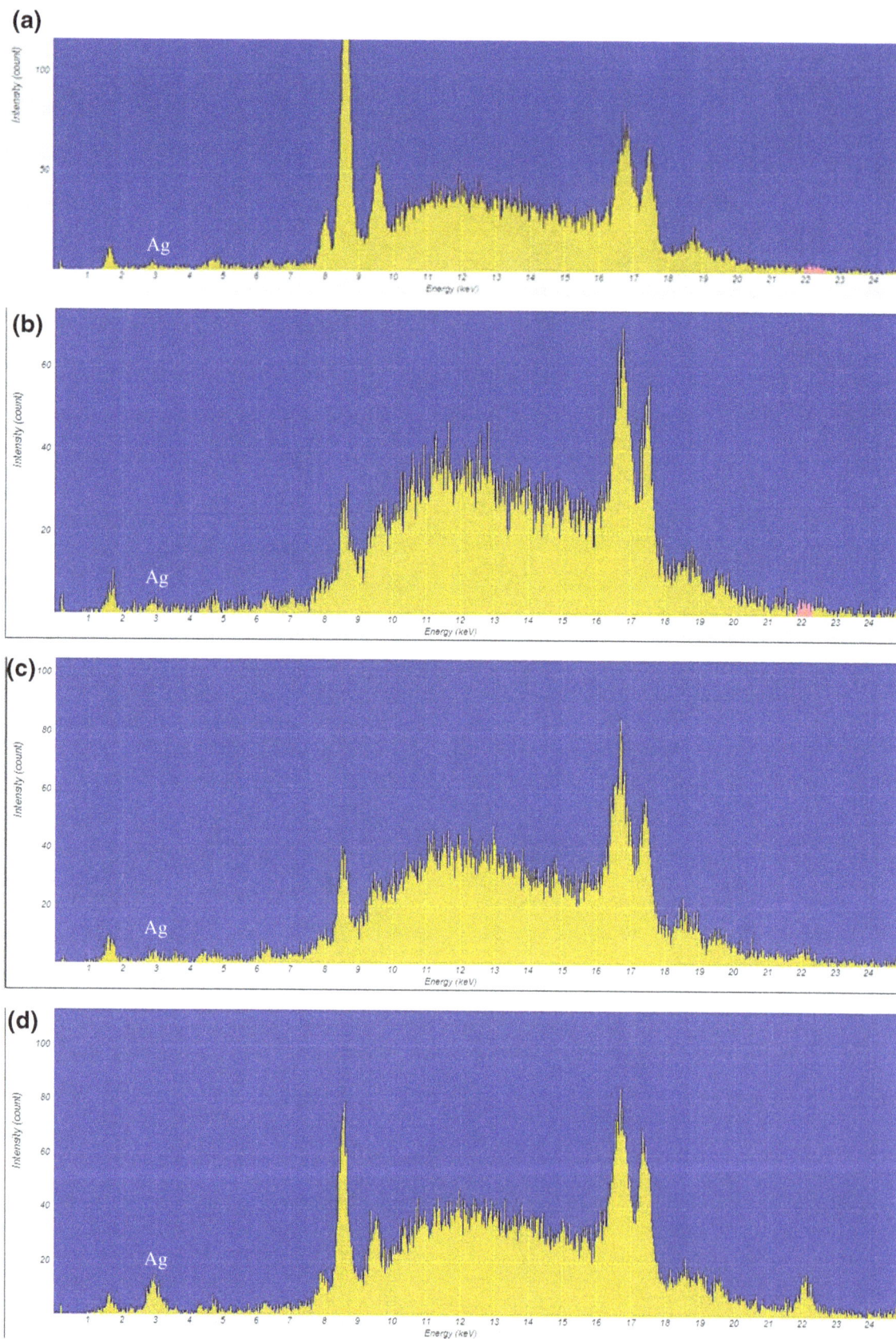

Fig. 3 The μ-EDXRF spectrum of composites, at deposition times of **a** 1 min and **b** 30 min at 100 °C and **c** 1 min and **d** 30 min at 300 °C

Fig. 4 The release of silver into deionized water after a 24 h period at 37 °C

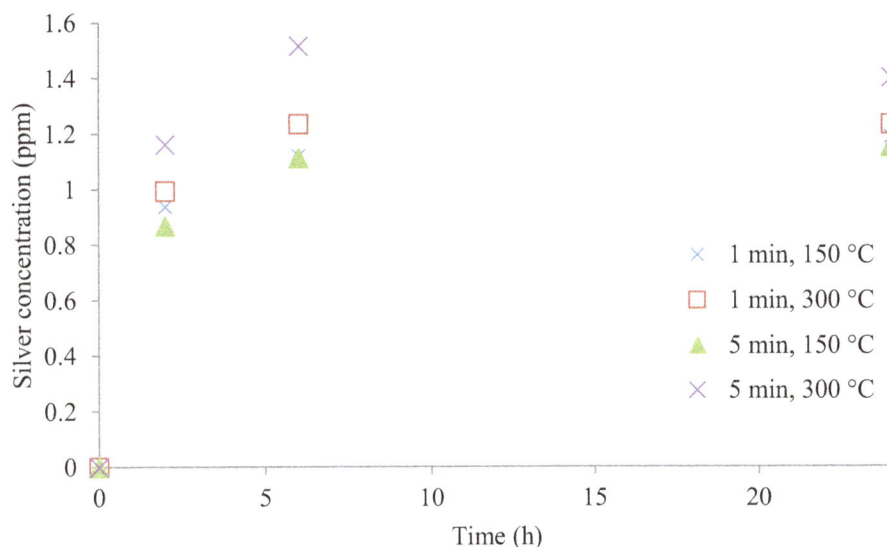

Legend:
× 1 min, 150 °C
□ 1 min, 300 °C
▲ 5 min, 150 °C
× 5 min, 300 °C

Table 1 Antibacterial activity of composites

Temperature of process (°C)	Time of process (min)	Inhibition zones (mm)	
		E. coli	S. aureus
150	1	0.7 ± 0.2	2.82 ± 0.6
	5	0.4 ± 0.2	2.05 ± 0.3
	10	0.9 ± 0.4	3.33 ± 0.4
	20	2.5 ± 1.1	3.10 ± 0.3
	30	4.1 ± 0.7	3.23 ± 0.8
300	1	6.0 ± 1.0	2.95 ± 1.1
	5	6.3 ± 1.4	7.07 ± 1.3
	10	5.85 ± 1.3	6.68 ± 1.5
	20	5.9 ± 1.7	6.49 ± 1.2
	30	5.7 ± 1.5	5.24 ± 2.0

Conclusions

An easy and new method was developed for the preparation of a bactericide nanocomposite. For the synthesis of the nanocomposite, SG was introduced into the molten salt of silver nitrate. After preparation of the nanocomposite, SEM and μ-EDXRF were utilized for its evaluation. The nanocomposite showed antibacterial properties. The prepared nanocomposite was stable during elution with water. There was negligible released silver during elution with water.

References

1. Girase, B., Depan, D., Shah, J.S., Xu, W., Misra, R.D.K.: Silver–clay nanohybrid structure for effective and diffusion-controlled antimicrobial activity. Mater. Sci. Eng. C **31**, 1759–1766 (2011)
2. Ghorbanpour, M.: Optimization of sensitivity and stability of gold/silver bi-layer thin films used in surface plasmon resonance chips. J. Nanostruct. **3**, 309–313 (2013)
3. Rai, M., Yadav, A., Gade, A.: Silver nanoparticles as a new generation of antimicrobials. Biotechnol. Adv. **27**, 76–83 (2009)
4. Hilonga, A., Kim, J.K., Sarawade, P.B., Quang, D.V., Shao, G., Elineema, G., Taik Kim, H.: Silver-doped silica powder with antibacterial properties. Powder Technol. **215–216**, 219–222 (2012)
5. Quang, D.V., Sarawade, P.B., Hilonga, A., Kim, J.K., Chai, Y.G., Kim, S.H., Ryu, J.Y., Taik Kim, H.: Preparation of silver nanoparticle containing silica micro beads and investigation of their antibacterial activity. Appl. Surf. Sci. **257**, 6963–6970 (2011)
6. Richardson, S., Postigo, C.: Drinking water disinfection by-products. In: The Handbook of Environmental Chemistry (2012)
7. Cao, G.F., Sun, Y., Chen, J.G., Song, L.P., Jiang, J.Q., Liu, Z.T., Liu, Z.W.: Sutures modified by silver-loaded montmorillonite with antibacterial properties. Appl. Clay Sci. **93–94**, 102–106 (2014)
8. Rivera-Garza, M., Olguõn, M.T., Garcõa-Sosa, I., Alcantara, D., Rodrõguez Fuentes, G.: Silver supported on natural Mexican zeolite as an antibacterial material. Microporous Mesoporous Mater. **39**, 431–444 (2000)
9. Ghorbanpour, M.: Stability modification of SPR silver nano-chips by alkaline condensation of aminopropyltriethoxysilane. J. Nanostruct. **5**, 105–110 (2015)
10. Varma, R.S., Kothari, D.C., Tewari, R.: Nano-composite soda lime silicate glass prepared using silver ion exchange. J. Non-Cryst. Solids **355**, 1246–1251 (2009)
11. Verne, E., Di Nunzio, S., Bosetti, M., Appendino, P., Vitale Brovarone, C., Maina, G., Cannas, M.: Surface characterization of silver-doped bioactive glass. Biomaterials **26**, 5111–5119 (2005)
12. Xu, K., Wang, J.X., Kang, X.L., Chen, J.F.: Fabrication of antibacterial monodispersed Ag-SiO$_2$ core-shell nanoparticles with high concentration. Mater. Lett. **63**, 31–33 (2009)

13. Zhang, X., Niu, H., Yan, J., Cai, Y.: Immobilizing silver nanoparticles onto the surface of magnetic silica composite to prepare magnetic disinfectant with enhanced stability and antibacterial activity. Colloids Surf. A **375**, 186–192 (2011)

14. Parandhaman, T., Das, A., Ramalingam, B., Samanta, D., Sastry, T.P., Mandal, A.B., Das, S.K.: Antimicrobial behavior of biosynthesized silica–silver nanocomposite for disinfection of water: a mechanistic perspective. J. Hazard. Mater. **290**, 117–126 (2015)

15. Ghorbanpour, M., Falamaki, C.: Micro energy dispersive X-ray fluorescence as a powerful complementary technique for the analysis of bimetallic Au/Ag/glass nanolayer composites used in surface plasmon resonance sensors. Appl. Opt. **51**, 7733–7738 (2012)

16. Ghorbanpour, M., Falamaki, C.: A novel method for the production of highly adherent Au layers on glass substrates used in surface plasmon resonance analysis: substitution of Cr or Ti intermediate layers with Ag layer followed by an optimal annealing treatment. J. Nanostruct. Chem. **3**, 661–667 (2013)

16

Synthesis and characterization of chitosan–copper nanocomposites and their fungicidal activity against two sclerotia-forming plant pathogenic fungi

Margarita S. Rubina[3] · Alexander Yu. Vasil'kov[3] · Alexander V. Naumkin[3] ·
Eleonora V. Shtykova[5] · Sergey S. Abramchuk[3] · Mousa A. Alghuthaymi[4] ·
Kamel A. Abd-Elsalam[1,2]

Abstract In this report, the metal-vapor synthesis (MVS) was used for the preparation of copper nanoparticles which was then used for the preparation of chitosan–copper nanocomposite. The antifungal activity of Cu@Chit NCs against two sclerotium-forming plant pathogenic fungi *Sclerotium rolfsii* (*S. rolfsii*) and *Rhizoctonia solani* (*R. solani*) AG-4 was evaluated in vitro and their effects on hyphal morphology, and sclerotia formation were observed for the first time. The NCs were prepared through impregnation of chitosan with colloid solution of copper nanoparticles in organic solvent (acetone or toluene). Transmission electron microscopy shows that the particles have predominantly spherical form, polydisperse character, the mean diameter about 2–3 nm and a rather uniform distribution in the chitosan matrice. Analysis of the small angle scattering curves suggests that the copper particles in the NCs with the size of ≤ 2 nm are mostly located in the chitosan pores with the same size. The effect of Cu@Chit NCs on fungal growth reveals some significant inhibitory activity against two tested fungi. The highest level of inhibition against *S. rolfsii* and *R. solani* AG-4 was observed using the high concentrations of Cu@Chit NC prepared using acetone as a solvent. A loss of the cytoplasm content, cytoplasmic coagulation, irregular shape of mycelia, or destruction in the hyphae was confirmed. The experiments demonstrate that the Cu@Chit NC synthesized via MVS using acetone was more effective than that of toluene in inhibiting fungal hyphae growth against *R. solani* AG-4 and *S. rolfsii*. The results show that the Cu@Chit NCs are fungicidal against both the tested fungus at high concentrations and the fungicidal or fungistatic activity is dependent on the tested fungus species.

Keywords Antifungal · Metallic nanocomposites · Copper nanoparticles · Metal-vapor synthesis · *Sclerotium rolfsii* · *Rhizoctonia solani*

Introduction

One of the most critical issues in the face of mankind is the solution search in the fight against pathogenic microorganisms. This issue now is well visible in the agriculture. A number of different pathogenic fungi and bacteria cause the plant diseases. To solve this problem, the synthetic organic or inorganic pesticides treatment of seeds, plants and crops is widely applied. However, other problems are often emerging. It is well known that chemical agriculture pesticides tend to accumulate in soil causing poisoning, gather in plants and crops, have toxic effect on ecology and sometimes has poor efficacy against some types of microorganisms. Copper compounds are well known as biocide from XVIII century. Nevertheless, the same problems with their usage exist. To enhance antifungal activity

✉ Kamel A. Abd-Elsalam
kamelabdelsalam@gmail.com

[1] Plant Pathology Research Institute, Agricultural Research Center (ARC), Giza, Egypt

[2] Unit of Excellence in Nano-Molecular Plant Pathology Research Center, Plant Pathology Research Institute, Giza, Egypt

[3] A.N. Nesmeyanov Institute of Organoelement Compounds (INEOS), Russian Academy of Sciences, Moscow, Russia

[4] Biology Department, Science and Humanities College, Shaqra University, Alquwayiyah, Saudi Arabia

[5] A.V. Shubnikov Institute of Crystallography, Russian Academy of Sciences, Moscow, Russia

and decrease environmental toxic effect, it is important to develop novel type of biocides [1, 2].

Chitosan is a biodegradable, non-toxic and biocompatible polymer obtained from partly deacytilation procedure of chitin sources. This polymer possesses unique number of biological properties such as antibacterial, antiviral and antifungal activities. Chitosan is established to exhibit the inhibition activity on different stages namely mycelia growth, sporulation, spore viability and germination, and the production of fungal virulence factors [3]. Chitosan and chitosan-based composites in the forms of powders, suspensions, film coatings, scaffolds, and capsules are utilized for crop protection and for increasing resistance of host plants [4]. The using of chitosan in agriculture is possible not only in crop protection of healthy plants but also in plant diseases control [5]. Although exact mechanisms of the chitosan inhibition have not been sufficiently studied, it is assumed that chitosan can act as chelating agent for minerals and nutrients of pathogens thus leading to theirs death [6]. It is known that chitosan antifungal activity depends on different factors like degree of acetylation, concentration, form of usage, type of target organism and molecular weight [7]. Numerous studies are dedicated to investigation of such correlations. Because of the large number of NH_2- and OH- in chitosans chemical structure, chitosan can be used as effective anchoring agent for metal nanoparticles (NPs). Cu NPs have pronounced bactericidal properties against various types of organisms [8]. The improved antimicrobial activity of Cu NPs as compared to copper salt is due to their unique property, i.e., large surface area to volume ratio [9]. Recently, the investigations showed that incorporation of Cu NPs into chitosan matrice significantly enhanced its antimicrobial and antifungal activity [10–13]. The Cu@Chit NCs can combine the properties of biopolymer and nanosized copper; therefore, their application can be more effective than separately.

Insertion of NPs into chitosan matrice can be done by different ways. The most versatile methods are based on reduction of metal salts with the presence of stabilizing and reducing agents in the water medium. The chemical reduction method is applied for synthesis of a huge number of different NPs including Cu NPs. Because of the high tendency of Cu NPs to be oxidized, it is a big issue to prepare copper in a zero-valent state.

One of the promising methods for preparing colloidal NPs solutions and their based metal–polymer NCs is metal-vapor synthesis (MVS). This cryochemical method is based on metal evaporating under $p = 10^{-4}$–10^{-6} Torr and $T = 77$ K and following co-condensation process with organic ligand/solvent on the walls of quartz reactor. The resulting cryomatrice upon being heated forms the colloidal solution of NPs in organic solvent called as organosol. To prepare composites with metal NPs, the different kinds of supports (organic/inorganic) are impregnated with organosol. MVS technique is well-proven and effective route to make composites with tailored magnetic, antibacterial, and catalytic properties. The MVS efficacy for preparing biopolymers and their hybrids with biologically active metal NPs has already been demonstrated [14–16].

This work is dedicated to the fabrication of Cu@Chit NCs and testing these materials against two sclerotium-forming plant pathogenic fungi *R. solani* AG-4 and *S. rolfsii*. To obtain metal-biopolymer NCs the original cryochemical technique will be employed.

Experimental part

Materials and reagents

Toluene and acetone with special purity grade, 99.5% were used as the solvents for MVS. Prior to synthesis toluene was dried over Na and acetone was dried under molecular sieves (4 Å) then solvents were degassed in vacuum of 10^{-1} Pa by freezing–thawing cycles. The metal source was Cu foils (99.99%) whose surface was pre-treated with concentrated HNO_3 and distilled H_2O to eliminate oxide films. Chitosan (CAS 9012-76-4) with the molecular weight of 310–375 kDa and degree of deacetylation ≥75% was purchased from Aldrich. Before the impregnation procedure chitosan powder was degassed for 12 h under vacuum of 10^{-1} Pa at 40 °C.

Preparation of Cu@Chit nanocomposites

The metal-vapor technique was used for the preparation of copper-carrying chitosan NCs. The full preparing procedure can be divided into the main three steps. In the first step, we obtained the NPs colloidal solution in organic solvent referred as organosol. In this work the copper-acetone and copper-toluene organosol were prepared. For this purpose 0.26 g of copper foil (about 200 μm thickness) were evaporated by resistively heating a tantalum boat ($l = 90$ mm, $b = 5$ mm) at a residual pressure of 10^{-5} Torr and co-condensed with 120 mL of solvent on liquid nitrogen-cooled walls of a 5 L reactor. The second step was the impregnation of chitosan powder with organosol. On this step the frozen matrix was melted and the organosol was infiltrated into chitosan powder in an evacuated Schlenk vessel. The black organosol was instantly discolored and the embedding of chitosan with metal NPs was occurred immediately. More detailed information about MVS set up and procedure reported elsewhere [17]. On the third step, the solvent was removed and the metal-carrying chitosan powder was dried in a vacuum of 1 Pa at 60 °C during 3 h. All manipulations were performed in a pure Ar atmosphere. The Cu@Chit NCs obtained by means of Cu-acetone and Cu-toluene impregnation are referred as Cu@Chit-1 and Cu@Chit-2, respectively.

Characterization of Cu@Chit nanocomposites

X-Ray fluorescence analysis (XRF)

Cu concentration (%wt) in the samples was measured with a VRA 30 X-ray fluorescent analyzer (Germany) using the M Kα line of the X-ray fluorescence spectrum. In order to excite XF, X-ray tube with a Mo anode was used in the 50 kV, 20 µA regime. Cu@Chit NCs were analyzed as pressed pills.

Transmission electron microscopy (TEM)

TEM microphotographs were performed using a transmission electron microscope LEO 912AB OMEGA, Zeiss (Germany) at acceleration voltage of 100 kV. The Cu-carrying chitosan powder was previously suspended in deionized water (18 MΩ) and sonicated in the ultrasonic bath during 15 min. Then, a small amount of the suspension was dripped onto a copper mesh covered with formvar film.

Small-angle X-ray scattering (SAXS)

Conventional small-angle X-ray scattering (SAXS) measurements were performed using a laboratory diffractometer "AMUR-K" (Institute of Crystallography, Moscow) at a wavelength $\lambda = 0.1542$ nm with a Kratky-type (infinitely long slit) geometry covered the range of momentum transfer $0.12 < s < 6.0$ nm^{-1} (here, $s = 4\pi \sin\theta/\lambda$, where 2θ is the scattering angle). The scattering curves were corrected for the background scattering and primarily processed using standard procedures and program PRIMUS [18].

Size distributions of heterogeneities presented in the specimens and those of Cu nanoparticles were computed using the indirect transform program GNOM [19]. Assuming the particles to be spherical, the program solves integral equation:

$$I(s) = \int_0^\infty D_V(R)\, m^2(R)\, i_0(sR)\, \mathrm{d}R,$$

where $I(s)$ is scattering intensity, R is the radius of a sphere, R_{min} and R_{max} are the minimum and maximum radii, respectively, $i_0(x) = \{[\sin(x) - x\cos(x)]/x^3\}^2$ is the sphere form factor and $m(R) = (4\pi/3)R^3\Delta\rho$, where $\Delta\rho$ is the particle contrast (difference between the scattering length

density of the particle and that of the matrix). The value of R_{min} was kept zero; that of R_{max} was selected for each individual data set by successive runs with different values of this parameter to obtain the minimal discrepancy between the experimental and calculated scattering curves.

X-ray photoelectron spectroscopy (XPS)

X-ray photoelectron spectra were acquired with an Axis Ultra DLD (Kratos, UK) spectrometer using monochromatized Al Kα (1486.6 eV) radiation at an operating power of 150 W of the X-ray tube. Survey and high-resolution spectra of appropriate core levels were recorded at pass energies of 160 and 40 eV and with step sizes of 1 and 0.1 eV, respectively. Sample area of 300 µm × 700 µm contributed to the spectra. The samples were mounted on a sample holder with a two-sided adhesive tape, and the spectra were collected at room temperature. The base pressure in the analytical UHV chamber of the spectrometer during measurements did not exceed 10^{-8} Torr. The energy scale of the spectrometer was calibrated to provide the following values for reference samples (i.e., metal surfaces freshly cleaned by ion bombardment): Au 4f$_{7/2}$–83.96 eV, Cu 2p$_{3/2}$–932.62 eV, Ag 3d$_{5/2}$–368.21 eV. The electrostatic charging effects were compensated by using an electron neutralizer. Sample charging was corrected by referencing to the C–C/C–H peak deconvoluted in the C 1s spectrum (284.8 eV). After charge referencing, a Shirley-type background with inelastic losses was subtracted from the high-resolution spectra. The Cu and Zn LMM Auger spectra were corrected using a linear background. The surface chemical composition was calculated using atomic sensitivity factors included in the software of the spectrometer corrected for the transfer function of the instrument.

In vitro investigation of antifungal activity of Cu@Chit nanocomposites

The tests were performed using the agar medium assay described by Tatsadjieu et al. [20]. To evaluate the in vitro antifungal effects of the Cu@Chit NCs against R. solani AG2 and AG4, three concentrations of the Cu@Chit NCs suspension (30, 60, and 100 mg/L) were added to Petri dishes before pouring plates with PDA. Each Petri dish was inoculated at the center with a mycelial disc (10 mm diameter) taken at the periphery of R. solani colony grown on PDA at 29 ± 1 °C for 3 days. Positive control (without chitosan nanocomposites) plates were inoculated following the same procedure. The growth inhibition percentage was calculated according to the formula described by Abd-El-Khair [21] which is mentioned below:

Growth inhibition $(\%) = (C - P/C) \times 100$,

where C is the diameter of mycelial growth in control plates, and P is the diameter of mycelial growth in treated plates. Three replicates were used per treatment, and the experiment was repeated three times. The Cu@Chit NCs with the highest levels of inhibition against the pathogens were selected for further experiments.

Morphology of mycelia expose to Cu@Chit nanocomposites

Hyphal morphology of *R. solani* anastomosis treated with the high concentration of the Cu@Chit NCs was investigated on microscopic slides after 1 h. Morphological changes resulting from Cu@Chit-1 and Cu@Chit-2 NCs on hyphal growth were examined under light microscope (LaboMed, Los Angeles, USA). This assay enables the observation of possible morphological changes displayed by the fungus when it was exposed to Cu@Chit NCs antifungal activity. Each assay was repeated at least three times.

Ability of *R. solani* and *S. rolfsii* isolates to produce sclerotia in vitro

One plug in diameter of 7 mm of *R. solani* and *S. rolfsii* isolates was taken for each isolate and placed in the center of a 60 mm Petri dish of PDA embedded with highest concentrations of Cu@Chit-1 NCs. Each plate was sealed with parafilm and incubated at 25 ± 2 °C for 18 days. The number of sclerotia was then measured using a transparent grid with subdivisions of 5 mm^2. Each subdivision with 50% or more of its area consisting of sclerotia was recorded and the number of sclerotia was calculated [22]. This procedure was replicated three times per isolate.

Statistical analyses

Data were subjected to analysis of variance (ANOVA). Least significant difference (LSD) was used to compare concentration means within genotypes. ANOVA was performed with MSTAT-C statistical package.

Results and discussion

In this work, the metal-vapor synthesis (MVS) was used for the preparation of copper nanoparticles which was then used for the preparation of chitosan–copper nanocomposite. According to XRF, the copper concentration (%wt) in the Cu@Chit-1 and Cu@Chit-2 is 3.7 and 5.1%, respectively. To analyze particles size and particles distribution in polymer matrice, TEM analysis was performed. According to microphotographs (Fig. 1a, d), the particles have predominantly spherical form, polydisperse character and slightly uniform distribution in the chitosan matrice. The average particle size of metal NPs in both Cu@Chit-1 and Cu@Chit-2 composites turned out to be almost equal and was about 2–3 nm. However, TEM microphotograph of Cu@Chit-2 in the black field showed a bunch of particles in the range of 4–8 nm whereas Cu@Chit-1 exhibited some amount of large particles in the range of 10–15 nm (Fig. 1,b,e). It is worth to point out that the fraction of the large particles is higher for copper nanocomposite obtained with Cu-toluene organosol. Selected area diffraction patterns (SAED) for both composites contained rings with lattice spacings corresponded to Cu_2O and Cu phase (Table 1). Thus, it can be suggested that the NCs contain copper nanoparticles covered by an overlay of copper (I) oxide which can perform protective function against further particles oxidation.

Experimental SAXS curves of Cu@Chit NCs and pristine chitosan are demonstrated in Fig. 2. The scattering curves are found to have a polydisperse character. The scattering curve for Cu@Chit-2 showed the maximal scattering intensity in very small angles (<2 nm^{-1}). This could be explained by a higher concentration of scattering metallic particles and possibly by the presence of the larger particles or their aggregates. The second suggestion was confirmed with TEM analysis.

It should be noticed that the curves coincide in the region of scattering vector higher than 5 nm^{-1}. Hence, the impregnation procedure of chitosan with organosol containing metallic particles has no influence on the pristine structure of polymer matrix. In this case, the scattering curve for pristine chitosan was subtracted from that of the chitosan with metal inclusions giving a curve for metallic NPs. The resulting difference curves are shown in Fig. 3.

The difference SAXS curves from Cu@Chit-1 and Cu@Chit-2 on the interval of scattering vectors from 0.4 up to about 3.0 nm^{-1} were used to calculate volume size distribution functions $D_V(R)$. The applying interval limitations were caused by the necessity to exclude the influence of both the primary beam and scattering from large particles and their aggregates, which are beyond the SAXS resolution. Figure 4 demonstrates the $D_V(R)$ for metallic particles along with the distribution function of the heterogeneities observed in pristine chitosan. The presence of heterogeneities in the pristine chitosan can be explained due to the pores or more dense regions of the polymer.

Small-angle X-ray scattering is caused by a difference in electronic densities of a scattering object and surrounding medium, i.e, by a contrast. The contrast may be either positive or negative. For the original chitosan, scattering may occur due to the pores (negative contrast) or more

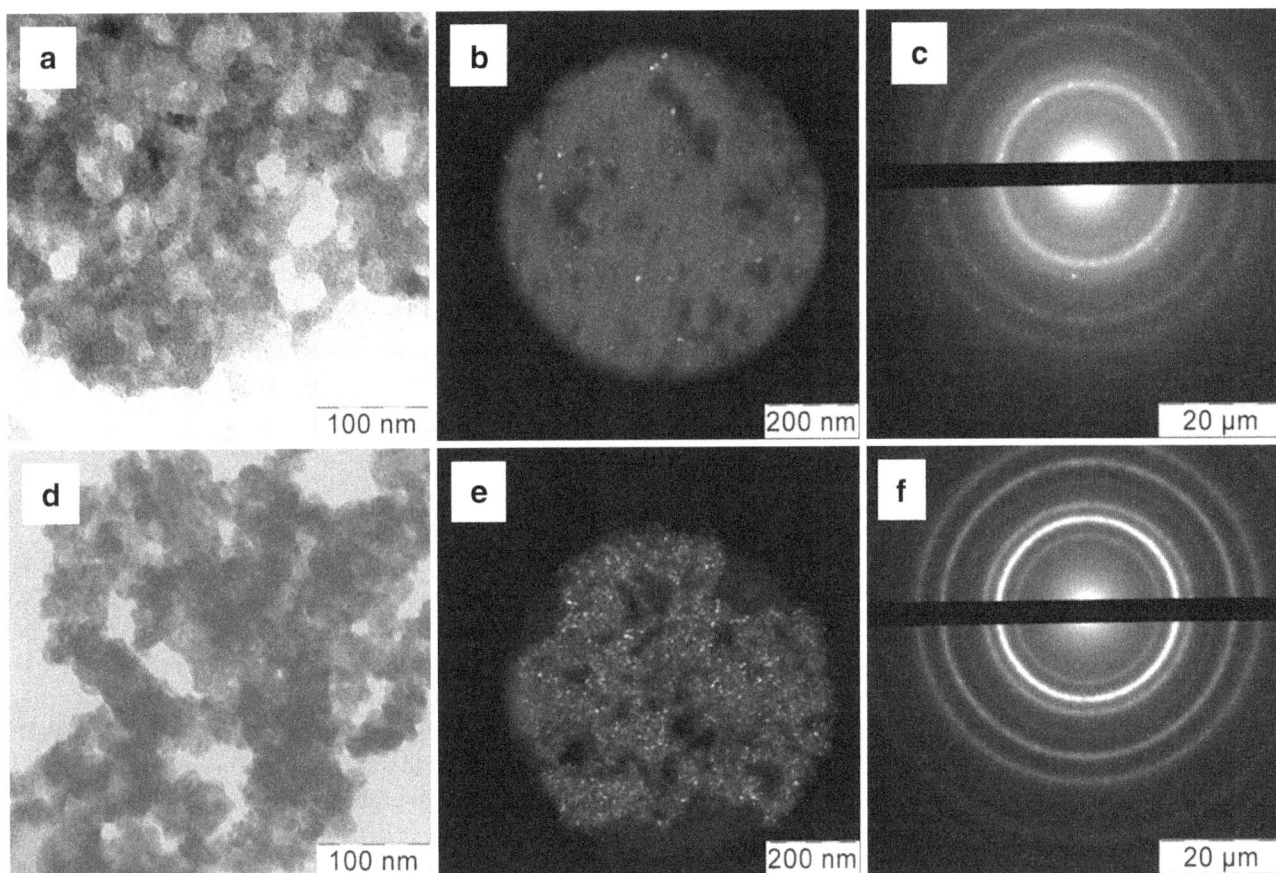

Fig. 1 TEM microphotographs in bright (**a**, **d**) and dark field (**b**, **e**) and SAED patterns (**c**, **f**) for Cu@Chit-1 and Cu@Chit-2 NCs, respectively

Table 1 The lattice spacings for Cu@Chit NCs determined from SAED patterns

	No.	d, Å	d, Å[a]	Phase
Cu@Chit-1	1	2.98	3.00	Cu_2O
	2	2.42	2.45	Cu_2O
	3	2.08	2.12/2.08	Cu_2O/Cu
	4	1.49	1.51	Cu_2O
	5	1.80	1.81	Cu
	6	1.27	1.28/1.28	Cu_2O/Cu
	7	1.21	1.23	Cu_2O
Cu@Chit-2	1	2.98	3.00	Cu_2O
	2	2.43	2.45	Cu_2O
	3	2.09	2.12/2.08	Cu_2O/Cu
	4	1.80	1.81	Cu
	5	1.47	1.51	Cu_2O
	6	1.26	1.28/1.28	Cu_2O/Cu

[a] d-data taken from [23]

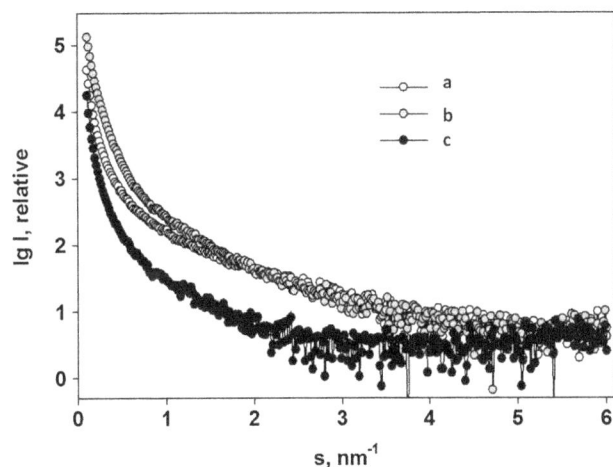

Fig. 2 Experimental *small-angle scattering curves* for Cu@Chit-1 (**a**) and Cu@Chit-2 (**b**) NCs and pristine non-modified chitosan (**c**)

dense regions of the polymer (positive contrast), but the profile of the size distribution function will be the same in both cases. The coincidence of the main fraction of all calculated size distribution functions reflects the fact that 2 nm metal nanoparticles most probably fill the chitosan pores with the same size. However, if for the Cu@Chit-1 fraction of the large particles is relatively small, the Cu@Chit-2 obtained with Cu-toluene organosol demonstrates much higher amount of larger NPs.

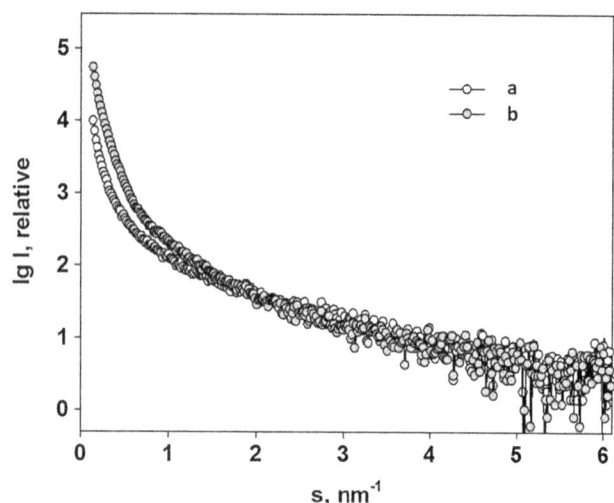

Fig. 3 Difference small-angle scattering curves for Cu@Chit-1 (**a**) and Cu@Chit-2 (**b**) NCs

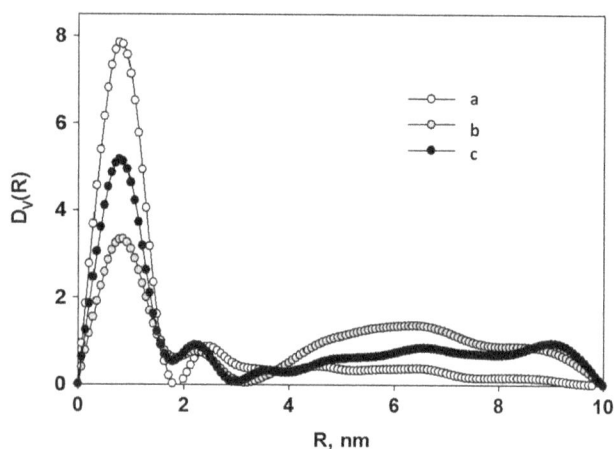

Fig. 4 Volume size distribution function for Cu@Chit-1 (**a**) and Cu@Chit-2 (**b**) NCs and pristine non-modified chitosan (**c**)

Table 2 XPS analysis of Cu@Chit NCs

Sample	Relative concentration, at. %						
	C	O	N	Cu	C/N	C/O	O/N
Chitosan	63.6	29.7	6.8	–	9.4	2.1	4.4
Cu@Chit-1	47.9	34.4	3.4	14.3	14.2	1.4	10.2
Cu@Chit-2	52.2	34.4	4.8	8.6	10.9	1.5	7.2

XPS analysis was applied for analysis of surface of the pristine and Cu@Chit NCs. It was revealed that surface composition after impregnation was changed due to slight increase of oxygen concentration while the carbon percentage decreases for the composites as compared with chitosan (Table 2). These changes are related to appearance of copper in different oxide states.

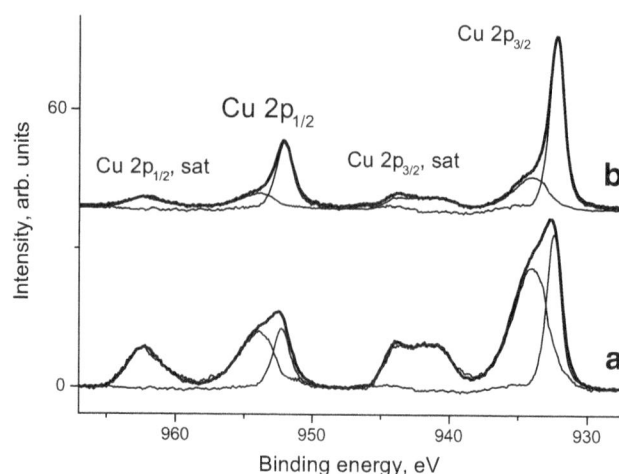

Fig. 5 The Cu 2p photoelectron spectra for the Cu@Chit-1 (**a**) and Cu@Chit-2 (**b**) NCs normalized by intensity of the main peak

Figure 5 presents the Cu 2p photoelectron high-resolution spectra of the Cu@Chit NCs and their fitting with two states. The binding energies and Auger parameters for Cu@Chit nanocomposites are closed to each other (Table 3). It evidences similar chemical states of copper in both samples. All the spectra contained satellite peaks at binding energies about 943 and 963 eV identifying Cu^{2+} state. Cu^{+} state was identified according to the binding energy of the Cu $2p_{3/2}$ peak and modified Auger parameter [24, 25]. Relative intensities of peaks determined form high resolution Cu 2p spectra are used to calculate the concentration of oxidation states of copper at surface. Thus, Cu^{2+} concentrations at the surface are 10.7 and 4.8%, whereas those of Cu^{+1} are 3.6 and 5.5% for Cu@Chit-1 and Cu@Chit-2, respectively.

Antifungal activity assay

The results of studies on the effect of Cu@Chit NCs on fungal growth revealed some significant inhibitory activity against two tested fungi. The effect of three concentrations of Cu@Chit NCs on mycelial growth of *R. solani* AG-4 and *S. rolfsii* is shown in Table 4.

The highest level of inhibition against *R. solani* AG-4 and *S. rolfsii* was observed using the high concentrations of Cu@Chit-1. While the lowest level of inhibition was observed using the high concentrations of Cu@Chit-2 with *R. solani* AG-4. The radial growth of *R. solani* AG-4 and *S. rolfsii* was reduced by high concentrations of Cu@Chit NCs in a dose dependent manner. The antifungal activity of the Cu@Chit NCs against *S. rolfsii* appears to be fungistatic rather than, fungicidal because growth of the fungus transferred from NCs-challenged plates to non-challenged condition was the same to growth of fungus started from untreated plates (Fig. 6). While, *R. solani* AG-4 has

Table 3 Binding energies of the Cu 2p photoelectron spectra and LMM Auger peaks

Sample	Cu 2p$_{3/2}$	Cu 2p$_{1/2}$	Cu 2p$_{3/2}$, sat1	Cu 2p$_{3/2}$, sat2	Cu 2p$_{1/2}$, sat1	Cu LMM	Auger parameter
Cu@Chit-1	932.44	952.33	940.66	943.93	962.28	916.42	1848.86
Cu@Chit-2	932.95	952.73	941.41	944.13	962.58	916.07	1849.02

Table 4 Mean's comparisons of in vitro antagonism of different concentrations of Cu@Chit NCs against linear growth of two sclerotium-forming plant pathogenic fungi

NCs concentrations	Sclerotium-forming fungi			
	R. solani		S. rolfsii	
	Cu@Chit-1	Cu@Chit-2	Cu@Chit-1	Cu@Chit-2
30 mg/L	1.500	0.967	4.100	0.000
60 mg/L	2.233	1.267	4.600	2.900
100 mg/L	3.267	2.467	7.033	3.100
Control	0.000	0.000	0.000	0.000
LSD ($p < 0.05$)	0.619	0.501	0.126	0.126

Fig. 6 Fungicidal effect of different concentrations of Cu@Chit-1 (**a**) and Cu-Chit-2 (**b**) on mycelial growth of *S. rolfsii*

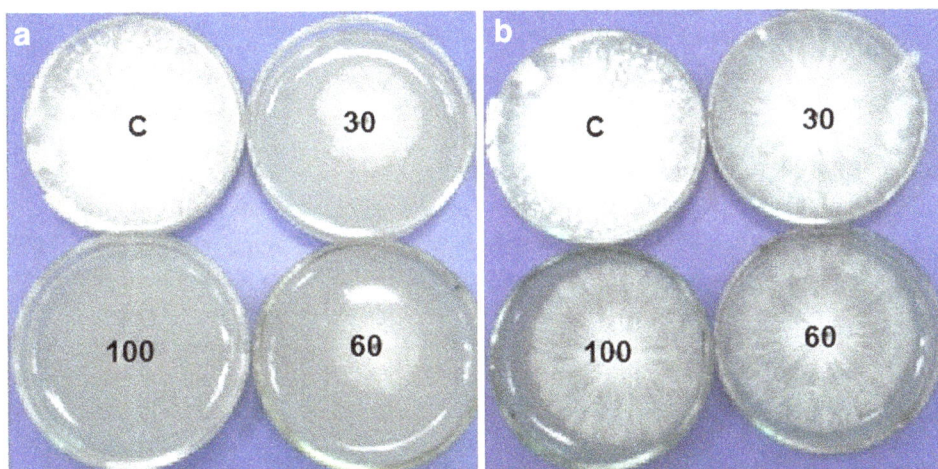

fungicidal effects in contrary with *S. rolfsii*. Since there were no significant differences in results between three concentrations of Cu@Chit NCs, more than 100 mg/L concentration was needed for antifungal activity (Table 4).

Cu–chitosan nanocomposite at 100 mg considerably hampered fungal mycelial growth and scolorotia formation in *R. solani* and *S. rolfsii*. In most cases, inhibition increased as the concentration of Chit NCs increased. This could be due to the high density at which the solution was able to saturate and cohere to fungal hyphae and to deactivate plant pathogenic fungi. The combined copper–chitosan colloids are used as a new generation of copper-based bio-pesticides [26]. Cu–chitosan has proven high antifungal activity by inhibiting spore and mycelium formation of *A. alternate*, *R. solani* and *Macrophomina phaseolina* [10, 12]. The phytopathogenic fungi usually have stronger resistance to reactive oxygen species (ROSs) attack than

bacteria or viruses, because of their much larger size and thicker cell wall/membrane. Thus, effective alternatives to chemical fungicides are difficult to develop [27].

Chitosan present in Cu–chitosan nanocomposite enhanced enzyme activities involved in plant defense. In addition, acidic environment created by the fungi in infected plant provokes the break-up of nanostructure and releases Cu ions [26]. Chitosan–Cu nanonetwork is evident through a higher Cu accumulation in porous area which supports the ion-exchange resins and surface chelating mechanism [12]. Cu–chitosan nanocomposite has a weighty effect on plant cells as they easily pass into the seeds along with encapsulated Cu and participate strongly in the metabolism of germinating seeds [13]. These studies strongly hypothesize a cumulative effect of chitosan elicitors to induce defense enzymes and direct Cu ions toxicity towards fungi.

Effect of Cu@Chit NCs on hyphal morphology

The fungal mycelia from the zone of inhibition produced by 100 mg of Cu@Chit-1 and Cu@Chit-2 NCs were observed under a light microscope (100×). A loss of the cytoplasm content, cytoplasmic coagulation, irregular shape of mycelia, or destruction in the hyphae was confirmed. In some cases, cell wall disruption, and consequent hyphal lysis or necrosis was detected. Upon observation, the hyphae showed mycelial deformations with swollen margins and broken mycelia, whereas the fungal mycelia from the control sample showed normal, highly branched intact morphology (Fig. 7).

However, there were other deformations such as structure of the cell membrane and inhibiting normal budding process of both *Rhizopus* sp. and *Aspergillus* sp., probably due to the destruction of the membrane integrity [28, 29]. Almost similar observation was reported by Ouda (2014) [30] who used copper and silver nanoparticles against two plant pathogens, *A. alternate* and *Botrytis cinerea*. The effects of tested Cu@Chit-1 and Cu@Chit-2 concentrations on sclerotia formation in *R. solani* AG-4 and *S. rolfsii* and was presented in Table 5. Significant differences ($p > 0.05$) were observed in the number of sclerotia produced in vitro between *R. solani* treated with Cu@Chit NCs. Plates treated with 100 mg Cu@Chit NCs revealed the minimum number of sclerotia compared with control plates (Fig. 8). The highest level of inhibition against *S. rolfsii* was observed using the high concentrations of Cu@Chit-2. These results indicate that Cu-NCs strongly suppressed sclerotia formation for *R. solani* and *S. rolfsii* under in vitro condition.

Conclusion

It is essential to develop an innovative and effective alternative nanocides such as nanometal polymer hybrids to develop new strategies for the control of sclerotium-forming plant pathogenic fungi. In this work, two Cu@Chit nanocomposites were fabricated through impregnation of chitosan with organosols Cu-acetone and Cu-toluene prepared via metal-vapor synthesis.

The particles in the nanocomposites have predominantly spherical form, polydisperse character, the mean diameter about 2–3 nm and a rather uniform distribution in the chitosan matrice. Analysis of the small angle scattering curves suggests that the copper particles in the nanocomposites with the size of ≤2 nm are mostly located in the chitosan pores with the same size. The binding energies of XPS spectra identified oxidation states of copper (I) and copper (II) for both composites; moreover, the surface of Cu@Chit nanocomposite obtained via Cu-acetone organosol is more oxidized than that of nanocomposite via Cu-toluene. Both Cu@Chit nanocomposites were examined against two plant pathogenic fungi *S. rolfsii* and *R. solani* AG-4. The results show that the Cu@Chit nanocomposites at concentration of 100 mg/L considerably hampered fungal mycelial growth and scolorotia formation in *R. solani* and *S. rolfsii*. In most cases, inhibition increased as the concentration of Cu@Chit nanocomposites increased. This could be due to the high density at which the solution was able to saturate and cohere to fungal hyphae and to deactivate plant pathogenic fungi. These results indicate that Cu@Chit nanocomposites strongly suppressed sclerotia formation for *R. solani* and *S. rolfsii* under in vitro

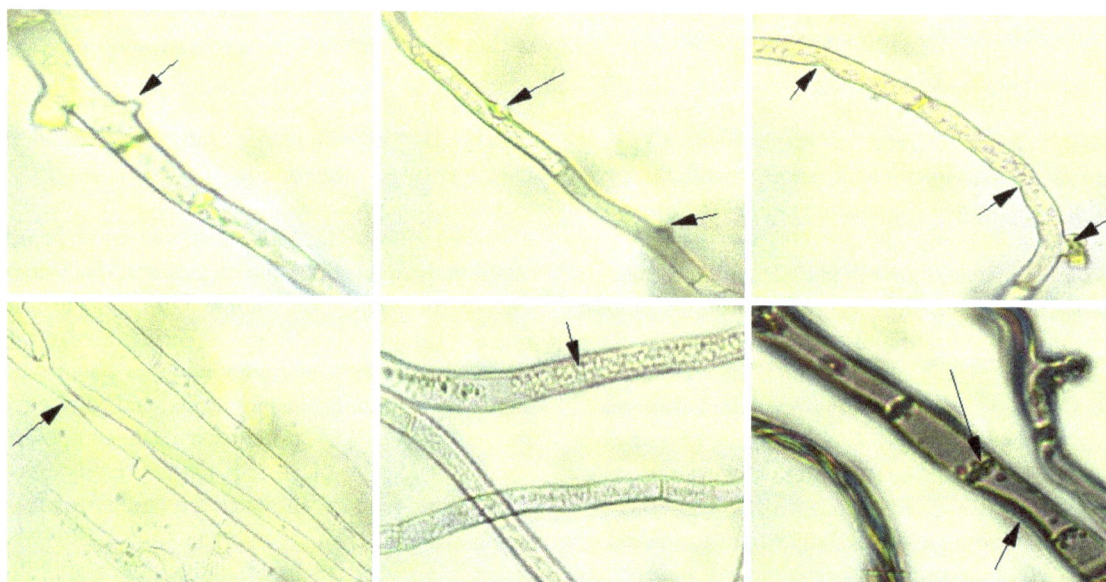

Fig. 7 Effect of Cu@Chit-1 and Cu@Chit-2 on hyphal morphology of *R. solani* grown in PDA media. *Small arrows mark* hyphal fragmentation and cytoplasmic coagulations. *Large arrows mark* cell wall disruption and necrosis on the fungal hyphae

Table 5 Effect of different concentrations of Cu@Chit NCs against scolorotia formation of two sclerotium-forming plant pathogenic fungi

NCs concentrations	Sclerotium-forming fungi			
	R. solani		S. rolfsii	
	Cu@Chit-1	Cu@Chit-2	Cu@Chit-1	Cu@Chit-2
100 mg/L	12.663	21.997	10.337	8.670
Control	19.100	18.997	19.100	18.997
LSD ($p < 0.05$)	4.228	4.854	4.228	4.854

Fig. 8 Effect of different concentrations of Cu@Chit-1 nanocomposites on sclerotia formation of *S. rolfsii* (**a**), and *R. solani* AG-4 incubated at 25 ± 2 °C for 18 days (**b**)

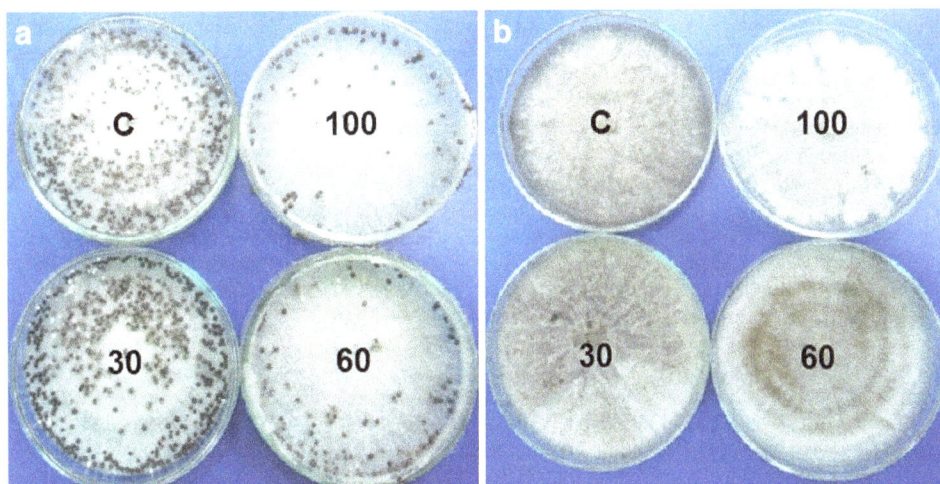

condition. The antifungal activity of the Cu@Chit nanocomposites against *S. rolfsii* appears to be fungistatic rather than, fungicidal because growth of the fungus transferred from NCs-challenged plates to non-challenged condition was the same to growth of fungus started from untreated plates. In the near future, production of nanoenabled fungicides with greater solubility, more stable dispersal, decreased persistence, and greater target specificity is clearly the most successful strategy for plant disease control. To the best of our knowledge, this is the first report concerning the evaluation of antifungal activity of Chit–Cu NCs developed by physico-chemical methods.

Acknowledegment The work is a part of the project Hybrid metal–chitosan nanocomposites: ecologically favorable synthesis, structure and plant pathogens protective properties supported by the Science and Technology Development Fund (STDF), Egypt (STDF- RFBR program) [Grant No. 13791]. Also, this work was partially funded by Russian Foundation for Basic Research grant (RFBR-15-53-61030).

References

1. Abd-Elsalam, K.A., Alghuthaymi, M.A.: Nanobiofungicides: are they the next-generation of fungicides? J. Nanotech. Mater. Sci. **2**, 1–3 (2015)
2. Abd-Elsalam, K.A., Khokhlov, A.R.: Eugenol oil nanoemulsion: antifungal activity against *Fusarium oxysporum* f. sp. *vasinfectum* and phytotoxicity on cottonseeds. Appl. Nanosci. **5**, 255–265 (2015)
3. Xu, J., Zhao, X., Han, X., Du, Y.: Antifungal activity of oligo-chitosan against *Phytophthora capsici* and other plant pathogenic fungi in vitro. Pest Biochem. Physiol. **87**, 220–228 (2007)
4. Reddy, M.V., Arul, J., Angers, P., Couture, L.: Chitosan treatment of wheat seeds induces resistance to *Fusarium graminearun* and improves seed quality. J. Agric. Food Chem. **47**, 1208–1216 (1999)
5. Muzzarelli, R.A.A., Muzzarelli, C., Tarsi, R., Miliani, M., Gabbanelli, F., Cartolari, M.: Fungistatic activity of modified chitosans against *Saprolegnia parasitica*. Biomacromol **2**, 165–169 (2001)
6. El Hadrami, A., Adam, L.R., El Hadrami, I., Daayf, F.: Chitosan in plant protection. Mar. Drugs. **8**, 968–987 (2010)
7. Liu, H., Bao, J., Du, Y., Zhou, X., Kennedy, J.F.: Effect of ultrasonic treatment on the biochemphysical properties of chitosan. Carbohydr. Polym. **64**, 553–559 (2005)
8. Kruk, T., Szczepanowicz, K., Stefańska, J., Socha, R.P., Warszyński, P.: Synthesis and antimicrobial activity of monodisperse copper nanoparticles. Colloids Surf. B **128**, 17–22 (2015)
9. Kanhed, P., Birla, S., Gaikwad, S., Gade, A., Seabra, A.B., Rubilar, O., Duran, N., Rai, M.: In vitro antifungal efficacy of copper nanoparticles against selected crop pathogenic fungi. Mater. Lett. **115**, 13–17 (2014)
10. Saharan, V., Mehrotra, A., Khatik, R., Rawal, P., Sharma, S.S., Pal, A.: Synthesis of chitosan based nanoparticles and their in vitro evaluation against phytopathogenic fungi. Int. J. Biol. Macromol. **62**, 677–683 (2013)

11. Arjunan, N., Singaravelu, C.M., Kulanthaivel, J., Kandasamy, J.: A potential photocatalytic, antimicrobial and anticancer activity of chitosan-copper nanocomposite. Int. J. Biol. Macromol. (2017). doi:10.1016/j.ijbiomac.2017.03.006

12. Saharan, V., Sharma, G., Yadav, M., Choudhary, M.K., Sharma, S.S., Pal, A., Raliya, R., Biswas, P.: Synthesis and in vitro antifungal efficacy of Cu–chitosan nanoparticles against pathogenic fungi of tomato. Int. J. Biol. Macromol. **75**, 346–353 (2015)

13. Saharan, V., Kumaraswamy, R.V., Choudhary, R.C., Kumari, S., Pal, A., Raliya, R., Biswas, P.: Cu-chitosan nanoparticles mediated sustainable approach to enhance seedling growth in maize by mobilizing reserved food. J. Agric. Food Chem. (2016). doi:10.1021/acs.jafc.6b02239

14. Belyakova, O.A., Shulenina, A.V., Zubavichus, Y.V., Veligzhanin, A.A., Naumkin, A.V., Vasil'kov, A.Y.: Diagnostics of gold-containing surgical-dressing materials with X-ray and synchrotron radiation. J. Surf. Invest. X-Ray, Synchrotron Neutron Techn. **7**, 509–514 (2013)

15. Rubina, M.S., Kamitov, E.E., Zubavichus, Y.V., Peters, G.S., Naumkin, A.V., Suzer, S., Vasil'kov, A.Y.: Collagen-Chitosan Scaffold modifying with Au and Ag nanopatrticles: synthesis, structure and properties. Appl. Surf. Sci. **366**, 365–371 (2016)

16. Nikitin, L.N., Vasil'kov, A.Y., Banchero, M., Manna, L., Naumkin, A.V., Podshibikhin, V.L., Abramchuk, S.S., Buzin, M.I., Korlyukov, A.A., Khokhlov, A.R.: Composite materials for medical purposes based on polyvinylpyrrolidone modified with ketoprofen and silver nanoparticles. Russ. J. Phys. Chem. A **85**, 1190–1195 (2011)

17. Vasil'kov, A.Y., Rubina, M.S., Naumkin, A.V., Zubavichus, Y.V., Belyakova, O.A., Maksimov, Y.V., Imshennik, V.K.: Metal-containing systems based on chitosan and a collagen-chitosan composite. Russ. Chem. Bull. **64**, 1663–1670 (2015)

18. Konarev, P.V., Volkov, V.V., Sokolova, A.V., Koch, M.H.J., Svergun, D.I.: PRIMUS: a Windows PC-based system for small-angle scattering data analysis. J. Appl. Crystallogr. **36**, 1277–1282 (2003)

19. Svergun, D.I.: Determination of the regularization parameter in indirect-transform methods using perceptual criteria. J. Appl. Cryst. **25**, 95–503 (1992)

20. Tatsadjieu, L., Dongmo Jazet, P.M., Ngassoum, M.B., Etoa, F.X., Mbofung, C.M.F.: Investigations on the essential oil of *Lippia rugosa* from Cameroun for its potential use as antifungal agent against *Aspergillus flavus* Link ex Fries. Food Cont. **20**, 161–166 (2009)

21. Abd-El-Khair, H., El-Gamal Nadia, G.: Effects of aqueous extracts of some plant species against *Fusarium solani* and *Rhizoctonia solani* in *Phaseolus vulgaris* plants. Arch. Phytopathol. Plant Protect. **44**, 1–16 (2011)

22. Woodhall, J.W., Lees, A.K., Edwards, S.G., Jenkinson, P.: Infection of potato by *Rhizoctonia solani*: effect of anastomosis group. Plant. Pathol. **57**, 897–905 (2008)

23. Hanawalt, J.D., Rinn, H.W., Frevel, L.K.: Chemical analysis by X-ray diffraction. Ind. Eng. Chem. Anal. Ed. **10**, 457–512 (1938)

24. Naumkin, A.V., Kraut-Vass, A., Gaarenstroom, S.W., Powell, C.J.: NIST X-ray Photoelectron Spectroscopy Database, Version 4.1 (National Institute of Standards and Technology, Gaithersburg; http://srdata.nist.gov/xps/. (2012)

25. Moretti, G.: The Wagner plot and the Auger parameter as tools to separate initial-and final-state contributions in X-ray photoemission spectroscopy. Surf. Sci. **618**, 3–11 (2013)

26. Brunel, F., Gueddari, N.E., Moerschbacher, B.M.: Complexation of copper (II) with chitosan nanogels: toward control of microbial growth. Carbohydr. Polym. **92**, 1348–1356 (2013)

27. Zhang, J., Liu, Y., Li, Q., Zhang, X., Shang, J.K.: Antifungal activity and mechanism of palladium-modified nitrogen doped titanium oxide photocatalyst on agricultural pathogenic fungi *Fusarium graminearum*. ACS Appl. Mater. Interfaces. **5**, 10953–10959 (2013). doi:10.1021/am4031196

28. Narayanan, K.B., Park, H.H.: Antifungal activity of silver nanoparticles synthesized using turnip leaf extract (*Brassica rapa* L.) against wood rotting pathogens. Eur. J. Plant Pathol. (2014). doi:10.1007/s10658-014-0399-4

29. Kim, K.J., Sung, W.S., Suh, B.K., Moon, S.K., Choi, J.S., Kim, J.G., Lee, D.G.: Antifungal activity and mode of action of silver nanoparticles on *Candida albicans*. Biometals **22**, 235–242 (2009)

30. Ouda, S.M.: Antifungal activity of silver and copper nanoparticles on two plant pathogens, *Alternaria alternate* and *Botrytis cinerea*. Res. J. Microbiol. **9**(1), 34–42 (2014). doi:10.3923/jm.2014.34.42

17

Synthesis, characterization, and application of nickel oxide/CNT nanocomposites to remove Pb^{2+} from aqueous solution

T. Navaei Diva[1] · K. Zare[1] · F. Taleshi[2] · M. Yousefi[1]

Abstract In this study, the efficiency of nickel oxide/carbon nanotube (NiO/CNT) nanocomposite to remove Pb^{2+} from aqueous solution is investigated. NiO/CNT nanocomposite was prepared using the direct coprecipitation method in an aqueous media in the presence of CNTs. Samples were characterized using simultaneous thermal analysis (STA), X-ray diffraction (XRD), filed emission scanning electron microscopy (FESEM), and Brunauer–Emmett–Teller (BET). To optimize the adsorption of Pb^{2+} ions on NiO/CNT nanocomposite, the effects of different parameters including pH, contact time, initial concentration of Pb^{2+}, and adsorbent mass—were also investigated. The optimum Pb^{2+} removal efficiency on NiO/CNT nanocomposite is achieved under experimental conditions of pH 7, contact time of 10 min, initial Pb^{2+} concentration of 20 ppm, and adsorbent mass of 0.1 g. The experimental data showed that the Pb^{2+} ions adsorption of NiO/CNT nanocomposite was through a Freundlich isotherm model rather than a Langmuir model. The kinetic data of adsorption of Pb^{2+} ions on the adsorbent was perfectly shown by a pseudo-second-order equation, to indicate their chemical adsorption. Thermodynamic parameters such as $\Delta G°$, $\Delta H°$, and $\Delta S°$ were also measured; the obtained values showed that the adsorption was basically spontaneous and endothermic.

Keywords Removal · Adsorption · Carbon nanotubes · Composite · Heavy metals

Introduction

Water, which is one of the main elements in the environment, is highly exposed to contamination. Among water pollutants, heavy metal cations, including Pb^{2+}, Cd^{2+}, Cr^{3+}, Cr^{6+}, Co^{2+}, Cu^{2+}, Fe^{3+}, Ni^{2+} and Zn^{2+}, are extremely toxic and non-biodegradable. These are commonly discharged into water sources and environment via either natural or industrial wastes. Today, elimination of water sources from such cations is among the main challenges faced by researchers and environmentalists. The accumulation of these metal cations in living organisms causes many physiological disorders [1]. The permitted level of Pb^{2+} in drinking water is 0.01 mg L^{-1} [2]. The exposure of human body organs to high concentrations of Pb^{2+} can result in anaemia, mental disorder, and renal and liver diseases.

Several different methods have been introduced and used for heavy metal removal [3], i.e. chemical precipitation, ion exchange, reverse osmosis, membrane-based processes, evaporation, solvent extraction, and adsorption. The efficiency of some of these methods is reduced due to major drawbacks including low removal efficiency and causing side effects that lead to new environmental issues.

In recent years, metal oxide nanoparticles especially metal oxide/CNT nanocomposites have attracted a good deal of persistent interest because of their unique chemical, physical, electrical, and thermal properties [4–6]. These materials are widely used in several areas, including chemistry, physics, material science, biology, medicine, and environment [7, 8].

✉ K. Zare
k-zare@sbu.ac.ir

1 Department of Chemistry, Science and Research Branch, Islamic Azad University, Tehran, Iran

2 Department of Physics, Qaemshahr Branch, Islamic Azad University, Qaemshahr, Iran

Metal oxide nanoparticle and metal oxide/CNT nanocomposite adsorbents have been applied for the removal of heavy metals from aqueous solutions. These adsorbents are economically more affordable and also more environment-friendly [9–12].

The amount of heavy metal uptake is directly associated with the total amount of active sites available on the adsorbent; a decrease in metallic nanoparticle dimensions increases their surface-to-volume ratio and consequently increases the active surface area for adsorption. In this regard, the utilization of appropriate substrate material in the synthesis procedure of metal oxide nanoparticles can prevent agglomeration, decrease the diameter of nanoparticles, and change the cluster-like morphology of nanoparticles to a powdered morphology. Therefore, the obtained nanocomposite with larger specific surface area and higher adsorption capacity can be employed as a suitable adsorbent for heavy metal removal.

Due to high surface-to-volume ratio, CNTs are considered as great substrates for the nucleation and growth of nanoparticles with control over diameter distribution [13–16]. One of the main challenges encountered when CNTs are used as the support material for nanoparticle synthesis is the hydrophobicity of their surface. In this regard, surface functionalization of CNTs for the generation of covalent bonds and further attachment of oxide nanoparticles seems to be of great significance [17–19]. The functionalization of CNTs is commonly performed using the chemical oxidation process introducing functional groups such as –COOH, C–O, C=O, and –OH on the surface of CNTs; such groups act as active surface sites for metal attachment [20, 21]. NiO nanostructure can be synthesized through various methods such as co-precipitation [22], sol gel [23], hydrothermal [24], spray pyrolysis method [25], and chemical precipitation [26]. In this study, NiO/CNT nanocomposite was synthesized by chemical precipitation method, which was simple and cost efficient.

NiO/CNT nanocomposite was applied as an adsorbent to remove Pb^{2+} from aqueous solution. However, pH, contact time, adsorbent dosage, and initial concentration of Pb^{2+} were the parameters, the effects of which on adsorption uptake have been investigated.

Experiment

Materials

Lead nitrate [$Pb(NO_3)_2$, 99.5%], nickel chloride hexa-hydrate ($NiCl_2 \cdot 6H_2O$, 98%), sodium hydroxide (NaOH, 99%), CNTs (MWCNTs, US4309, $20 < d < 30$ nm,

SSA = 264 m^2 g^{-1}, 95%), and sulphuric (95–97%), nitric (60%), and hydrochloric (37%) acids were applied without further purification.

To synthesize NiO/CNT nanocomposite, the surface of CNTs was functionalized as follows: initially, the desired amount of CNTs was added to the mixture of sulphuric/nitric/hydrochloric acids (6 M) and ultrasonicated for 30 min. The obtained mixture was stirred for 2 h at temperature of 80 °C and was then filtered and washed with deionized water until the pH reached 7. Finally, the functionalized CNTs were dried in an oven at 120 °C.

For NiO/CNT nanocomposite preparation with 1:1 weight ratio, 3.2 g of $NiCl_2 \cdot 6H_2O$ was dissolved in 50 ml of deionized water containing 1 g of functionalized CNTs. This was then ultrasonicated for 10 min and magnetically stirred for 15 min at 80 °C. The addition of 1.1 g of NaOH to the mixture and stirring for 30 min completed the precipitation of $Ni(OH)_2$/CNTs nanocomposite. The obtained black mixture was filtered, washed with absolute ethanol and deionized water, dried at 120 °C for 24 h, and finally calcinated at 300 °C for 2 h in static air.

Methods of analysis

To determine the optimum calcination temperature of NiO/CNT nanocomposite powder, simultaneous thermal analysis (STA 1500) was applied in a static air atmospheric (10 °C min^{-1}). The residual concentration of Pb^{2+} ions in aqueous media was analysed with the aid of Buck Scientifics 210 VGP flame atomic adsorption spectroscopy. The crystallinity of samples was determined using Xpertpr Pananalytical X-ray diffraction apparatus (Holland) with $Cu(K_\alpha)$ source and wavelength of $\lambda = 1.5405$ Å. The morphology of powders was recorded using Field Emission Scanning Electron Microscopy on a Mira3-XMU system. The BET specific surface area and porosity were determined using nitrogen adsorption–desorption porosimetry (77 K) by a porosimeter (Bel Japan, Inc.).

Adsorption studies

In this section, 1000 ppm of Pb^{2+} stock solution was initially provided through dissolving the desired volume of ($Pb(NO_3)_2$) in deionized water; the corresponding concentrations were then obtained from the dilution of the stock solution. For batch adsorption experiments, 0.1 g of NiO/CNT was added to 50 ml of 20 ppm solution of Pb^{2+} and then stirred. After adsorption, the nanocomposite was

Fig. 1 STA curves of Ni(OH)$_2$/CNT in air

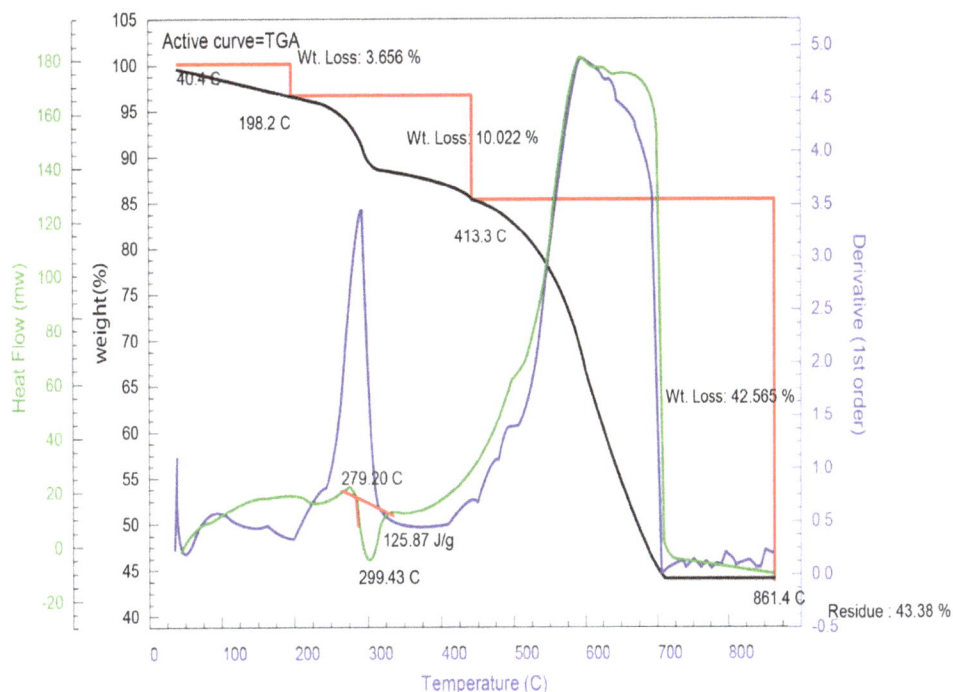

taken from the solution and the residual Pb^{2+} concentration was measured using flame atomic adsorption spectroscopy. The effects of the different parameters namely pH, initial Pb^{2+} concentration, adsorbent dosage, and contact time on the amount of adsorption were investigated through the removal percentage (R) of Pb^{2+} by the equation mentioned below:

$$\%R = \frac{C_0 - C_e}{C_0} \times 100,$$

here, C_0 and C_e are the initial and equilibrium concentrations of Pb^{2+} (mg L^{-1}).

Result and discussion

NiO/CNT characterization

To study the temperature-dependent behaviour of nanocomposite, STA was performed on Ni(OH)$_2$/CNT synthesized precursor in static air. According to the obtained spectra (Fig. 1), in the temperature range of 40–800 °C, three endothermic reactions occur in Ni(OH)$_2$/CNT. In the range of 40–400 °C, two stages of weight loss in the sample are observed. These can be attributed to two reactions in the initial stage at 40–200 °C, attributing to

Fig. 2 FESEM of **a** pure NiO and **b** NiO/CNT nanocomposite

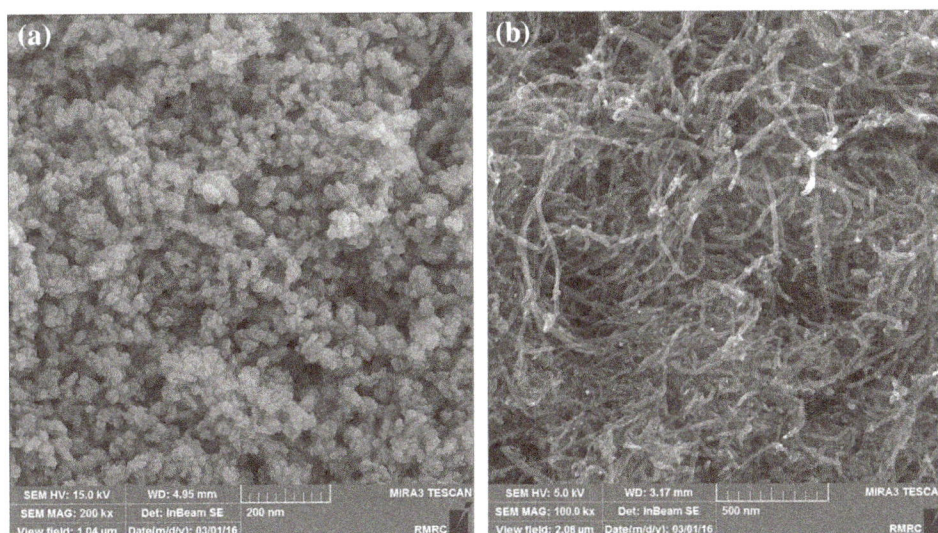

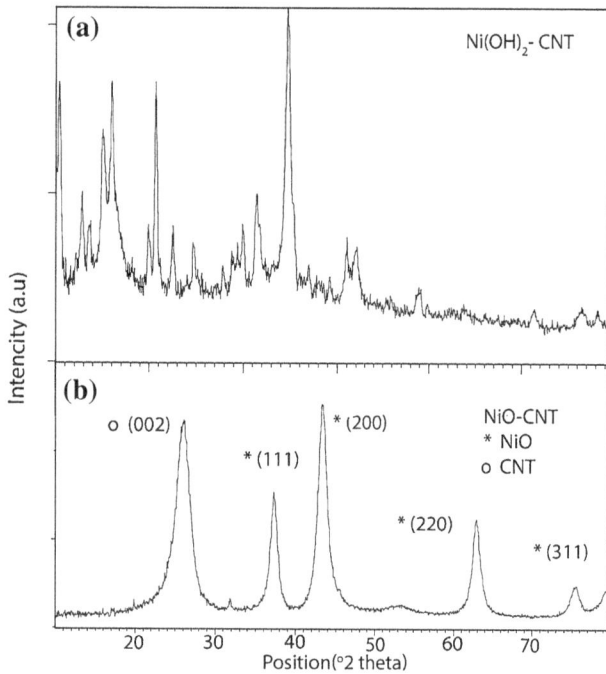

Fig. 3 XRD pattern of a Ni(OH)$_2$/CNT and b NiO/CNT nanocomposite

Fig. 4 BET plot of NiO/CNT nanocomposite

The slow decrease in weight at a temperature above 300 °C is related to the removal of small amounts of hydroxyl-groups remaining during the development of the NiO phase [27].

The weight percentage of the sample at 400 °C was 86%, which is related to the NiO/CNT nanocomposite. At 800 °C, the weight percentage of the sample reached 43%, due to the oxidation of CNTs into CO$_2$ [28]. Following the specifications of STA, 300 °C was considered as the reaction temperature in the study.

The morphologies of NiO/CNT and pure NiO were analysed using FESEM images, as shown in Fig. 2. It can be seen in Fig. 2a that NiO nano-crystallites are aggregated, forming clusters with larger grains. The formation of this agglomerated structure has unfavourable impacts on

surface moisture evaporation and the second one at 200–400 °C is the result of water release and the formation of NiO chemical structure with the following reactions:

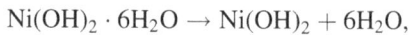

$$Ni(OH)_2 \cdot 6H_2O \rightarrow Ni(OH)_2 + 6H_2O,$$

$$Ni(OH)_2 \rightarrow NiO + H_2O.$$

Fig. 5 a Effect of pH, b contact time, c adsorbent mass and d initial concentration on Pb^{2+} removal by NiO/CNT nanocomposite

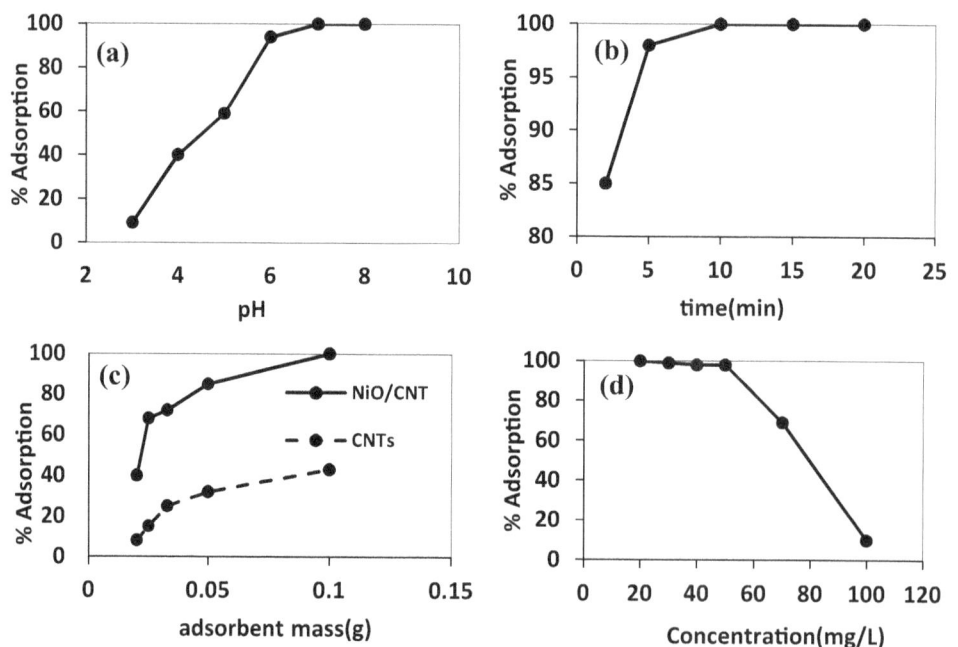

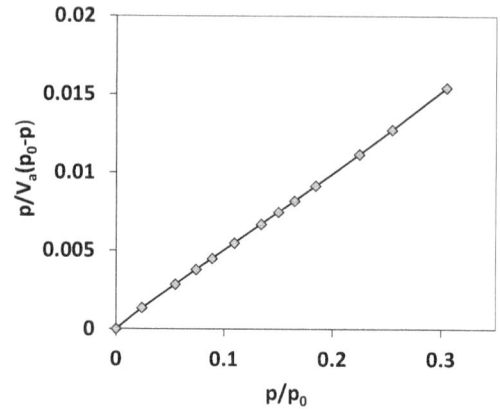

Fig. 6 **a** Langmuir,
b Freundlich adsorption
isotherm, **c** pseudo-first-order,
d pseudo-second-order kinetic
models and **e** Van't Hoff plot
for Pb^{2+} removal on NiO/CNT
nanocomposite

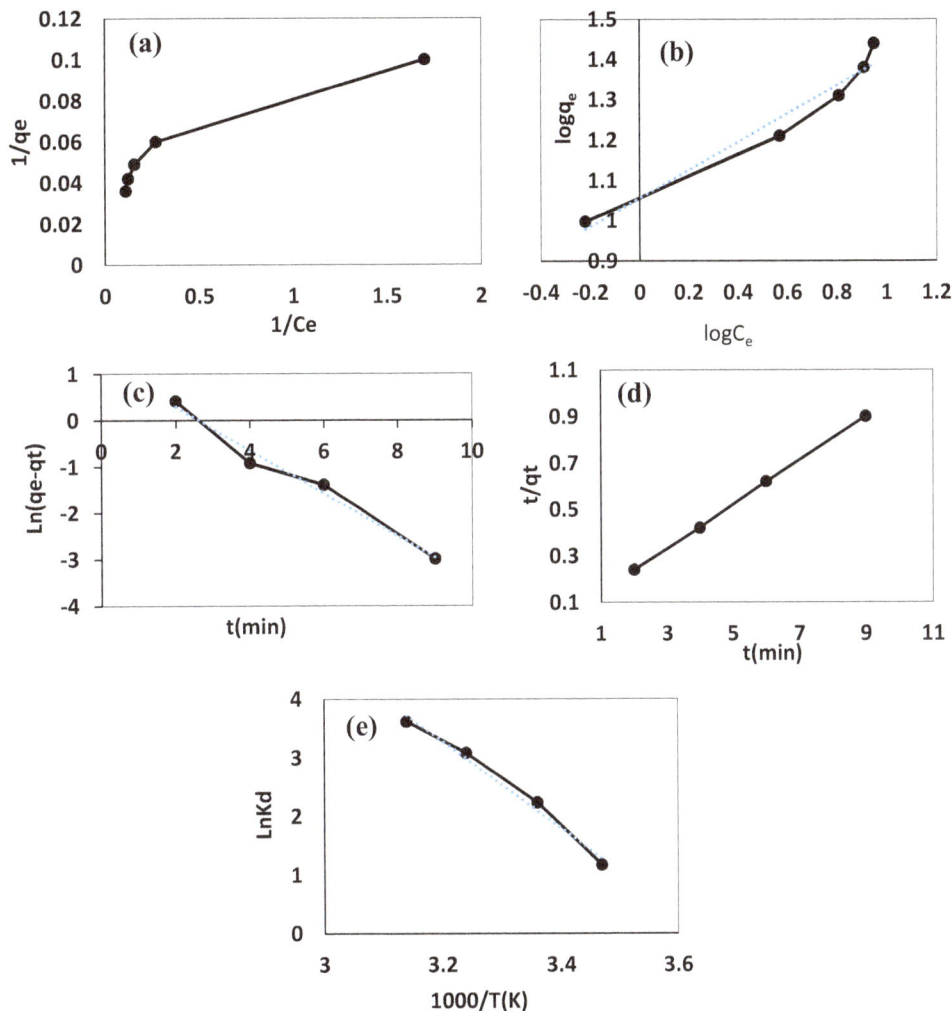

the physical and chemical features of nanoparticles. The utilization of CNTs as the support material for the growth of nanoparticles reduces the agglomeration and changes the powder morphology from a cluster-like to a filamentous structure (Fig. 2b). It is observed that the nanoparticles on the surface of CNTs possess smaller dimensions in comparison with pure NiO nanoparticles [27].

Figure 3 indicates the X-ray diffraction pattern of Ni(OH)$_2$/CNT and NiO/CNT nanocomposites. Characteristic peaks at $2\theta = 37.3°$, $43.4°$, $62.9°$ and $75.4°$ are assigned to (111), (200), (220), and (311) crystallographic orientations in NiO nanoparticles, respectively (Fig. 3a), the strong diffraction peak at $2\theta = 26°$ reflected from (002) lattice plane in CNTs [29]. The peaks of Ni(OH)$_2$/CNT cannot be seen at NiO/CNT, which is due to the crystallization and change in the structure of crystal. The size of NiO/CNT nanoparticles was calculated using the Scherrer equation $D = \frac{k\lambda}{\beta \cos\theta}$. The main reflection peak of the XRD pattern at $2\theta = 43°$ can be attributed to (200) plane. The size of NiO/CNT nano-crystallites was 18.5 nm.

N_2 adsorption–desorption isotherms was utilized to assess the specific surface area, mean pore diameter and the total volume of NiO/CNT nanocomposite. The specific surface area of NiO/CNT nanocomposite was measured by the Brunauer–Emmett–Teller (BET) method (Fig. 4). The specific surface area, mean pore diameter and the total pore volume of NiO/CNT nanocomposite were determined to be 90 m^2 g^{-1}, 4.16 nm and 9.26×10^{-2} cm^3 g^{-1}, respectively.

Effect of pH

The pH is among the most remarkable parameters that should be controlled in metal ions adsorption process of aqueous solutions. To evaluate the effect of pH, the initial pH varied from 3 to 8 in the solution. It was observed (Fig. 5a) that the rate of removal of Pb^{2+} increased rapidly, from about 9–94% as pH increased from 3 to 6, and then remained constant following the increase of pH value from 7 to 8. Based on the authors' best knowledge, lead species are available in Pb^{2+} and $Pb(OH)^+$ forms at pH 3–8 [30].

Table 1 Langmuir, Freunlich parameters and correlation coefficients of Pb^{2+} adsorption onto NiO/CNT

Langmuir	$\frac{1}{q_e} = \left(\frac{1}{q_m K_L}\right) + \left(\frac{1}{C_e}\right) + \frac{1}{q_m}$		Freundlich	$\log q_e = \log K_F + \frac{1}{n}\log C_e$	
q_m (mg g^{-1})	K_L (L mg^{-1})	R^2	K_F (L mg^{-1})	$1/n$	R^2
24.63	1.14	0.9279	11.39	0.3493	0.9502

Table 2 The equilibrium capacities of Pb(II) on various adsorbents

Adsorbents	q_{max} (mg g^{-1})	References
MWCNTs	1	[35]
MWCNTs	4	[36]
Activated carbon	13.05	[37]
MWCNTs	15.6	[35]
Oxidized MWCNTs	2.06	[38]
Oxidized N-doped MWCNTs	29	[39]
MWCNT-f	14.56	[40]
Tiol/CNT	65.52	[41]
SiO$_2$/MWCNT	13	[42]
Iron oxid/MWCNT	12	[43]
Al$_2$O$_3$/CNTs	67.5	[44]
PAAM/MWCNTs	29.71	[45]
Zeolites: chabazite	6	[46]
NiO/CNTs	24.63	This study

When the pH value is low (<6), the predominant lead specie is Pb^{2+} and Pb^{2+} is removed by sorption reaction. Therefore, the lower adsorption percentage of Pb^{2+} on NiO/CNT at lower pH values partly depends on the competition between H^+ and Pb^{2+} on the surface sites Pb^{2+} removal rate, which is maximized in pH values 6 to 8 and then remains constant. In this range, the main species are Pb^{2+} and $Pb(OH)^+$; hence, Pb^{2+} removal is regulated by the sorption of Pb^{2+} and $Pb(OH)^+$. According to the results of the current study, to eliminate Pb^{2+} from the solution by NiO/CNT, pH 7 is the most optimum value for the employed system.

Effect of contact time

The contact time and its effect on Pb^{2+} elimination rate using NiO/CNT is depicted in Fig. 5b. The highest amount of adsorption approximately 100%, was achieved after 10 min of adsorption at pH 7. Therefore, a period of 10 min was considered as the optimum contact time for Pb^{2+} adsorption on the NiO/CNT nanocomposite.

Effect of adsorbent mass

Figure 5c illustrates the effect of adsorbent mass on the removal rate for a 20 ppm solution of Pb^{2+} by NiO/CNT

nanocomposite and CNTs. In both adsorbents, the percentage of adsorbed lead increased as the mass of adsorbent was increased over the range of 0.02–0.1 g. A removal rate of approximately 100% was obtained when 0.1 g of NiO/CNT nanocomposite was drowned in 50 mL of the 20 ppm Pb^{2+} solution, compared to 43% removal when 0.1 g of CNTs was used. Based on the given results, it is obvious that the mechanism of nanocomposite adsorption towards Pb^{2+} may be attributed to three factors: the van der Waals interaction between the hexagonally arrayed carbon atoms in the graphite sheets of the CNTs and the positively charged lead ions, the electrostatic attraction between the negatively charged adsorbent surface of the CNT and Pb^{2+} cations, and the electrostatic attraction between Pb^{2+} and the electron pairs on the oxygen atoms of nickel oxide [30].

Effect of Pb^{2+} initial concentration

Figure 5d indicates the variation in removal rate in the initial concentration of Pb^{2+}. It is evident that the initial concentration increase of Pb^{2+} from 20 to 100 ppm reduces the removal efficiency from almost 100–10%. At lower initial concentrations of Pb^{2+}, an adequate number of active sites are available for adsorption, leading to a higher adsorption capacity. It is also noteworthy that at higher initial concentrations, the number of metal ions is higher compared with active adsorption sites; therefore, the removal efficiency decreases [31].

Adsorption isotherms

Adsorption isotherms represent the amount of adsorbed metal ion per unit mass of the adsorbent. Among all the isotherm models, the Langmuir and Freundlich models are the most popular ones describing adsorption in these systems [32–34]. The equilibrium adsorption capacity (q_e) is calculated using:

$$q_e = \frac{C_0 - C_e}{m}V,$$

here q_e is the adsorbed Pb^{2+} amount (mg g^{-1}), m is the composite mass (g), and V is the solution volume (L).

Figure 6a, b, respectively, illustrate Langmuir and Freundlich isotherms for Pb^{2+} adsorption on NiO/CNT nanocomposite with the obtained parameters summarized in Table 1. According to the data shown in Table 1, both the Langmuir and Freundlich models are appropriate for

Table 3 Adsorption kinetic parameters of Pb^{2+} adsorption onto NiO/CNT

Psedo-first-order $Ln(q_e - q_t) = Lnq_e - k_1 t$			Psedo-second-order $\frac{t}{q_t} = \frac{1}{k_2 q_e^2} + \frac{1}{q_e} t$			Experimental data
k_1 (min^{-1})	$q_{e,cal}$ (g mg^{-1} min^{-1})	R^2	k_2 (g mg^{-1} min^{-1})	$q_{e,cal}$ (mg g^{-1})	R^2	$q_{e,exp}$ (mg g^{-1})
0.46	16.54	0.9787	0.189	10.55	0.9997	10

Table 4 Values of thermodynamic parameters of adsorption of Pb^{2+} onto NiO/CNT

$LnK_d = \frac{\Delta S^\circ}{R} - \frac{\Delta H^\circ}{RT}$		ΔG° (temperature, K)			
ΔS°	ΔH°	$\Delta G^\circ = -RT \ln K_d$			$K_d = \frac{q_e}{C_e}$
224.26	61.55	2.80 (288)	−5.52 (298)	−7.89 (308)	−9.57 (318)

interpreting the obtained experimental data; however, the Freundlich model, with a higher correlation coefficient (R^2), showed higher compatibility with the data, compared with the Langmuir model.

To get the synthesized absorbent efficiency approved, its maximum amount of absorption was compared with other adsorbents. As shown in Table 2, the maximum absorption capacity obtained in this study is significant compared to other adsorbents [35–46], which indicates the efficiency of the synthetic adsorbent.

Adsorption kinetics

One of the most important features of designing an adsorption system is the prediction of the adsorption rate, i.e., how fast the adsorption is taking place, which is controlled by adsorption kinetics. The adsorption kinetics depends on the adsorbent physical and chemical properties; it also affects the adsorption mechanism. To evaluate the adsorption mechanism, adsorption kinetic constants, Lagergren pseudo first-order and Ho, pseudo second-order rate equations can be used [47].

The obtained results are summarized in Fig. 6c, d, and Table 3. According to Table 3, the pseudo second-order model, with higher values of correlation coefficient (R^2) and closer values of calculated adsorption capacity ($q_{e,cal}$) to experimental one ($q_{e,exp}$) is a better model to describe the adsorption capacity of ions on NiO/CNT nanocomposite. This indicates that the adsorption of Pb^{2+} onto NiO/CNT may be the chemisorption involving valence forces through the exchange or sharing of electrons between adsorbate and adsorbent [48].

Adsorption thermodynamics

Another feature of any adsorption process with high significance is the determination of thermodynamic parameters, including changes in Gibbs free energy (ΔG°, KJ mol^{-1}), enthalpy (ΔH°, KJ mol^{-1}), and entropy (ΔS°,

J mol^{-1} K^{-1}). Adsorption thermodynamics were evaluated at different temperatures (288, 298, 308, and 318 K). The results are depicted in Fig. 6e. To measure the thermodynamic parameters and K_d values, Van't Hoff equation was applied [49].

The results of calculation are shown in Table 4. The negative values of ΔG° indicate that the adsorption was a spontaneous reaction. The ΔH° and ΔS° positive values are attributed to the endothermic nature of adsorption and the increase in the randomness at the solid/liquid interface during the adsorption of Pb^{2+} on NiO/CNT nanocomposite, respectively.

Conclusion

In this study, NiO/CNT nanocomposite was utilized as an adsorbent to remove Pb^{2+} from an aqueous solution. The NiO/CNT nanocomposite powder was prepared using direct coprecipitation process in the aqueous media. The results indicate that the optimum efficiency for Pb^{2+} removal is achieved under experimental conditions of pH 7, contact time of 10 min, and the initial Pb^{2+} concentration of 20 ppm. Moreover, increasing the initial concentration of Pb^{2+} and reducing the adsorbent dosage diminishes the removal efficiency. Therefore, NiO/CNT nanocomposite can be considered as a useful adsorbent for the elimination of aqueous solutions from Pb^{2+}. In addition, results obtained from Pb^{2+} adsorption on NiO/CNT nanocomposite follow the pseudo second-order rate equation and Freundlich isotherm.

References

1. Bedelean, H., Maicaneanu, A., Burca, S., Stanca, M.: Removal of heavy metal ions from wastewaters using natural clays. J. Clay Miner. **44**, 487–495 (2009)
2. Chand, P., Pakade, Y.B.: Synthesis and characterization of

hydroxyapatite nanoparticles impregnated on apple pomace to enhanced adsorption of Pb(II), Cd(II) and Ni(II) ions from aqueous solution. Environ. Sci. Pollut. Res. **22**, 10919–10929 (2015)

3. Shafaei, A., Ashtiani, F.Z., Kaghazchi, T.: Equilibrium studies of the sorption of Hg(II) ions onto chitosan. Chem. Eng. J. **133**(1–3), 311–316 (2007)

4. Kumar, R., Singh, R.K., Dubey, P.K., Singh, D.P., Yadav, R.M., Tiwari, R.S.: Freestanding 3D graphene–nickel encapsulated nitrogen-rich aligned bamboo like carbon nanotubes for high-performance supercapacitors with robust cycle stability. Adv. Mater. Interfaces **2**, 1500191 (2015)

5. Kumar, R., Singh, R.K., Singh, D.P., Vaz, A.R., Yadav, R.R., Rout, C.S., Moshkalev, S.A.: Synthesis of self-assembled and hierarchical palladium-CNTs-reduced graphene oxide composites for enhanced field emission properties. Mater. Des. **122**, 110–117 (2017)

6. Kumar, R., Singh, R.K., Dubey, P.K., Singh, D.P., Yadav, R.M.: Self-assembled Hierarchical formation of conjugated 3D cobalt oxide nanobeads–CNTs–graphene nanostructure using microwave for high performance supercapacitor electrode. ACS Appl. Mater. Interfaces **7**, 15042–15051 (2015)

7. Zare, K., Gupta, V.K., Moradi, O., Makhlouf, A.S.H., Sillanpää, M., Nadagouda, M.N., Sadegh, H., Shahryari-ghoshekandi, R., Pal, A., Wang, Z., Tyagi, I., Kazemi, M.: A comparative study on the basis of adsorption capacity between CNTs and activated carbon as adsorbents for removal of noxious synthetic dyes: a review. J. Nanostruct. Chem. **5**(2), 227–236 (2015)

8. Moradi, O., Fakhri, A., Adami, S., Adami, S.: Isotherm, thermodynamic, kinetics, and adsorption mechanism studies of Ethidium bromide by single-walled carbon nanotube and carboxylate group functionalized single-walled carbon nanotube. J. Colloid Interface Sci. **395**, 224–229 (2013)

9. Abd El Fatah, Ossman, M.E.: Removal of heavy metal by nickel oxide nano powder. Int. J. Environ. Res. **8**(3), 741–750 (2014)

10. Coston, J.A., Fuller, C.C., Davis, J.A.: Pb^{2+} and Zn^{2+} adsorption by a natural aluminum-and iron-bearing surface coating on an aquifer sand. Geochim. Cosmochim. Acta **59**, 3535–3547 (1995)

11. Agrawal, A., Sahu, K.K.: Kinetic and isotherm studies of cadmium adsorption on manganese nodule residue. J. Hazard. Mater. **137**, 915–924 (2006)

12. Srivastava, N.K., Jha, M.K., Sreekrishnan, T.R.: Removal of Cr(VI) from waste water using NiO nanoparticles. Int. Sci. Environ. Technol. **3**(2), 395–402 (2014)

13. Awasthi, S., Awasthi, K., Kumar, R., Srivastava, O.N.: Functionalization effects on the electrical properties of multi-walled carbon nanotube-polyacrylamide composites. J. Nanosci. Nanotechnol. **9**, 5455–5460 (2009)

14. Kumar, R., Singh, R.K., Ghosh, A.K., Sen, R., Srivastava, S.K., Tiwari, R.S., Srivastava, O.N.: Synthesis of coal-derived single-walled carbon nanotube from coal by varying the ratio of Zr/Ni as bimetallic catalyst. J. Nanopart. Res. **15**, 1406 (2013)

15. Kumar, R., Singh, R.K., Singh, D.P.: Natural and waste hydrocarbon precursors for the synthesis of carbon based nanomaterials: graphene and CNTs. Renew. Sustain. Energy Rev. **58**, 976–1006 (2016)

16. Li, Y.H., Di, Z., Ding, J., Wu, D., Luan, Z., Zhu, Y.: Adsorption thermodynamic, kinetic and desorption studies of Pb^{2+} on carbon nanotubes. Water Res. **39**, 605–609 (2005)

17. Taleshi, F., Hosseini, A.A.: Synthesis of uniform MgO/CNT nanorods by precipitation method. J. Nanostruct. Chem. **3**(1), 1–5 (2012)

18. Jiang, S., Storr Handberg, E., Liu, F., Liao, Y., Wang, H., Li, Z., Song, S.: Effect of doping the nitrogen into carbon nanotubes on the activity of NiO catalysts for the oxidation removal of toluene. Appl. Catal. B **160–161**, 716–721 (2014)

19. Songa, S., Meng, A., Jiang, S., Cheng, B., Jiang, C.: Construction of Z-scheme Ag_2CO_3/N-doped graphene photocatalysts with enhanced visible-light photocatalytic activity by tuning the nitrogen species. Appl. Surf. Sci. **396**, 1368–1374 (2017)

20. Taleshi, F., Porkia, M., Shakeri Chenari, I., Pahlavan, A., Ahmadi Tarsi, M., Zabihi, F., Dehghan-niarostami, N.: Morphology of $CuFe_2O_4$/CNT composites prepared by precipitation, plastics, rubber and composites. Macromol. Eng. **43**, 240–244 (2014)

21. Han, W.Q., Zettl, A.: Coating single-walled carbon nanotubes with tin oxide. Nano Lett. **3**, 681–683 (2003)

22. Mallick, P., Chandanarath Biswal, R., Mishra, N.C.: Structural and magnetic properties of Fe doped NiO. Indian J. Phys. **83**, 517–523 (2009)

23. Mallick, P., Sahoo, C.S., Mishra, N.C.: Structural and optical characterization of NiO nanoparticles synthesized by sol-gel route. AIP Conf. Proc. **1461**, 229–232 (2012)

24. Takami, S., Hayakawa, R., Wakayama, Y., Chikyow, T.: Continuous hydrothermal synthesis of nickel oxide nanoplates and their use as nanoinks for p-type channel material in a bottom-gate field-effect transistor. Nanotechnology **21**, 134009 (2010)

25. Yudin, A., Shatrova, N., Khaydarov, B., Kuznetsov, D., Dzidziguri, E., Issi, J.P.: Synthesis of hollow nanostructured nickel oxide microspheres by ultrasonic spray atomization. J. Aerosol Sci. **98**, 30–40 (2016)

26. Chakrabarty, S., Chatterjee, K.: Synthesis and characterization of nanodimensional NiO semiconductor. J. Phys. Sci. **13**, 245–250 (2008)

27. Lin, P., She, Q., Hong, B., Liu, X., Shi, Y., Shi, Z., Zheng, M., Donga, Q.: The nickel oxide/CNT composites with high capacitance for supercapacitor. J. Electrochem. Soc. **157**(7), A818–A823 (2010)

28. Kumar, R., Singh, R.K., Vaz, A.R., Savu, R., Moshkalev, S.A.: Self-assembled and one-step synthesis of interconnected 3D network of Fe_3O_4/reduced graphene oxide nanosheets hybrid for high-performance supercapacitor electrode. Appl. Mater. Interfaces **9**, 8880–8890 (2017)

29. Taleshi, F.: A new strategy for increasing the yield of carbon nanotubes by the CVD method. Fuller. Nanotubes Carbon Nanostruct. **22**, 921–927 (2014)

30. Gupta, V.K., Agarwal, S., Saleh, T.A.: Synthesis and characterization of alumina-coated carbon nanotubes and their application for lead removal. J. Hazard. Mater. **185**, 17–23 (2011)

31. Rao, M.M., Ramesh, A., Rao, G.P.C., Seshaiah, K.: Removal of copper and cadmium from the aqueous solutions by activated carbon derived from *Ceiba pentandra* hulls. J. Hazard. Mater. B **129**, 123–129 (2006)

32. Tahermansouri, H., Dehghan, Z., Kiani, F.: Phenol adsorption from aqueous solutions by functionalized multiwalled carbon nanotubes with a pyrazoline derivative in the presence of ultrasound. RSC Adv. **5**, 44263–44273 (2015)

33. Langmuir, I.: The adsorption of gases on plane surfaces of glass, mica and platinum. J. Am. Chem. Soc. **40**, 1361–1403 (1918)

34. Freundlich, H.: Over the adsorption in solution. J. Phys. Chem. **57**, 385–470 (1906)

35. Li, Y.H., Wang, S., Wei, J., Zhang, X., Xu, C., Luan, Z., Wei, B.: Lead adsorption on carbon nanotubes. Chem. Phys. Lett. **357**, 263–266 (2002)

36. Chen, Y., Haddon, R.C., Fang, S., Rao, A.M., Eklund, P.C., Lee, W.H., Smalley, R.E.: Chemical attachment of organic functional groups to single-walled carbon nanotube material. J. Mater. Res. **13**, 2423–2431 (1998)

37. Imamoglu, M., Tekir, O.: Removal of copper(II) and lead(II) ions from aqueous solutions by adsorption on activated carbon from a new precursor hazelnut husks. Desalination **228**, 108–113 (2008)

38. Xu, D., Tan, X., Chen, C., Wang, X.: Removal of Pb(II) from

aqueous solution by oxidized multiwalled carbon nanotubes. J. Hazard. Mater. **154**, 407–416 (2008)

39. Perez-Aguilar, N., Muñoz-Sandoval, E., Diaz-Flores, P., Rangel-Mendez, J.: Adsorption of cadmium and lead onto oxidized nitrogendoped multiwall carbon nanotubes in aqueous solution: equilibrium and kinetics. J. Nanopart. Res. **12**, 467–480 (2010)

40. Jahangiri, M., Kiani, F., Tahermansouri, H., Rajabalinezhad, A.: The removal of lead ions from aqueous solutions by modified multi-walled carbon nanotubes with 1-isatin-3-thiosemicarbazone. J. Mol. Liq. **212**, 219–226 (2015)

41. Zhang, C., Sui, J., Li, J., Tang, Y., Cai, W.: Efficient removal of heavy metal ions by thiol functionalized superparamagnetic carbon nanotubes. Chem. Eng. J. **210**, 45–52 (2012)

42. Saleh, T.A.: Nanocomposite of carbon nanotubes/silica nanoparticles and their use for adsorption of Pb(II): from surface properties to sorption mechanism. Desalin. Water Treat. **57**, 10730–10744 (2016)

43. Hu, J., Shao, D., Chen, C., Sheng, G., Li, J., Wang, X., Nagatsu, M.: Plasma-induced grafting of cyclodextrin onto multiwall carbon nanotube/iron oxides for adsorbent application. J. Phys. Chem. B **114**, 6779–6785 (2010)

44. Hsieh, S.H., Horng, J.J.: Adsorption behavior of heavy metal ions by carbon nanotubes grown on microsized Al_2O_3 particles. J. Univ. Sci. Technol. **14**, 77–84 (2007)

45. Yang, S., Hu, J., Chen, C., Shao, D., Wang, X.: Mutual effects of Pb(II) and humic acid adsorption on multiwalled carbon nanotubes/polyacrylamide composites from aqueous solutions. Environ. Sci. Technol. **45**, 3621–3627 (2011)

46. Ouki, S.K., Kavannagh, M.: Performance of natural zeolites for the treatment of mixed metal-contaminated effluents. Waste Manag. Res. **15**, 383–394 (1997)

47. Benguell, B., Benaissa, H.: Cadmium removal from aqueous solutions by chitin: kinetic and equilibrium studies. Water Res. **36**, 2463–2474 (2002)

48. Gu, H., Lou, H., Tian, J., Liu, S., Tang, Y.: Reproducible magnetic carbon nanocomposites derived from polystyrene with superior tetrabromobisphenol A adsorption performance. J. Mater. Chem. A **4**, 10174–10185 (2016)

49. Gu, H., Lou, H., Ling, D., Xiang, B., Guo, Z.: Polystyrene controlled growth of zerovalent nanoiron/magnetite on a sponge-like carbon matrix towards effective Cr(VI) removal from polluted water. RSC Adv. **6**, 110134–110145 (2016)

18

Functionalization of acidified multi-walled carbon nanotubes for removal of heavy metals in aqueous solutions

A. A. Farghali[1] · H. A. Abdel Tawab[2] · S. A. Abdel Moaty[2] · Rehab Khaled[3]

Abstract Water pollution is a worldwide issue for the eco-environment and human society. Removal of various pollutants including heavy metals from the environment is a big challenge. Techniques of adsorption are usually simple and work effectively. In the current study, MWCNTs were prepared by chemical vapor deposition (CVD) of acetylene at 600 °C. Fe–Co/CaCO$_3$ catalyst/support was prepared by wet impregnation method. The crystal size of the catalyst was identified using XRD. Acidified functionalized multi-walled carbon nanotubes (MWCNT) were produced from oxidation of multi-walled carbon nanotubes by mixture of H$_2$O$_2$ + HNO$_3$ in a ratio of 1:3 (v/v) at 25 °C. The structure and purity of synthesized functionalized CNTs were examined by TEM, N$_2$-BET method and thermogravimetric analysis. The functional groups produced at CNTs surface were investigated using FTIR spectroscopy. Acidified functionalized MWCNTs with a high surface area of 194 m^{2}g^{-1} and porous structure (17.19 nm) were used for water treatment from harmful cations (Pb^{2+}, Cu^{2+}, Ni^{2+} and Cd^{2+}), single cation solutions and quaternary solution at different pH values and different times. The results were interesting because in single solutions the catalyst removed Pb^{2+}, Ni^{2+}, Cu^{2+} and Cd^{2+} with percentages of 93, 83, 78 and 15%, respectively, in 6 h. While in quaternary solution, adsorption was more complex and the order of the adsorbed metals was as following: Pb^{2+} (aq) > Cu^{2+}(aq) > Cd^{2+} (aq) > Ni^{2+} (aq).

Keywords Carbon nanotubes · Chemical vapor deposition · Functionalization · Adsorption

Introduction

Heavy metals are the most important pollutants in water due to their strong toxicity to plants, animals and human beings. The most heavy metals in polluted waters include Hg, Pb, Ag, Cu, Cd, Cr, Zn, Ni, Co and Mn [1]. The heavy metals cannot be degraded or destroyed as they tend to bioaccumulate in food chains. Moreover, their natural process of mineralization is very slow. Many methods are used for removing them from polluted water. The best way to get rid of these heavy metals is by immobilization of good sorbents in solutions leading to adsorption of these heavy metals on the sorbents. Many adsorbents are used for that purpose, such as activated carbon (AC) [2–4], fly ash [5], chitin [6], activated carbon cloth [7] and resins [8]. Among all of these adsorbents, the carbon nanotubes, which are one of the carbon family that possess a great potential for removing many kinds of pollutants, such as dioxin from air [9], lead [4], cadmium, zinc, fluoride [10], 1,2 dichlorobenzene [11] from water.

Studies on adsorption of heavy metals with CNTs presented in the literature are limited to few examples as shown in Table 1. The poor solubility of CNTs in most solvents limits their applications. Their poor solubility in aqueous and organic solvents and limited compatibility with polymer matrices are major drawbacks, rendering these materials incapable of achieving their full potential.

✉ S. A. Abdel Moaty
samah.mohamed@science.bsu.edu.eg

[1] Materials Science and Nanotechnology Department, Faculty of Postgraduate Studies for Advanced Sciences, Beni-Suef University, Beni Suef, Egypt

[2] Materials Science Lab, Chemistry Department, Faculty of Science, Beni-Suef University, Beni Suef, Egypt

[3] Department of Chemistry, Faculty of Science, Beni-Suef University, Beni Suef, Egypt

Table 1 Reported adsorption capacities of some CNT adsorbents for some metal ion

Type of CNT	Metal ion adsorbed	Adsorptivity (%)	References
CNTs	Pb^{2+}	17.44	[40]
MWCNTs	Cd^{2+}	7.4	[41]
	Mn^{2+}	4.8	
	Ni^{2+}	6.8	
MnO_2/CNTs	Pb^{2+}	78.7	[42]
MWCNTs	Cd^{2+}	10.8	[43]
	Cu^{2+}	24.4	
	Pb^{2+}	97.06	
CNTs/Al_2O_3	Pb^{2+}	67.5	[44]
	Cu^{2+}	26.3	
	Cd^{2+}	8.8	
MWCNTs	Co^{2+}	2.77	[45]
MWCNTs	Pb^{2+}	8.7	[46]
Functionalized MWCNTs	Pb^{2+}	93 and 68	This study
	Ni^{2+}	82 and zero	
	Cu^{2+}	78 and 5	
	Cd^{2+}	15 and 4	

Hence, the functionalisation of nanotubes is extremely important, as it increases their solubility and process ability. Several modification approaches like physical, chemical or combined modifications have been exploited for their homogeneous dispersion in common solvents to improve their solubility and applications.

In the present study, MWCNTs were produced by CVD method by Fe–Co/$CaCO_3$ as a catalyst at 600 °C. The catalyst was characterized by XRD. Acidified functionalization of MWCNTs was done using H_2O_2 + HNO_3 in a ratio of 1:3 (v/v) at 25 °C. Functionalized MWCNTs were characterized by thermal analysis, FT-IR, SEM and TEM. Their capabilities to adsorb different heavy metals in single and quaternary aqueous solutions (Pb^{2+}, Cu^{2+}, Cd^{2+} and Ni^{2+}) in different pH [5–9] solutions and different contact time were studied as shown in Scheme 1. Our work showed high % removal efficiency compared with previous studies as shown in Table 1. Langmuir and Freundlich isotherm models of adsorption were applied to fit the experimental data.

Experimental

Materials

Table 2 shows all materials which were used: iron nitrate, Fe $(NO_3)_3 \cdot 9H_2O$, SDFCL, India; cobalt nitrate, Co $(NO_3)_2 \cdot 4H_2O$, and cadmium nitrate, Cd $(NO_3)_2 \cdot 6H_2O$, Oxford laboratory reagent, India; calcium carbonate, $CaCO_3$, sodium hydroxide, NaOH, PioChem Manufacturer

of Laboratory Chemicals, Egypt; hydrochloric acid HCl; hydrogen peroxide, H_2O_2; nitric acid, HNO_3. All used chemicals were of analytical reagent grade and were not more purified; besides, all solutions were prepared using bi-distilled water.

Preparation of the catalyst

The catalyst/support (Fe–Co/$CaCO_3$) was prepared by impregnation method [12]. In the first step, commercial $CaCO_3$ was milled for 10 h to decrease the crystallite size and increase the surface area (see Table 3). Iron and cobalt nitrates, (Fe$(NO_3)_3 \cdot 9H_2O$) and (Co$(NO_3)_2 \cdot 6H_2O$), were added to support ($CaCO_3$) with certain weight ratio (2.5:2.5:95), respectively. Milling was continued for another 2 h. Few drops of distilled water were added to the produced catalyst/support mix making a paste just to ensure the homogeneity. The paste was then dried overnight at 120 °C for 12 h and then ground well to obtain a fine powder of Fe–Co/$CaCO_3$ catalyst/support mixture [13].

Preparation and acidification of CNTs

CNT preparation and oxidized MWCNTs were prepared by catalytic chemical vapor deposition (CVD). The produced supported catalysts were stored in a sealed vessel and CNTs were synthesized over the catalyst. About 2 g of catalyst was packed in a cylindrical alumina cell. The catalyst was preheated to 600 °C in a flow of nitrogen gas (70 ml/min) for 10 min. Then acetylene gas was allowed to pass over the catalyst bed with a rate of 10 ml/min for

Scheme 1 The application of Fe–Co/CaCO₃ nanoparticles to produce MWCNTs for removing heavy metals from wastewater

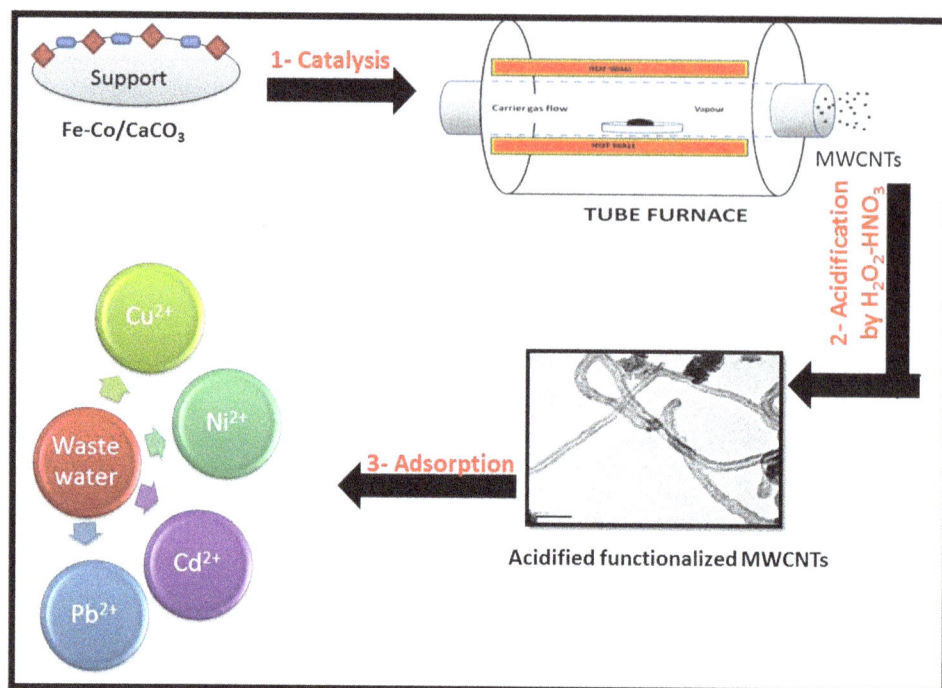

Table 2 The specifications of the chemicals used

Cobalt (II) nitrates (Co(NO₃)₂·6H₂O)	(Oxford Laboratory Reagent, India)
Ferric (III) nitrates (Fe(NO₃)₃·9H₂O)	(SDFCL, India)
Calcium carbonate powder (CaCO₃)	(PioChem)
Acetylene gas (C₂H₂) pure	(Commercial)
Nitrogen (N₂) pure gas	(Commercial)
Nitric acid (HNO₃) conc.	(Commercial)
Hydrogen peroxide (H₂O₂)	(PioChem)
Sodium hydroxide (NaOH)	(PioChem)
Hydrochloric acid (HCl) conc.	(Commercial)
Nickl(II) nitrates (Ni(NO₃)₂·6H₂O	(WINLAB, UK)
Lead (II) nitrates (Pb(NO₃)₂)	(WINLAB, UK)
Cadmium chloride (CdCl₂)	(Oxford Laboratory Reagent, India)
Copper nitrate (Cu(NO₃)₂)	(WINLAB, UK)

Table 3 Ball milling conditions for preparing Fe–Co/CaCO₃ catalyst/support

Condition	Description
Vessel size	7.5 cm diameter
Balls diameters	Ranged from 1.11 to 1.75 cm diameter
Materials of vessels	Stainless steel
Materials of balls	Porcelain
Ball/precipitate mass ratio	8:1 mass ratio
Speed	300 rpm
Time	10 h

60 min. The acetylene gas flow was stopped and the product on the alumina cell was cooled to room temperature in a flow of nitrogen gas.

MWCNTs purification process was achieved using chemical acidified oxidation method. Approximately 0.5 g of MWCNTs was sonicated at 25 °C in 200 ml of mixture of the H₂O₂ + HNO₃ in a ratio of 1:3 (v/v). After 3 h of sonication, the acid-treated MWCNTs was diluted with 200 ml of distilled water and filtered through a filter paper with 3 μm porosity. The acid-treated MWCNTs were then washed thoroughly with distilled water until a neutral pH is reached and dried at 100 ± 0.5 °C for 3 h [14].

Adsorption experiments

Analytical grade lead nitrate, nickel nitrate, cadmium chloride and copper nitrate were employed to prepare a stock solution containing 1000 mg/l for the four metal ions, which were further diluted to the required concentrations before usage. The adsorption of all cations was studied by a batch operation at 25 ± 0.5 °C [15].

Single metal ion adsorption experiments

In single metal ion experiments, 0.05 g of acidified functionalized MWCNTs was placed in 100 ml solutions of concentration 100 mg/l. Each single cation solution was adjusted at different pH values [5–9]. If necessary, an appropriate volume of 0.1 M HNO_3 or 0.1 M NaOH solutions was used to adjust the pH of the solution. The prepared samples were shaken with an orbital shaker at a shaking speed of 200 rpm at room temperature for 6 h. Then solid/liquid phases were separated by filtration. The concentration of the different cations before and after adsorption was determined using atomic absorption spectrometry (Agilent Technologies 200 Series AA). The adsorbed amounts of metal ions onto the acidified functionalized MWCNTs were determined according to the following equations [1]:

$$Q = \frac{(C_o - C_t)}{C_o} \times 100, \tag{1}$$

where Q is the adsorptivity (%), C_o represents the initial concentration of metal ion and C_t is the concentration of metal ions in (mg/l) after adsorption at time t (min). The amount of metal ion adsorption at equilibrium q_e (mg/g) was determined by the following equation:

$$q_e = \frac{V(C_o - C_e)}{W}. \tag{2}$$

In the equation, the equilibrium adsorption capacity of adsorbent in mg (metal)/g (adsorbent) represented by q_e, C_o stands for the initial concentration of metal ions before adsorption in mg/l and C_e is the equilibrium concentration of metal ions in mg/l. The metal ion solution volume in l is represented by V, and W stands for the adsorbent weight in g [16].

Competitive adsorption experiments

To investigate the competitive adsorption of the four cations (Pb^{2+}, Cu^{2+}, Ni^{2+} and Cd^{2+}) on MWCNTs at different pH values [5–9], 0.05 g of acidified functionalized MWCNTs was added to 100 ml solution with equal initial concentrations of the four heavy metals (100 mg/l) and the experiment was completed as previously described.

The same experimental conditions were carried out to study the effect of time by shaking the solution for 10 h and the filtrate concentration was measured every 2 h by atomic adsorption spectroscopy. The absorptivity percentage was calculated by equation [1].

Characterization

($Fe–Co/CaCO_3$) was characterized by X-ray diffraction technique using JSX-60P JEOL diffractometer. The morphology of acidified functionalized MWCNTs was investigated by transmission electron microscope (JEOL JEM-1230).The physical properties of sorbents were determined by nitrogen adsorption at 77 K using ASAP-2010 surface area analyzers. N_2 adsorption isotherms were measured at a relative pressure range 0.0001–0.99. The adsorption data were then employed to determine surface area using Brunauer–Emmett–Teller equation and pore size distribution (including average pore diameter and pore volume) using Barrett–Johner–Halenda equation. The functional groups on the surface sites of MWCNTs were detected by a Fourier transform infrared spectrum (model FT/IR-6100 type A). The carbon content of the sorbents was determined by a thermogravimetric analyzer (model Labsys TG-DSC 50H). Heavy metals' concentrations were determined by atomic adsorption spectrometer (model ZEISS-AA55, Germany).

Results and discussion

Characterizations of synthesized materials

Figure 1 shows the XRD patterns of the catalyst (Fe–Co/$CaCO_3$) used for the preparation of MWCNTs by CVD method and the resulting peaks indicate the presence of the following phases (1: $CaCO_3$, 2: Fe_2O_3, 3: CoO). $CaCO_3$ is a non-porous support material by which the formation of amorphous carbon is suppressed during nanotubes growth and, therefore, selective formation of CNTs is promoted. Purification can be achieved in one step in which both metallic particles and catalyst support can be dissolved in $H_2O_2 + HNO_3$. After purification, CNTs were produced with high yield, high purity and less damage of graphitic walls.

Figure 2a, b shows the TEM images of acidified functionalized MWCNTs. The tubes are highly long (~ 3 μm) and curved with some open tips. The images show clearly a hollow inner tube with a diameter of 3 nm and an outer diameter of 19 nm. Catalyst nanoparticles were encapsulated at the ends of nanotubes, confirming a tip-growth mechanism [17]. The mixed oxide particles seem to be necessary for the growth because they are often found at

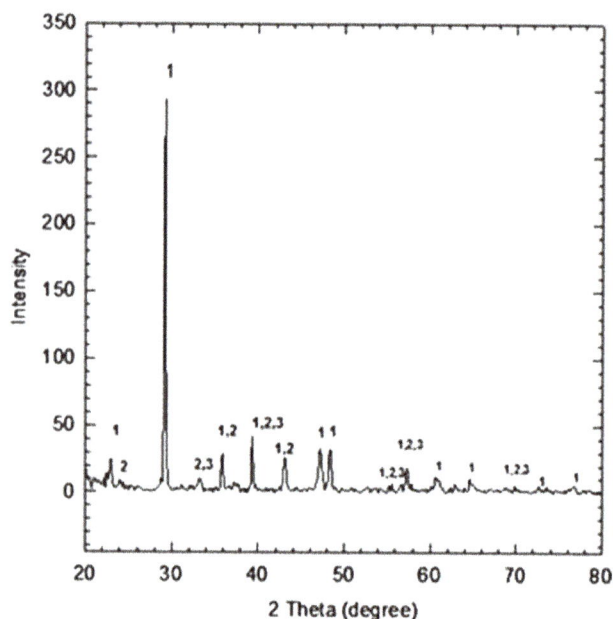

Fig. 1 X-ray diffraction patterns for Fe–Co/CaCO$_3$ catalyst/support, Fe$_2$O$_3$ [1], CoO [2] supported on CaCO$_3$ [3]

the tip inside the nanotube or also somewhere else in the middle of the tube as shown in Fig. 2c (marked with black spot). It is supposed that acetylene decomposes at 600 °C on the top of a supported catalyst as shown in Fig. 2d. The dissolved carbon diffuses in the catalyst, precipitates on the rear side and forms nanotubes. The carbon diffuses through

the catalyst due to a thermal gradient formed by the heat release of the exothermic decomposition of acetylene [18].

FTIR spectra of acidified functionalized MWCNTs are shown in Fig. 3, indicating that the acid treatment generated functional groups on the surface of MWCNTs. The corresponding band close to 3438 cm^{-1} could be attributed to free hydroxyl groups on acidified functionalized MWCNTs surface. On the other hand, the characterized peak that appeared at 2925 cm^{-1} could be attributed to the stretching vibration of C–H, while the peak at 1704 cm^{-1} could be attributed to carboxyl groups. The peak observed at 1628 cm^{-1} is the C=C stretch of the MWCNTs, while asymmetric carboxylate anions' stretch mode was shown at 1575 cm^{-1}. The peak located at 1462 cm^{-1} could be attributed to carbonyl groups [19]. The absorptions at 1384 cm^{-1} were associated with symmetric COO$^-$ stretching [2, 20, 21]. These produced functional groups abundantly on the external and internal surfaces of acidified functionalized MWCNTs, which can provide numerous chemical sorption sites and thus increase the ion exchange capacity for the metal ion, in other words, the hydrophilic properties of these functional groups improve the dispersity of MWCNTs in aqueous solution.

To give a further insight into the specific surface area and porosity of the as-prepared acidified functionalized MWCNTs, the BET surface area, average pore diameter and pore volume were calculated using the BJH method as shown in Table 4. The specific surface area value of the

Fig. 2 TEM images of **a**, **b** acidified functionalized MWCNTs, **c** MWCNTs after adsorption of Pb(II) and **d** growth model of vapor grown carbon nanotubes

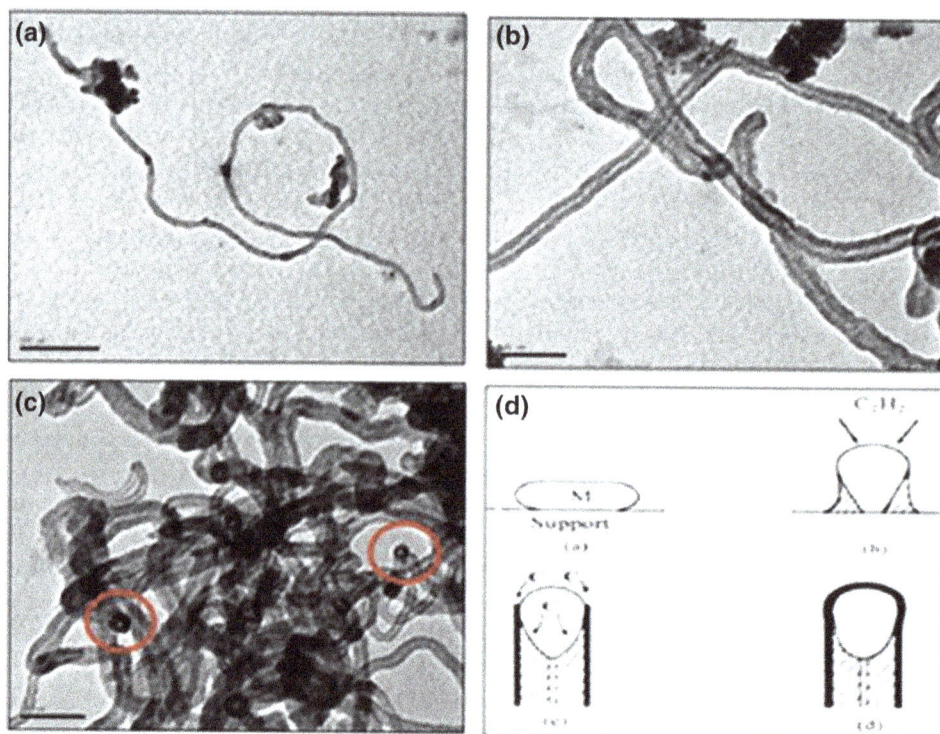

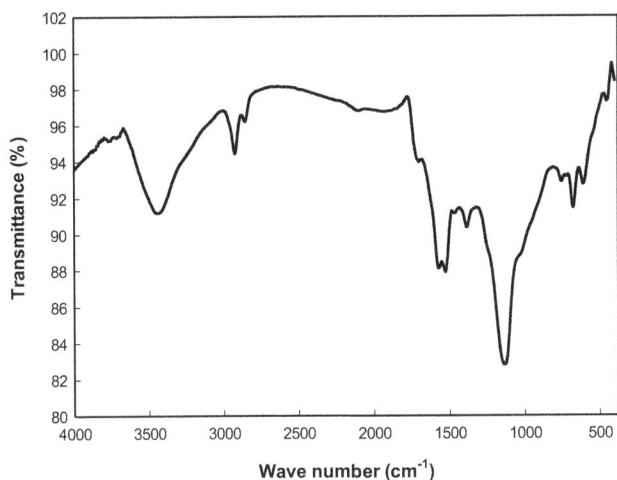

Fig. 3 FT-IR spectra of acidified functionalized MWCNTs

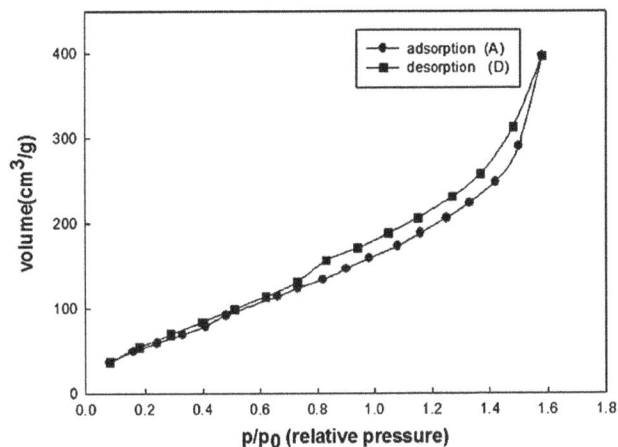

Fig. 4 N_2 adsorption–desorption isotherms of acidified functionalized MWCNTs

acidified functionalized MWCNTs is found to be 194.36 m^2/g, which is greater than that of prepared MWCNTs adsorbent [17], and thus adsorption capability and adsorption active sites will be increased. Removing amorphous carbon, carbon black and carbon nanoparticles introduced by CVD method leads to better dispersion of CNTs, breaks the inner tube spaces and even opens the tips partially. Figure 4 presents the adsorption–desorption isotherms of N_2 on the acidified functionalized MWCNTs, and it is apparent from the adsorption and desorption curves that it exhibits a type II shape. It was observed that there is a small closed adsorption–desorption hysteresis loop with relative pressure above 0.4, which is suggested to be due to mesopores and capillary condensation [22].

Figure 5 reveals the TGA results of acidified functionalized MWCNTs, which show temperature range for weight loss and exhibit two main weight loss regions. The first weight loss region ($\sim 10\%$) can be attributed to the loss of various kinds of functional groups that were produced on the surface of MWCNTs due to acidification treatment. The second region ($\sim 35\%$) may be attributed to the gasification of MWCNTs at which its decomposition begins at 405 °C and ends at 610 °C.

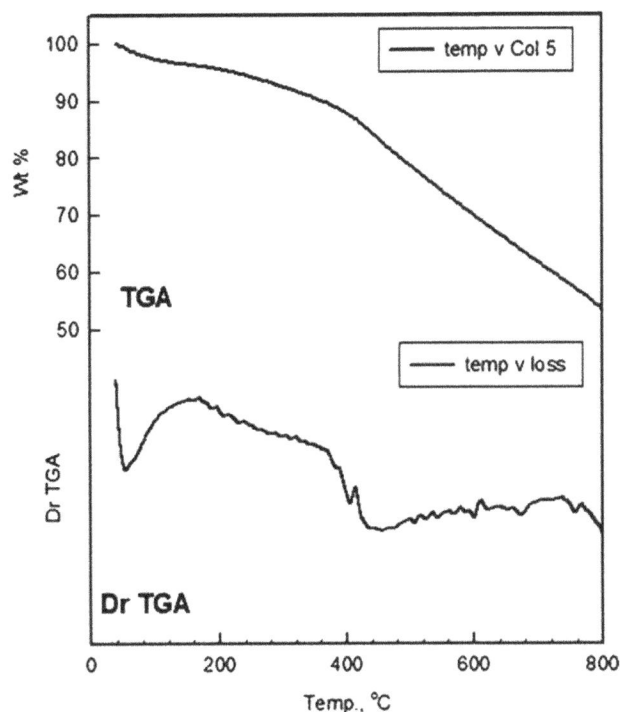

Fig. 5 TGA of acidified functionalized MWCNTs

Adsorption analysis of different metal ions (Pb^{2+}, Ni^{2+}, Cu^{2+}, Cd^{2+})

Adsorption isotherm of single metal ions solutions

The adsorption equilibrium isotherm is important for describing how the adsorbate molecules distribute between the liquid and the solid phases when the adsorption process reaches an equilibrium state. The adsorption isotherms of single cation solutions of Pb^{2+}, Cu^{2+}, Cd^{2+} and Ni^{2+} on the acidified functionalized MWCNTs are shown in

Table 4 Surface area measurements for MWCNTs

Surface area (m^2/g)	194.4
Total pore volume (cc/g)	0.0835
Average pore diameter (nm)	17.19
Micro pore volume (cc/g)	0.17

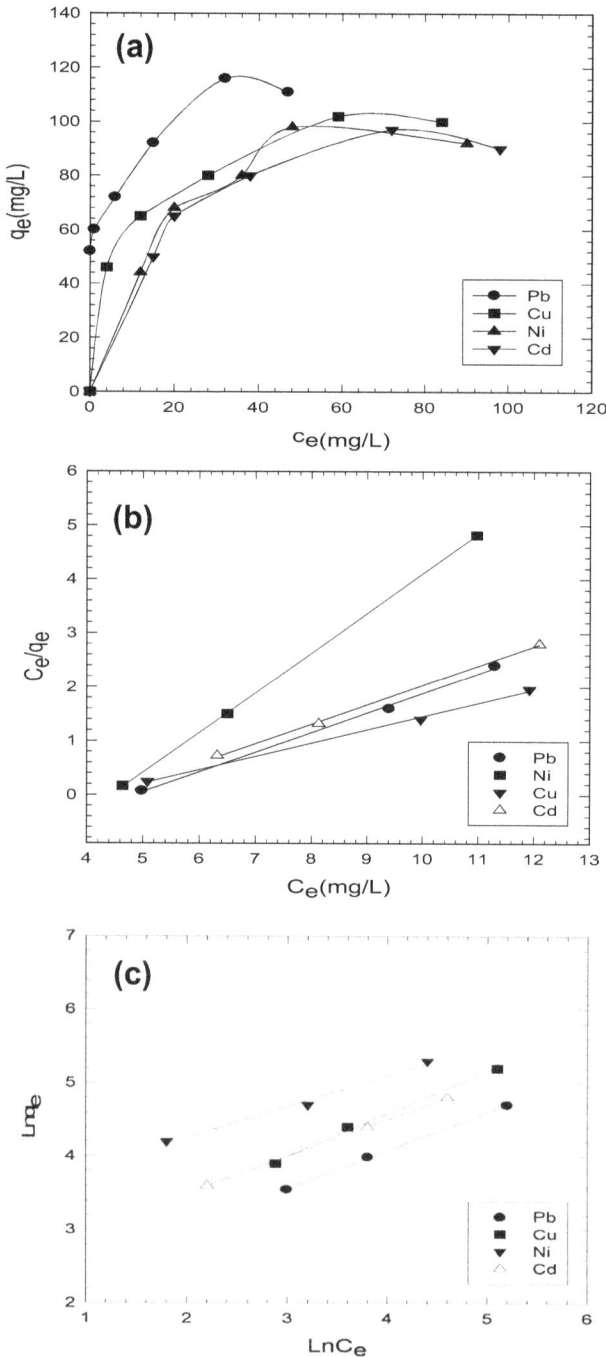

Fig. 6 Langmuir (**a**) and Freundlich (**b**) isotherms for adsorption of Pb^{2+}, Cu^{2+}, Ni^{2+} and Cd^{2+} onto acidified functionalized MWCNTs

Fig. 6a. Equilibrium uptake increased with heavy metal concentrations. This is a result of the increase in the driving force from the concentration gradient. In the same conditions, if the concentrations of heavy metals in solutions are higher, the active sites of the CNTs are surrounded by many more heavy metal ions and the process of adsorption would be carried out sufficiently. The experimental data for single component solutions containing P^{2+}, Cu^{2+}, Cd^{2+}

and Ni^{2+} ions could be approximated by Langmuir and Freundlich isotherm models.

The Langmuir model assumes that there is no interaction between the adsorbate molecules and the adsorption is localized in a monolayer. The Langmuir isotherm [23] is represented by the following linear equation [3]:

$$\frac{C_e}{q_e} = \frac{1}{(q_o K_L)} + \frac{1}{(q_o)} C_e, \tag{3}$$

where C_e (mg/l) is the equilibrium concentration, q_e (mg/g) is the amount of adsorbate adsorbed per unit mass of adsorbate, and q_o and K_L are the Langmuir constants related to the adsorption capacity and the rate of adsorption, respectively. When C_e/q_e was plotted against C_e, a straight line with a slope of $1/q_o$ was obtained (Fig. 6b), indicating that the adsorption of the four heavy metals (Pb^{2+}, Cu^{2+}, Ni^{2+} and Cd^{2+}) on acidified functionalized MWCNTs follows the Langmuir isotherm. The Langmuir constants K_L and q_o were calculated from this isotherm and their values are listed in Table 3. Another important parameter, R_L, called the separation factor or the equilibrium parameter, is evaluated in this study and determined from the following relation [4, 24]:

$$R_L = \frac{1}{[1 + K_L C_o]}, \tag{4}$$

where K_L is the Langmuir constant (l/mg) and C_o (mg/l) is the highest metal ion concentration. The value of R_L indicates whether adsorption onto the MWCNTs will be unfavorable $(R_L > 1)$, linear $(R_L = 1)$, favorable $(0 < R_L < 1)$ or irreversible $(R_L = 0)$. R_L values for Pb^{2+}, Cu^{2+}, Cd^{2+} and Ni^{2+} were less than 1 and greater than zero indicating favorable adsorption (Table 5).

The Freundlich isotherm model is an empirical relationship describing the adsorption of solutes from a liquid to a solid surface and assumes that different sites with several adsorption energies are involved. The linear form of the Freundlich equation [5] is

$$\ln q_e = \ln K_F + \left(\frac{1}{n}\right) \ln C_e, \tag{5}$$

where q_e is the amount adsorbed at equilibrium (mg/g) and C_e is the equilibrium concentration of the four metal ions. K_F and n are Freundlich constants, where K_F (mg/g (l/mg)$^{1/n}$) is the adsorption capacity of the adsorbent and n gives an indication of how favorable the adsorption process is. The slope $1/n$ ranging between 0 and 1 is a measure of adsorption intensity or surface heterogeneity. The surface becomes more heterogeneous as its value gets closer to 0 [25]. Figure 9c shows straight lines with slope $1/n$. The adsorption of Pb^{2+}, Cu^{2+}, Cd^{2+} and Ni^{2+} also follows the Freundlich isotherm. Accordingly (Fig. 6c) Freundlich

Table 5 Isotherm parameters for removal of Pb^{2+}, Cu^{2+}, Cd^{2+} and Ni^{2+} by acidified MWCNTs

Isotherms	Parameters	Heavy metal ions			
		Pb^{2+}	Cu^{2+}	Ni^{2+}	Cd^{2+}
Langmuir	q_o (mg/g)	166	123	95	101
	K_L (l/mg)	0.485	0.384	0.061	0.560
	R_L	0.017	0.19	0.020	0.018
	R^2	0.997	0.997	0.998	0.999
Freundlich	K_F (mg/g (l/mg)$^{1/n}$)	42.23	15.12	37.58	31.96
	N	3.89	2.065	3.012	2.95
	R^2	0.999	0.995	0.990	0.971

constants (K_F and n) were calculated and are listed in Table 5.

Effect of contact time in Pb^{2+} solution

The importance of contact time comes from the need for identification of the possible rapidness of binding and removal processes of the tested metal ions by the synthesized adsorbents and obtaining the optimum time for complete removal of target metal ion. Figure 7 illustrates the effect of contact time on the adsorption of Pb^{2+} ions onto acidified MWCNTs at initial concentration of 100 mg/l and pH 9. In agreement with previous studies [26, 27] there is a significant increase in the adsorption capacity of Pb^{2+} ions as the contact time increases. The analysis of batch adsorption of metal ions was carried out in 2 h and the concentration of each sample was measured by atomic absorption spectroscopy; first after 2 h, the adsorption percentage reached 84%, which may be due to the fact that the initial adsorbent sites were vacant and the solute concentration gradient was high. Later, equilibrium

was achieved at a time period ranging from 6 to 8 h and the adsorption percentage reached to 89.8 and 90.2%, respectively. Finally, a sharp increase after 8 h occurred and complete adsorption was achieved at 10 h as adsorption percentage reached 99.6% and was considered the optimum condition.

Effect of MWCNTs dose

To clarify the function of MWCNTs on adsorption capacity of lead ions, eight weights of MWCNTs (0.1, 0.15, 0.2, 0.25, 0.3, 0.4, 0.5 and 0.6 g) were used to adsorb Pb^{2+} ions from aqueous solution as shown in Fig. 8. After shaking time, adsorption percentage (%) of lead ions onto MWCNTs reached sharply 13.63, 23.55, 39.13, 50.96, 72.23, 73.1, 73 and 72.9 as the CNTs weight increased from 0.1, 0.15, 0.2, 0.25, 0.3, 0.4, 0.5 and 0.6 g, respectively. Therefore, the increase of MWCNTs can obviously increase the adsorption percent of Pb^{2+} [28–30]. Furthermore, the increase in percentage of removed lead ions with an adsorbent dosage can be attributed to an increase in the adsorbent surface which increased the availability of

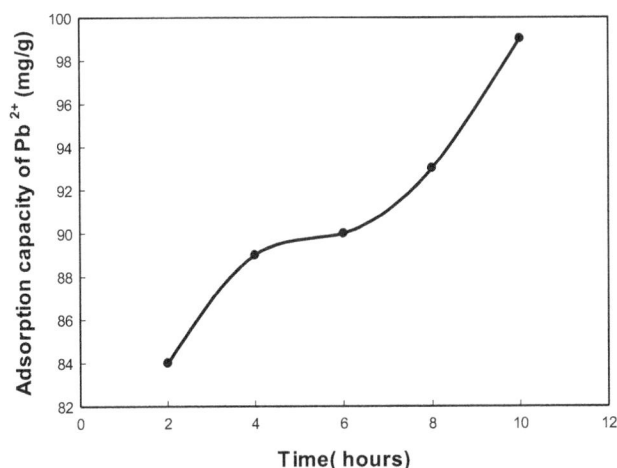

Fig. 7 The effect of contact time on adsorption of Pb^{2+} ions for acidified functionalized MWCNTs

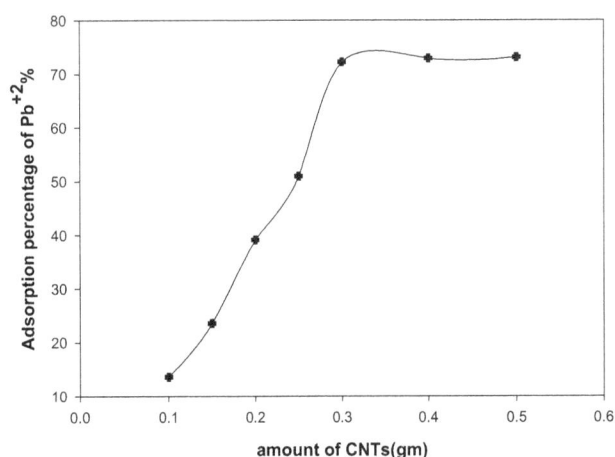

Fig. 8 Shows the effect of MWCNTs dose on the adsorption of Pb^{2+}

adsorption sites; while further increase in CNT dose did not show any increase in adsorption percentage [31]. As indicated from data, there is no more adsorption up to 0.5 g of used CNTs.

Effect of different pH for different metal ions (pb^{2+}, Ni^{2+}, Cu^{2+}, Cd^{2+})

Previous studies [32] showed that the pH of solution is the most important variable governing heavy metal ions adsorption. Figure 9 shows that at low pH value, the surface of the adsorbent would be closely associated with hydronium ions (H$_3$O$^+$) and hold mainly protonated sites. As a result, the surface maintains a net positive charge. So it hinders the access of the metal ions to the surface functional group. Consequently, the percentage removal of metal ions decreases as the pH values decreased. The positive charge on adsorbent surface, however, gradually decreases as pH increases, thus reducing the electrical repulsion between sorbing surface and cations. Moreover, lower H concentration also favors cation sorption by mass action. For example, the adsorption of bivalent cations such as M^{2+} on iron oxide can be written as:

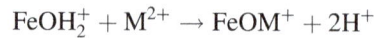

$$FeOH_2^+ + M^{2+} \rightarrow FeOM^+ + 2H^+$$

Lowering H$^+$ concentration will drive this reaction toward the right-hand side and favor the sorption of M^{2+} by increasing pH [33]. That shows good agreement with Surface Complex Formation Theory (SCF), which states the following: by increasing pH, the competition for adsorption sites between protons and metal species decreases [19]. As a result, by increasing pH value there are more and more attractive forces due to the presence of surface negative charges.

Effect of different metal ions on the adsorption of Pb^{2+} at different pH

In a quaternary aqueous solution containing equal concentrations of Pb^{2+}, Cu^{2+}, Cd^{2+} and Ni^{2+} using serial pH values [5–9], the effect of pH was presented in Fig. 10 which shows that as the pH increase, the adsorption capacities for Pb^{2+}, Cu^{2+} and Cd^{2+} increase, while Ni^{2+} did not show any adsorption capacity in all pH values. The maximum adsorption capacities reached at pH 9 for three metals Pb^{2+}, Cu^{2+} and Cd^{2+}. The experimental data indicate that the adsorption affinities for the four metals take the sequence, Pb^{2+}> Cu^{2+}> Cd^{2+}> Ni^{2+}, which agrees with Li et al. [34]

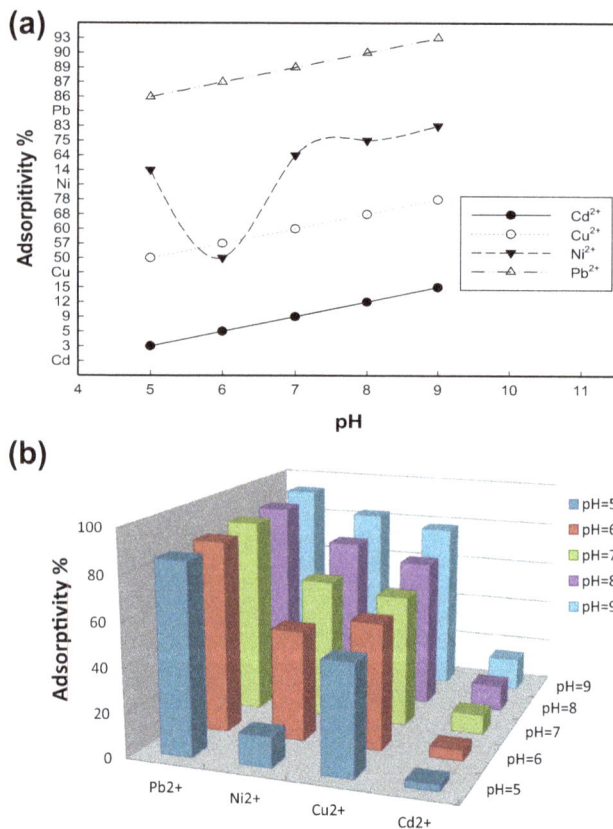

Fig. 9 The effect of different pH [5–9] on the adsorption uptake of metals Pb^{2+}, Ni^{2+}, Cu^{2+} and Cd^{2+}, in single solution

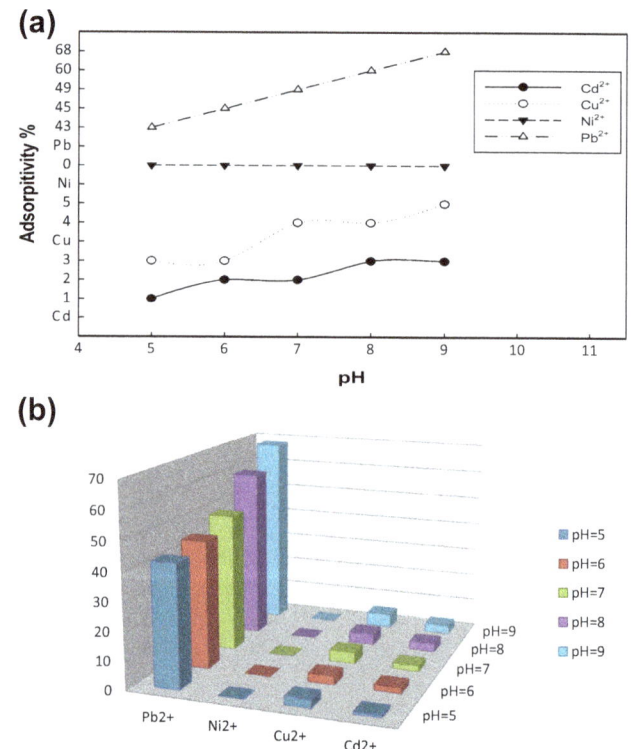

Fig. 10 The effect of different pH [5–9] on the adsorption uptake of metals Pb^{2+}, Ni^{2+}, Cu^{2+} and Cd^{2+}, in multi-solution

Ni^{2+} has no ability to compete the other three metal ions as it is considered to be the most weakly adsorbed species. Many studies attributed the behavior of the multi-component systems to the CNT adsorption sites [34–36] and found that Pb^{2+} and Cu^{2+} both have higher affinities more than Cd^{2+} and Ni^{2+}. The competitive adsorption may be related to ion exchange and electrochemical properties of the cations. The higher the complex redox properties and ion exchange processes, the higher is the polarizability of the ion, the smaller is the value of N/R_H (N is the number of water of hydration and R_H is the hydrated radius] and greater the affinity between the ions and functional groups [37]. However, there is no consensus among researchers regarding the competitive adsorption of metal ions [38, 39].

Conclusions

Fe–Co/CaCO$_3$ catalyst/support was successfully prepared by wet impregnation method and characterized by XRD. The MWCNTs was synthesized in a homogeneous form by chemical vapor deposition (CVD) of acetylene on the catalyst surface at 600 °C with high density and high purity. The synthesized MWCNTs were purified by a mixture of $H_2O_2 + HNO_3$ in a ratio of 1:3 (v/v) at 25 °C leading to formation of functional groups on its surface. The functionalized MWCNTs were characterized using FTIR, TEM, BET and TGA. FTIR spectra of functionalized MWCNTs confirmed the presence of several functional groups. These produced functional groups can provide numerous chemical sorption sites on MWCNTs surface. From single and quaternary solutions of the four heavy metals (Pb^{2+}, Ni^{2+}, Cu^{2+} and Cd^{2+}), the following conclusions could be obtained:

1. The adsorption of Pb^{2+}, Ni^{2+}, Cu^{2+} and Cd^{2+} in both single and quaternary solutions increases with increase of the solution pH.
2. The large adsorption capacity of acidified MWCNTs for the metals is mainly due to the oxygenous functional groups on its surface which could react with metals to form salt or complex deposits on the surface of MWCNTs.
3. The removal process of the metals is related to the surface chemistry of oxidized MWCNTs.

The adsorption experimental results of these heavy metals are in a good correspondence with the Langmuir and Freundlich isotherms.

Acknowledgements Special acknowledgment for Mr. Khaled Samir Hemida (MA in EFL) for helpful discussions and valuable support in English and grammar revision. The work was supported by the Faculty of Science, Materials Science Lab, Chemistry Department and Faculty of Postgraduate Studies for Advanced Sciences, Materials Science and Nanotechnology Department, Beni-Suef University, Egypt.

References

1. Wang, S., et al.: Adsorptive remediation of environmental pollutants using novel graphene-based nanomaterials. Chem. Eng. J. **226**, 336–347 (2013)
2. Gabaldón, C., et al.: Single and competitive adsorption of Cd and Zn onto a granular activated carbon. Water Res. **30**, 3050–3060 (1996)
3. Faur-Brasquet, C., et al.: Modeling the adsorption of metal ions (Cu^{2+}, Ni^{2+}, Pb^{2+}) onto ACCs using surface complexation models. Appl. Surf. Sci. **196**, 356–365 (2002)
4. Mubarak, N., et al.: Removal of heavy metals from wastewater using carbon nanotubes. Sep. Purif. Rev. **43**, 311–338 (2014)
5. Kadirvelu, K., Faur-Brasquet, C., Cloirec, P.L.: Removal of Cu (II), Pb(II), and Ni (II) by adsorption onto activated carbon cloths. Langmuir **16**, 8404–8409 (2000)
6. Benguella, B., Benaissa, H.: Cadmium removal from aqueous solutions by chitin: kinetic and equilibrium studies. Water Res. **36**, 2463–2474 (2002)
7. Bayat, B.: Comparative study of adsorption properties of Turkish fly ashes: I. The case of nickel (II), copper (II) and zinc (II). J. Hazard. Mater. **95**, 251–273 (2002)
8. Diniz, C.V., Doyle, F.M., Ciminelli, V.S.: Effect of pH on the adsorption of selected heavy metal ions from concentrated chloride solutions by the chelating resin Dowex M-4195. Sep. Sci. Technol. **37**, 3169–3185 (2002)
9. Long, R.Q., Yang, R.T.: Carbon nanotubes as superior sorbent for dioxin removal. J. Am. Chem. Soc. **123**, 2058–2059 (2001)
10. Ren, X., et al.: Carbon nanotubes as adsorbents in environmental pollution management: a review. Chem. Eng. J. **170**, 395–410 (2011)
11. Peng, X., et al.: Adsorption of 1,2-dichlorobenzene from water to carbon nanotubes. Chem. Phys. Lett. **376**, 154–158 (2003)
12. Schwarz, J.A., Contescu, C., Contescu, A.: Methods for preparation of catalytic materials. Chem. Rev. **95**, 477–510 (1995)
13. Bahgat, M., et al.: Synthesis and modification of multi-walled carbon nano-tubes (MWCNTs) for water treatment applications. J. Anal. Appl. Pyrol. **92**, 307–313 (2011)
14. Mkhondo, N., Magadzu, T.: Effects of different acid-treatment on the nanostructure and performance of carbon nanotubes in electrochemical hydrogen storage. Dig. J. Nanomater. Biostruct. (DJNB) **9**, 1331–1338 (2014)
15. Yuan, F., et al.: Adsorption of Cd(II) from aqueous solution by biogenic selenium nanoparticles. RSC Adv. **6**, 15201–15209 (2016)
16. Kandah, M.I., Meunier, J.-L.: Removal of nickel ions from water by multi-walled carbon nanotubes. J. Hazard. Mater. **146**, 283–288 (2007)
17. Abdi, Y., et al.: PECVD-grown carbon nanotubes on silicon substrates with a nickel-seeded tip-growth structure. Mater. Sci. Eng. C **26**, 1219–1223 (2006)
18. Khedr, M., Bahgat, M., Abdel-Moaty, S.: Catalytic decomposition of acetylene over CoFe$_2$O$_4$/BaFe$_{12}$O$_{19}$ core shell nanoparticles for the production of carbon nanotubes. J. Anal. Appl. Pyrolysis **84**, 117–123 (2009)
19. Xu, D., et al.: Removal of Pb(II) from aqueous solution by oxidized multiwalled carbon nanotubes. J. Hazard. Mater. **154**, 407–416 (2008)
20. Üçer, A., Uyanik, A., Aygün, Ş.: Adsorption of Cu(II), Cd(II), Zn(II), Mn(II) and Fe(III) ions by tannic acid immobilised activated carbon. Sep. Purif. Technol. **47**, 113–118 (2006)

21. Cho, H.-H., et al.: Influence of surface oxides on the adsorption of naphthalene onto multiwalled carbon nanotubes. Environ. Sci. Technol. **42**, 2899–2905 (2008)

22. Bansal, R.C., Goyal, M.: Activated carbon adsorption. CRC Press, Boca Raton (2005)

23. Farghali, A., et al.: Decoration of MWCNTs with $CoFe_2O_4$ nanoparticles for methylene blue dye adsorption. J. Solut. Chem. **41**, 2209–2225 (2012)

24. Hall, K., et al.: Pore-and solid-diffusion kinetics in fixed-bed adsorption under constant-pattern conditions. Ind. Eng. Chem. Fundam. **5**, 212–223 (1966)

25. Haghseresht, F., Lu, G.: Adsorption characteristics of phenolic compounds onto coal-reject-derived adsorbents. Energy Fuels **12**, 1100–1107 (1998)

26. Ernhart, C.B.: A critical review of low-level prenatal lead exposure in the human: 1. Effects on the fetus and newborn. Reprod. Toxicol. **6**, 9–19 (1992)

27. Ernhart, C.B.: A critical review of low-level prenatal lead exposure in the human: 2. Effects on the developing child. Reprod. Toxicol. **6**, 21–40 (1992)

28. Li, Y.H., et al.: Competitive adsorption of Pb^{2+}, Cu^{2+} and Cd^{2+} ions from aqueous solutions by multiwalled carbon nanotubes. Carbon **41**, 2787–2792 (2003)

29. Lu, C., Liu, C.: Removal of nickelII from aqueous solution by carbon nanotubes. J. Chem. Technol. Biotechnol. **81**, 1932–1940 (2006)

30. Kabbashi, N.A., et al.: Kinetic adsorption of application of carbon nanotubes for Pb(II) removal from aqueous solution. J. Environ. Sci. **21**, 539–544 (2009)

31. Rao, G.P., Lu, C., Su, F.: Sorption of divalent metal ions from aqueous solution by carbon nanotubes: a review. Sep. Purif. Technol. **58**, 224–231 (2007)

32. Moaty, S.A., Farghali, A., Khaled, R.: Preparation, characterization and antimicrobial applications of Zn-Fe LDH against MRSA. Mater. Sci. Eng. C **68**, 184–193 (2016)

33. Mohapatra, M., et al.: A comparative study on Pb(II), Cd (II), Cu(II), Co(II) adsorption from single and binary aqueous solutions on additive assisted nano-structured goethite. Int. J. Eng. Sci. Technol. **2**, 89–103 (2012)

34. Li, Y.H., et al.: Adsorption of cadmium(II) from aqueous solution by surface oxidized carbon nanotubes. Carbon **41**, 1057–1062 (2003)

35. Sun, Y.-P., et al.: Functionalized Carbon Nanotubes: properties and Applications. Acc. Chem. Res. **35**, 1096–1104 (2002)

36. Gao, Z., et al.: Investigation of factors affecting adsorption of transition metals on oxidized carbon nanotubes. J. Hazard. Mater. **167**, 357–365 (2009)

37. Trivedi, P., Axe, L., Dyer, J.: Adsorption of metal ions onto goethite: single-adsorbate and competitive systems. Colloids Surf. A **191**, 107–121 (2001)

38. Gabaldon, C., et al.: Single and competitive adsorption of Cd and Zn onto a granular activated carbon. Water Res. **30**, 3050–3060 (1996)

39. Ûçer, A., Uyanik, A., Aygün, A.: Adsorption of Cu(II), Cd(II), Zn(II), Mn(II) and Fe(III) ions by tannic acid immobilised activated carbon. Sep. Purif. Technol. **47**, 113–118 (2006)

40. Li, Y.-H., et al.: Lead adsorption on carbon nanotubes. Chem. Phys. Lett. **357**, 263–266 (2002)

41. Liang, P., et al.: Multiwalled carbon nanotubes as solid-phase extraction adsorbent for the preconcentration of trace metal ions and their determination by inductively coupled plasma atomic emission spectrometry. J. Anal. Atom. Spectrom. **19**, 1489–1492 (2004)

42. Wang, H., et al.: Adsorption characteristic of acidified carbon nanotubes for heavy metal Pb (II) in aqueous solution. Mater. Sci. Eng. A **466**, 201–206 (2007)

43. Li, Y.-H., et al.: Competitive adsorption of Pb^{2+}, Cu^{2+} and Cd^{2+} ions from aqueous solutions by multiwalled carbon nanotubes. Carbon **41**, 2787–2792 (2003)

44. Hsieh, S.-H., Horng, J.-J.: Adsorption behavior of heavy metal ions by carbon nanotubes grown on microsized Al_2O_3 particles. J. Univ. Sci. Technol. **14**, 77–84 (2007)

45. Pyrzyńska, K., Bystrzejewski, M.: Comparative study of heavy metal ions sorption onto activated carbon, carbon nanotubes, and carbon-encapsulated magnetic nanoparticles. Colloids Surf. A. **362**, 102–109 (2010)

46. Mamba, G., et al.: Application of multiwalled carbon nanotube-cyclodextrin polymers in the removal of heavy metals from water. J. App. Sci. (Faisalabad) **10**, 940–949 (2010)

19

Synthesis, characterization and electrochemical-sensor applications of zinc oxide/graphene oxide nanocomposite

Ehab Salih[1] · Moataz Mekawy[1] · Rabeay Y. A. Hassan[2]⊙ · Ibrahim M. El-Sherbiny[1]

Abstract Nanostructured metal oxides received considerable research attention due to their unique properties that can be used for designing advanced nanodevices. Thus, in the present study, zinc oxide/graphene oxide (ZnO/GO) nanocomposite was synthesized, characterized and implemented in an electrochemical system. The formation of a compacted ZnO/GO nanocomposite was confirmed by field emission scanning electron microscopy, high-resolution transmission electron microscopy (HRTEM), X-ray diffraction (XRD), and attenuated total reflectance spectroscopy. HRTEM showed that ZnO nanocrystals (NCs) are well formed on the GO surface and are interconnected via GO functional groups. From the XRD patterns, the average size of ZnO NCs was found to be about 21.7 ± 2.3 nm which is in agreement with the HRTEM results. The newly developed nanocomposite-based electrochemical system showed a significant improvement in both electrical conductivity and the electrocatalytic activity as noted from the cyclic voltammetry measurements. Consequently, direct electron transfer efficiency was confirmed and used for the amperometric detection of hydrogen peroxide (H_2O_2). Fast and sensitive electrochemical responses for the detection of H_2O_2 at 1.1 V in the linear response range from 1 to 15 mM with the detection limit (S/N = 3) of 0.8 mM were obtained. These results demonstrated that the prepared ZnO/GO/CPE displayed a good performance along with high sensitivity and long-term stability.

Keywords Electrochemical biosensors · Nanocomposite · Zinc oxide/graphene oxide composites · Hydrogen peroxide detection

Introduction

Electrochemical techniques have recently showed many advantages in medical and biological analysis such as high sensitivity, low cost, rapid response, and simplicity [1]. In the era of nanomaterials, several electrochemical systems have been developed using various nanomaterials such as nanostructured metal oxides [2]. Amongst the nanostructured metal oxides, zinc oxide semiconductor nanocrystals (ZnO NCs) have been widely used in photocatalytic [3], photonic [4] spintronic [5], and many other optoelectronic applications [6]. This could be attributed to their wide band gap (3.37 eV) and large excitonic binding energy (60 meV) [7, 8]. However, the use of ZnO NCs as a single electrode modifier in electrochemical biosensors is limited since it behaves as n-type semiconductor. This causes fast recombination of the generated electron-hall pairs and low operating speed, and thus, the capability of direct electron transfer is rather difficult [9]. Instead, implementation of artificial electron shuttles [10] or using hydride substances [11] to liberate the captured-(stored)-electrons is highly recommended.

To avoid the utilization of artificial redox mediators, the direct electrochemical communication is more preferable.

✉ Rabeay Y. A. Hassan
rabeayy@yahoo.com; rabeayy@gmail.com

[1] Center for Materials Science, Zewail City of Science and Technology, 6th October City, Giza 12588, Egypt

[2] Microanalysis Lab, Applied Organic Chemistry Department, National Research Centre (NRC), El-Buhouth St., Dokki, Cairo 12622, Egypt

Thus, the integration (or hybridization) of ZnO with other carbon-based materials has been reported [12, 13]. Due to its interesting mechanical and electrochemical properties, and the ease of its mass production in addition to the abundance of several function groups on its surface [12, 14], graphene oxide (GO) has attracted our attention to build up sensitive and reliable electrochemical sensors and biosensors. In fact the interesting catalytic activity of the ZnO/GO nano-structured has been shown in several electrochemical applications [15–20]. However, better understanding for the mechanisms behind the use of either the composite or its elemental composition (i.e. individual use of ZnO NPs, or GO NPs) for the direct electron transfer is still unclear.

Therefore, the main concern of this study is to identify the possibility of direct electrochemical uses of ZnO/GO nanocomposite. To reach that goal, fabrication with full characterization of ZnO/GO nanocomposite was performed. Electrocatalytic activity of the target nanostructure was investigated using the redox functions of ferricyanide (FCN). In addition, direct electrochemical detections were performed after testing the role of ZnO alone or upon combination with GO in form of a nanocomposite. Consequently, the modified electrode with the nanocomposite was successfully used for the direct amperometric determination of hydrogen peroxide, as one of the most important biomarkers for several biochemical process and enzymatic functions [21, 22]. The developed ZnO/GO modified CPE presents high sensitivity, low potential and long-term stability towards the detection of H_2O_2, which could be a promising approach for the development of non-enzymatic H_2O_2 sensor.

Materials and methods

Materials

Graphite flakes were purchased from Fisher (UK), chemical reagents such as sulfuric acid (H_2SO_4), phosphoric acid (H_3PO_4), potassium permanganate ($KMnO_4$), hydrogen peroxide (H_2O_2), dimethylformamide (DMF), Zinc acetate dihydrate $Zn(CH_3COO)_2 \cdot 2H_2O$, phosphate buffer saline (PBS) were purchased from Sigma-Aldrich (Germany).

Methods

Preparation of GO

GO was prepared using improved Hummer method with a slight modification [23]. Typically, 3 g of graphite flakes was added to a 4:1 mixture of concentrated H_2SO_4/H_3PO_4 (160:40 mL), 15 g of $KMnO_4$ was added to the mixture

with keeping the temperature below 5 °C using ice bath. The reaction was heated to 60 °C with stirring for 24 h. Finally, the reaction was cooled to room temperature and poured onto ice (~ 400 mL) with 30 % H_2O_2 (5 mL). The resulting solution was maintained at room temperature overnight and then the supernatant was decanted away. This process was repeated for 4 days. Afterwards, the dispersion was washed repeatedly with water in a cycle of centrifugation and decantation, and finally washed with ethanol. The product was allowed to dry at 50 °C for 12 h.

Preparation of ZnO/GO nanocomposite

In a typical procedure, 10 mg GO was first dispersed in 10 mL of DMF by sonication for 4 min. Then, the GO suspension was added with stirring to 50 mL of zinc acetate dihydrate $[Zn(CH_3COO)_2 \cdot 2H_2O]$ dissolved in DMF (0.02 M). Subsequently, the mixture was heated to 90 °C and maintained at that temperature for 5 h. The color of the resulting ZnO/GO powder was then changed into grayish-white. The product was subjected to repeated washing with ethanol followed by centrifugation, and finally washing with water. The pure ZnO/GO powder was obtained after drying the product overnight at 50 °C. ZnO NCs were prepared using the same procedure and used as the control sample.

Morphological characterization

Attenuated total reflectance (ATR) spectroscopy of the produced ZnO/GO nanocomposite was performed using FTIR spectrometer (NECOLET iS10). The morphology of ZnO/GO nanocomposites was investigated by HRTEM (JEOL -JEM- 2100) at an accelerating voltage of 200 kV. Field emission scanning electron microscope FESEM images were carried out using a (Nova Nano SEM 450) at an accelerating applied potential of 15 keV. The X-ray diffraction patterns of the samples were recorded by X-ray diffractometer (Philips PW 1390) using Cu Kα1 as an X-ray source. A thermogravimetric analyzer (TGA Q50) was used to study the thermal behavior of ZnO NCs and the ZnO/GO nanocomposites in the temperature range from 0 to 700 °C with a heating rate of 10 °C min^{-1} in a 40 mL min^{-1} nitrogen flow.

Sensor preparation

Nano-sensors were prepared by thoroughly mixing 75 mg ZnO/GO and 675 mg synthetic carbon powder with 0.2 mL paraffin oil in a small agate mortar. A portion of the paste was then packed into the tip of the electrode assembly with a surface area of (0.3 cm^2) [24]. Electrode surface regeneration was performed before each experiment by polishing

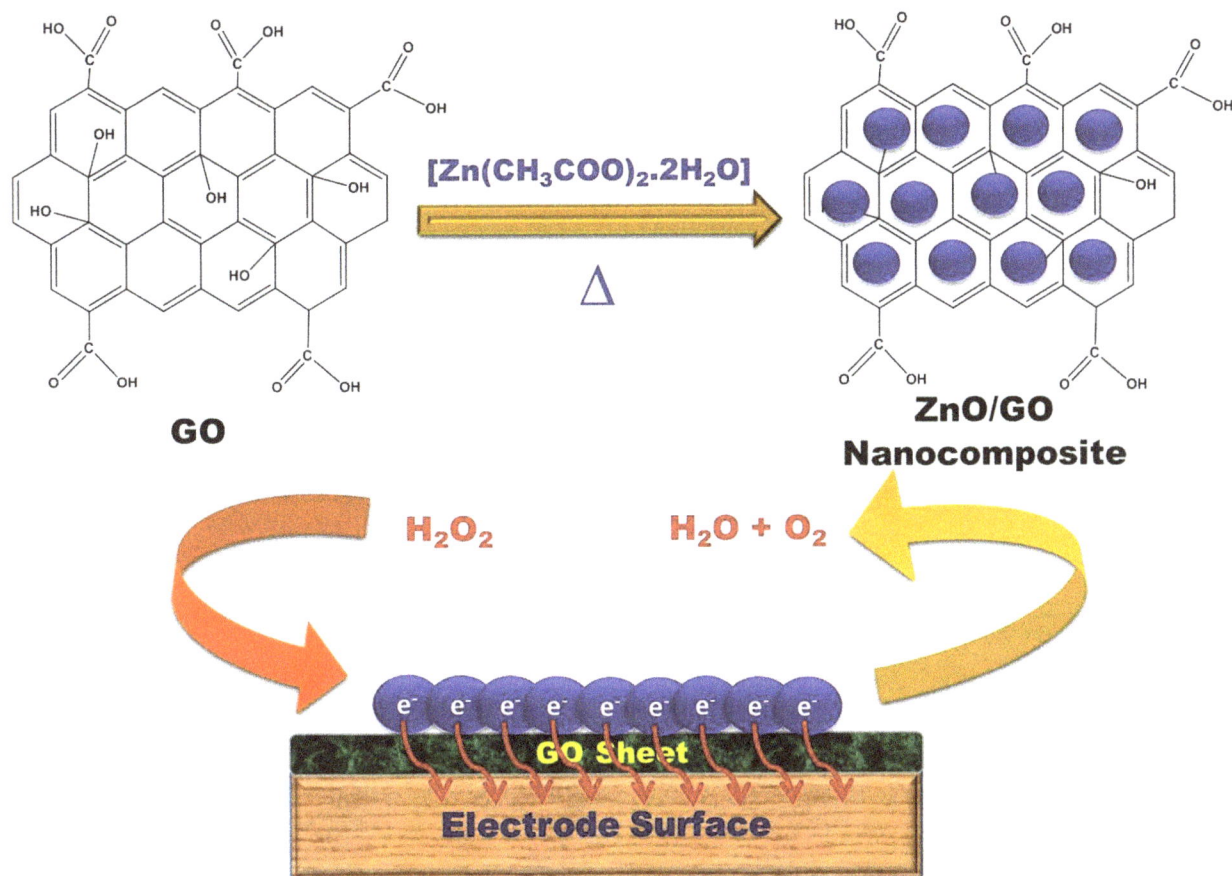

Scheme 1 A schematic illustration of the preparation of the ZnO/GO nanocomposite and the fabrication of the sensor's electrode

it with a smooth wet filter paper until a shiny and clean electrode surface was obtained. The fabrication of the ZnO/GO-based nano-sensor is illustrated in Scheme 1.

Electrochemical characterization of ZnO/GO nanocomposite

All electrochemical measurements were performed using a computer controlled Gamry Potentiostat/Galvanostat/ZRA G750, which was connected to a three electrode system comprising of a CPE working electrode, a Pt disc auxiliary electrode, and an Ag/AgCl reference electrode. Prior to the measurements, the working electrode was electrochemically activated in 0.1 M KCl by 10-cyclic scans from -0.2 to 1.0 V with scan rate of 100 mVs^{-1}. Aliquots of the FCN were introduced into the electrochemical cell containing 30 mL of KCl.

Direct electrochemical detection of H_2O_2

The surface of the nanocomposite sensor was activated in phosphate buffer (pH 7.4) by 10-cyclic scans from -0.2 to 1.0 V with a scan rate of 100 mVs^{-1}. Then, certain

amounts of H_2O_2 were introduced into the electrochemical cell containing 30 mL of phosphate buffer. All the experiments were carried out at room temperature.

Results and discussion

Structural features of ZnO/GO nanocomposite

The morphological characterizations of the synthesized GO, and ZnO/GO nanocomposite were examined by FESEM. Figure 1a shows the basic shape of GO sheet that was significantly exfoliated, and looks like pieces of leaves with a dimension ranging from several hundred nm to several microns. Figure 1b shows the top view of ZnO/GO nanocomposite which indicates that ZnO NCs are closely anchored at the surface of GO.

From the TEM images, the GO surface looks smooth and integrated (Fig. 2a). In the case of ZnO/GO nanocomposite (Fig. 2b), a large number of ZnO NCs with average diameters 21.7 ± 2.3 nm were observed uniformly on the surface of the GO. The arrows clearly show the edges of the GO sheet. The high magnification TEM image

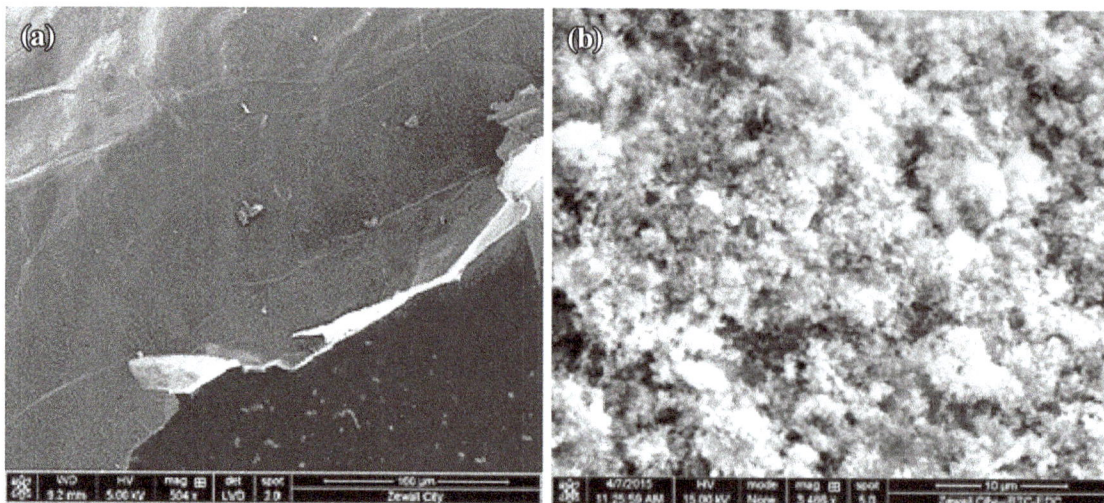

Fig. 1 FESEM of **a** GO, and **b** ZnO/GO nanocomposite

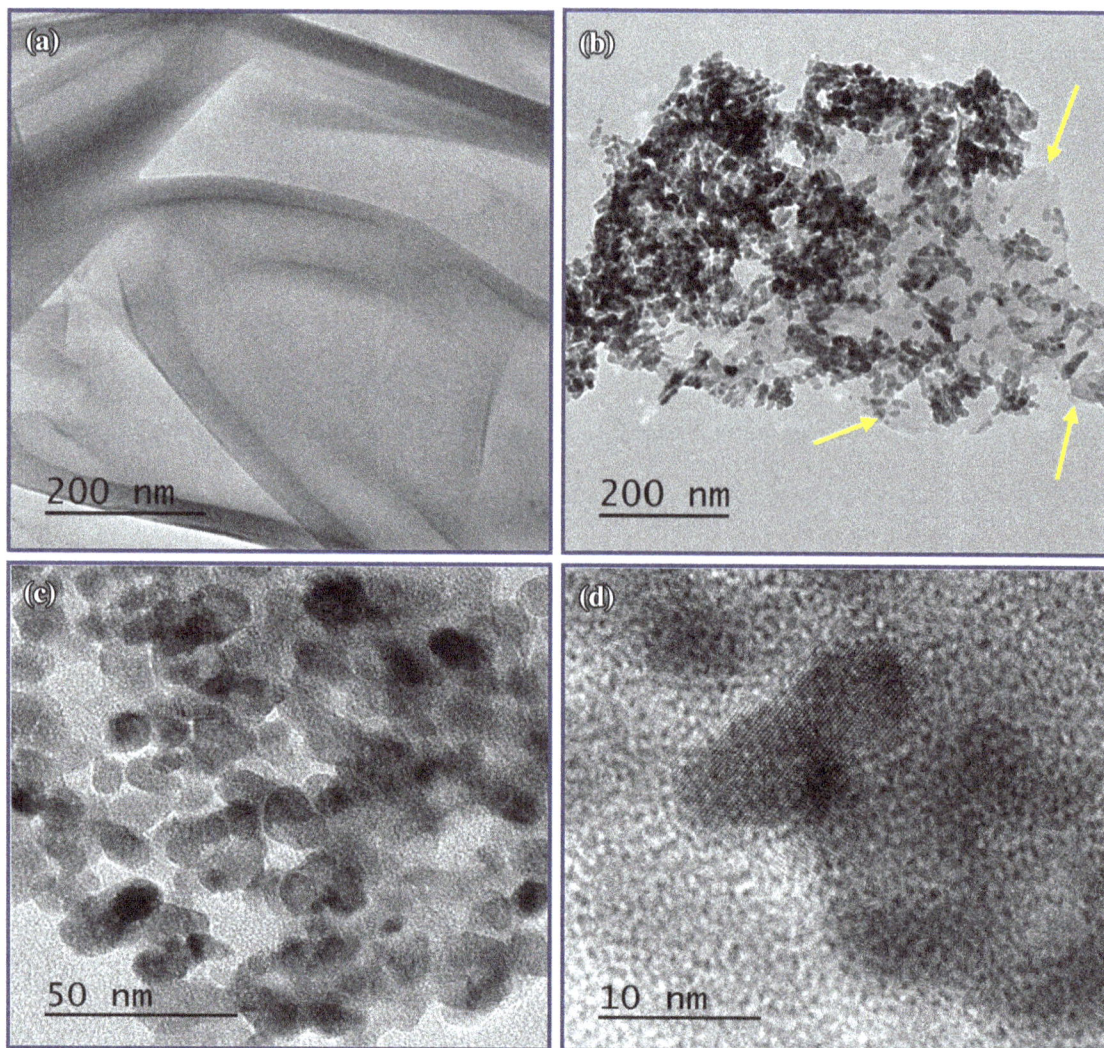

Fig. 2 TEM images of **a** GO, **b**, **c** ZnO/GO nanocomposite with different magnifications, and **d** HRTEM image of ZnO NCs anchored GO surface

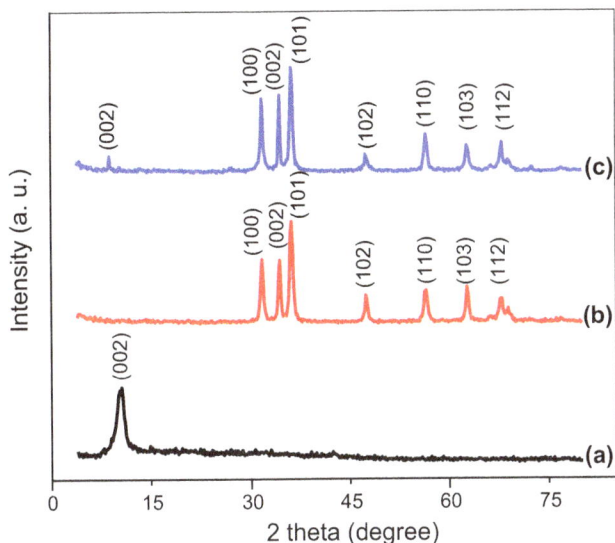

Fig. 3 XRD patterns of *a* GO, *b* ZnO NCs, and *c* ZnO/GO nanocomposite

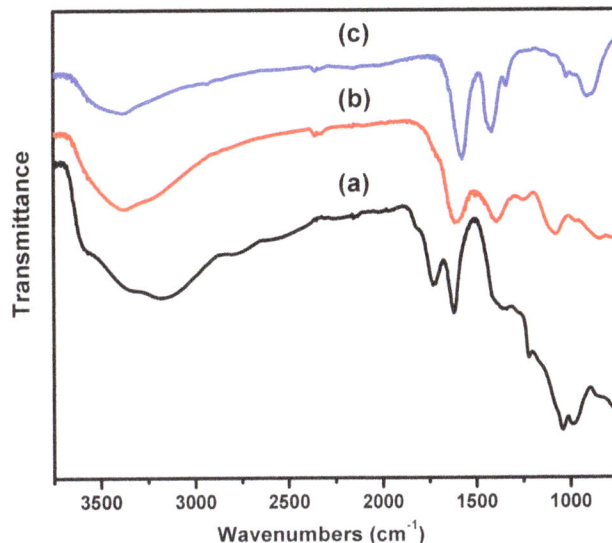

Fig. 4 FTIR-ATR spectra of *a* GO, *b* ZnO/GO nanocomposite, and *c* ZnO NCs

(Fig. 2c) further reveals that ZnO NCs are almost spherical in shape. The high-resolution transmission electron microscopy (HRTEM) image of the ZnO NCs (Fig. 2d) shows a clear lattice fringe which indicates a high degree of crystallinity of the ZnO.

Furthermore, the XR diffraction patterns of GO, ZnO NCs and ZnO/GO nanocomposite are illustrated in (Fig. 3). In the diffractogram of GO (Fig. 3a), the peak appearing at 10.6° is due to the presence of oxygen carrying groups on the GO-surface. This peak was significantly decreased in ZnO/GO nanocomposite (Fig. 3c) due to exfoliation of GO sheet as a result of the ZnO NCs surface loading. As obvious from (Figs. 3b, c), a coincidence of diffraction peaks for ZnO NCs and ZnO/GO nanocomposite is highly remarkable indicating the formation of well crystalline structure of ZnO NCs onto the GO surface. Peaks at 31.7°, 34.4°, 36.2°, 47.4°, 56.6° 62.9°, 65.5°, 68.0° and 69.1° that are corresponding to (100), (002), (101), (102), (110), (103), (200), (112) and (201) lattice planes, respectively, indicating the formation of wurtzite structure with 2D hexagonal *P6₃mc* space group [25, 26].

For the functional analysis, FTIR-ATR spectra were used as shown in Fig. 4. The following functional groups were identified; O–H stretching vibrations (3240–3300 cm^{-1}), C=O stretching vibration (1720–1740 cm^{-1}), C=C from un-oxidized sp^2 C–C bonds (1590–1620 cm^{-1}), and C–O vibrations (1250 cm^{-1}) [23]. Comparing the spectra of ZnO/GO nanocomposite with that of GO demonstrates a slight shift with a reduction in the intensity of the O–H peak (at 3240 cm^{-1}). Besides, the peak at about 1740 cm^{-1} corresponding to the C=O was disappeared in ZnO/GO nanocomposite. This confirms the formation of ZnO NCs onto the surface of GO accompanied by a partial reduction of the GO. For further physical investigations, thermal gravimetric analysis has been performed (see Fig. 1 in the supplementary information).

Electrocatalytic activity of ZnO/GO

Carbon paste electrode was used in this study due to its ease of fabrication, modification and renewability of its active surface area. For achieving better direct electrochemical detection and identifying the role of the nanocomposite, different electrodes, ZnO, ZnO/GO nanocomposite or unmodified electrode, were prepared and their electrochemical responses were measured. In this regard, Ferricyanide (FCN) was used for testing the electrocatalytic performance of the prepared sensors. Therefore, the redox reactions of FCN using the modified and unmodified electrodes were obtained. As shown in Fig. 5a, the electrochemical peak currents, either the oxidation or the reduction peaks, increased after incorporating the nanomaterials, either ZnO or ZnO/GO, into the electrode matrix. However, the addition of ZnO/GO nanocomposite exhibited the highest electrochemical signals.

Due to the catalytic properties of graphene oxides, the conjugation of GO to the ZnO nanoparticles enhanced the electron transfer. GO role in the composite could be demonstrated as a wire-transfer for bridging the electron between the ZnO (that accepts the electrons from the redox molecules) and the electrode surface.

Consequently, the effect of ZnO/GO nanocomposite concentrations, within the electrode matrix, on the

(a)

(b)

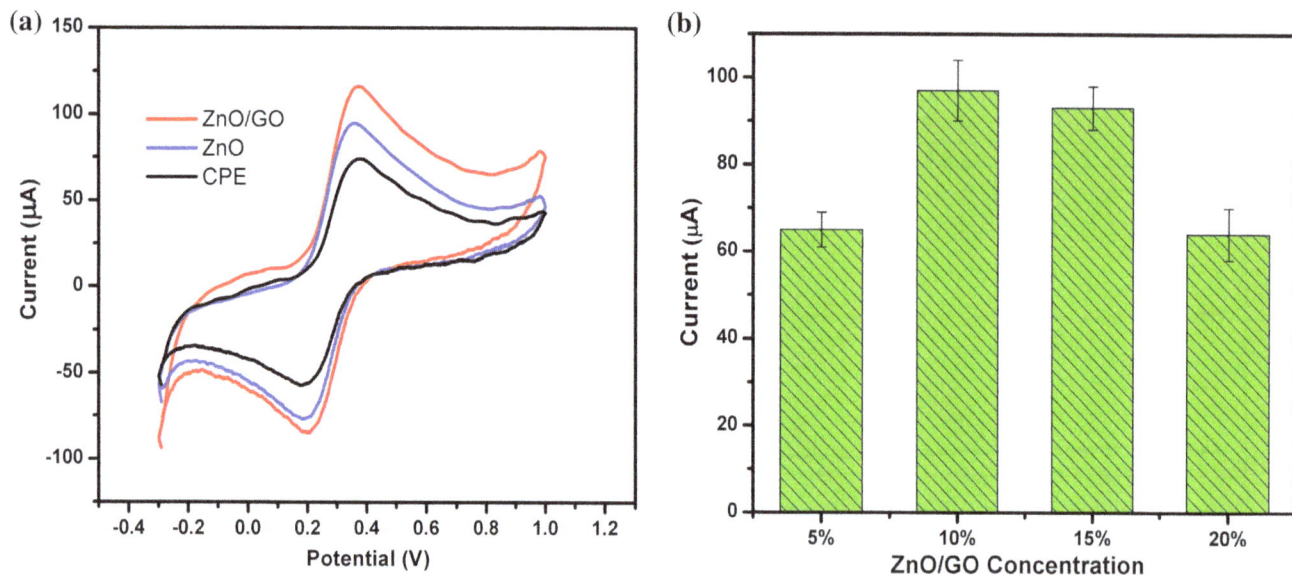

Fig. 5 **a** Cyclic voltammograms of different electrodes against FCN (2 mM) in KCl (0.1 M), **b** effect of ZnO/GO concentration on the electrochemical performance of FCN

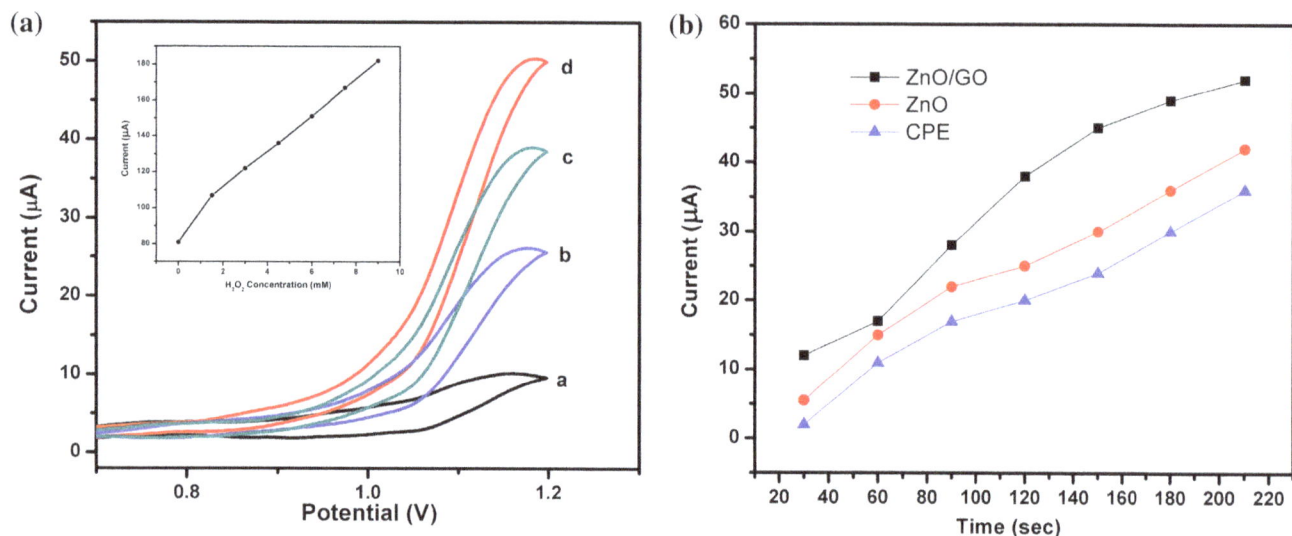

(a)

(b)

Fig. 6 **a** Cyclic voltammograms obtained at ZnO/GO/CPE in the *a* absence, and presence of *b* 1.5 mM, *c* 3 mM, and *d* 4.5 mM H_2O_2, The *inset* is the corresponding calibration curve between the current response and concentration of H_2O_2. **b** Amperometric response of different electrodes against H_2O_2

measurable oxidation current of FCN was investigated (Fig. 5b). Specifically, different concentrations ranging from 5 to 20 % (w/w) were selected. As shown in (Fig. 5b), the electrochemical signal of the 10 % w/w was found to be the highest. To that end, 10 % w/w of ZnO/GO nanocomposite was selected as optimal concentration. Moreover, the impact of FCN concentration was examined against 10 % w/w of ZnO/GO nanocomposite (see Fig. 2 in the supplementary information) and the results showed increase in electrochemical response with the increase of FCN concentration.

The use of ZnO/GO-based sensor for H_2O_2 detection

As the main concern of this study is to enable the direct detection of H_2O_2 using the developed nano-structured electrode, the capability of direct electron transfer was tested. In this regards, different concentrations of peroxides were injected into the electrochemical cell and the possibility of direct oxidation was measured. As shown in Fig. 6, the oxidation peak currents of H_2O_2 at ~ 1.1 V were obtained without the addition of artificial redox mediator. However, the chronoamperometric responses of the nano-composite were much higher than that obtained by either the ZnO-based electrode or the bare-CPE.

Therefore, the used nanostructured electrode has the enough electrocatalytic activity to enable the direct detections of peroxide. In addition, the oxidation currents were correlated with the concentrations of H_2O_2 which reflects the sensitivity as well as the reliability of the proposed sensor.

Effect of pH

Figure 7 shows the influence of pH on the performance of the ZnO/GO modified electrode surface towards the oxidation of peroxides. It was found that the oxidation current

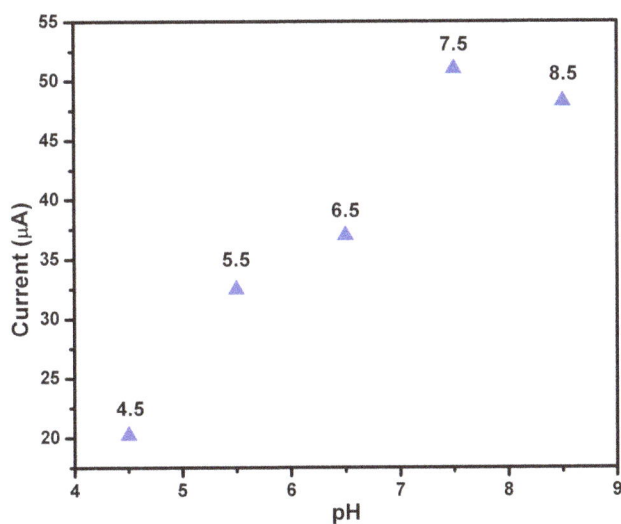

increased by increasing the pH values until reaching 7.4 which have the highest peak, after that the peak height decreased. The decreasing of oxidation current at lower pH may be attributed to the protonation of the electrode surface which resists the electron transfer. Therefore, PBS at pH 7.4 was selected as the supporting electrolyte in all subsequent experiments, given that the biosensors would be used under normal physiological conditions [27].

Chronoamperometric determinations of peroxide

Using the CVs measurements, the capability of direct oxidation of peroxide using the nanostructured electrode was confirmed. Consequently, the chronoamperometric measurements were performed at 1.1 V versus Ag/AgCl. A standard addition was done by adding a certain concentration of H_2O_2 at fixed time intervals (30 s). From the time/current curve (Fig. 8a), a fast response towards each addition was noted, which reflects the rapid electron transfer as well as the electrocatalytic power of the developed nanostructured electrode. Figure 8b shows the corresponding calibration curve that has a linear response range from 1 to 15 mM with the detection limit (S/N = 3) of 0.8 mM. Method sensitivity for the real physiological concentration of peroxides in the biological samples is sufficient.

Conclusion

In this study, we have reported a facile and cost-effective approach for the fabrication of ZnO NCs on the GO surface. The resulting ZnO/GO nanocomposite was

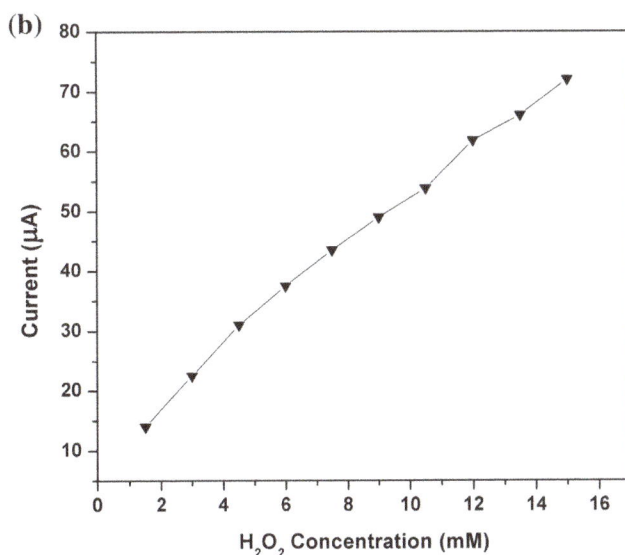

Fig. 7 Effect of pH on the oxidation peak of the composite-modified electrode in 0.1 M PBS

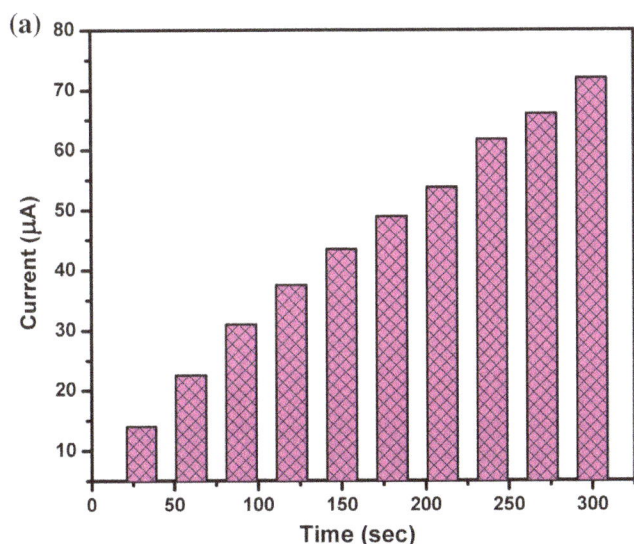

Fig. 8 a Amperometric response of the ZnO/GO electrode upon addition of H_2O_2 at 1.1 V. **b** The corresponding calibration curve between the current response and concentration of H_2O_2

incorporated into the carbon paste electrode matrix producing a significant improvement in the electrical conductivity, electrocatalytic activity which provides a direct electrochemical detection of superoxide. The effect of ZnO/GO nanocomposite concentrations incorporated into the electrode was investigated and it was found that 10 % (w/w) has the highest electrochemical signal. These results revealed that the developed nanocomposite with the high surface area and electrocatalytic activity offers great promise for a non-enzymatic biosensor.

Acknowledgments   The authors are grateful for the group leader of Biological Systems Analysis, Prof. Dr. Ursula (Bilitewski, Helmholtz Centre for Infection Research, HZI, and Braunschweig, Germany) for presenting the Potentiostat (Gamry Potentiostat/Galvanostat/ZRA G750).

Compliance with ethical standards

Conflict of interest   The authors declare that there are no conflicts of interest.

References

1. Thenmozhi, K., Narayanan, S.S.: Electrochemical sensor for H_2O_2 based on thionin immobilized 3-aminopropyltrimethoxy silane derived sol–gel thin film electrode. Sens. Actuators B Chem. **125**, 195–201 (2007)

2. Shen, G., Chen, P.C., Ryu, K., Zhou, C.: Devices and chemical sensing applications of metal oxide nanowires. J. Mater. Chem. **19**, 828–839 (2009)

3. Chu, D., Masuda, Y., Ohji, T., Kato, K.: Formation and photocatalytic application of ZnO nanotubes using aqueous solution. Langmuir **26**, 2811–2815 (2009)

4. Willander, M., Nur, O., Zhao, Q., Yang, L., Lorenz, M., Cao, B., et al.: Zinc oxide nanorod based photonic devices: recent progress in growth, light emitting diodes and lasers. Nanotechnology **20**, 332001 (2009)

5. Pearton, S., Norton, D., Heo, Y., Tien, L., Ivill, M., Li, Y., et al.: ZnO spintronics and nanowire devices. J. Electron. Mater. **35**, 862–868 (2006)

6. Godlewski, M., Guziewicz, E., Kopalko, K., Łuka, G., Łukasiewicz, M., Krajewski, T., et al.: Zinc oxide for electronic, photovoltaic and optoelectronic applications. Low Temp. Phys. **37**, 235–240 (2011)

7. Özgür, Ü., Alivov, Y.I., Liu, C., Teke, A., Reshchikov, M., Doğan, S., et al.: A comprehensive review of ZnO materials and devices. J. Appl. Phys. **98**, 041301 (2005)

8. Janotti, A., Van de Walle, C.G.: Native point defects in ZnO. Phys. Rev. B **76**, 165202 (2007)

9. Schulz, P., Kelly, L.L., Winget, P., Li, H., Kim, H., Ndione, P.F., et al.: Tailoring electron-transfer barriers for zinc oxide/C60 fullerene interfaces. Adv. Funct. Mater. **24**, 7381–7389 (2014)

10. Xie, L., Xu, Y., Cao, X.: Hydrogen peroxide biosensor based on hemoglobin immobilized at graphene, flower-like zinc oxide, and gold nanoparticles nanocomposite modified glassy carbon electrode. Colloids Surf. B Biointerfaces **107**, 245–250 (2013)

11. Chawla, S., Pundir, C.S.: An amperometric hemoglobin A1c biosensor based on immobilization of fructosyl amino acid oxidase onto zinc oxide nanoparticles-polypyrrole film. Anal. Biochem. **430**, 156–162 (2012)

12. Zhou, F., Zhao, X., Zheng, H., Shen, T., Tang, C.: Synthesis and electrochemical properties of ZnO 3D nanostructures. Chem. Lett. **34**, 1114–1115 (2005)

13. Palanisamy, S., Cheemalapati, S., Chen, S.M.: Highly sensitive and selective hydrogen peroxide biosensor based on hemoglobin immobilized at multiwalled carbon nanotubes-zinc oxide composite electrode. Anal. Biochem. **429**, 108–115 (2012)

14. Wang, H., Pan, Q., Cheng, Y., Zhao, J., Yin, G.: Evaluation of ZnO nanorod arrays with dandelion-like morphology as negative electrodes for lithium-ion batteries. Electrochim. Acta **54**, 2851–2855 (2009)

15. Kang, C.G., Kang, J.W., Lee, S.K., Lee, S.Y., Cho, C.H., Hwang, H.J., et al.: Characteristics of CVD graphene nanoribbon formed by a ZnO nanowire hardmask. Nanotechnology **22**, 295201 (2011)

16. Wan, Y., Wang, Y., Wu, J., Zhang, D.: Graphene oxide sheet-mediated silver enhancement for application to electrochemical biosensors. Anal. Chem. **83**, 648–653 (2011)

17. Wang, J., Li, Y., Ge, J., Zhang, B.P., Wan, W.: Improving photocatalytic performance of ZnO via synergistic effects of Ag nanoparticles and graphene quantum dots. Phys. Chem. Chem. Phys. **17**, 18645–18652 (2015)

18. Yang, K., Xu, C., Huang, L., Zou, L., Wang, H.: Hybrid nanostructure heterojunction solar cells fabricated using vertically aligned ZnO nanotubes grown on reduced graphene oxide. Nanotechnology **22**, 405401 (2011)

19. Geng, W., Zhao, X., Zan, W., Liu, H., Yao, X.: Effects of the electric field on the properties of ZnO-graphene composites: a density functional theory study. Phys. Chem. Chem. Phys. **16**, 3542–3548 (2014)

20. Chen, J., Li, C., Eda, G., Zhang, Y., Lei, W., Chhowalla, M., et al.: Incorporation of graphene in quantum dot sensitized solar cells based on ZnO nanorods. Chem. Commun. **47**, 6084–6086 (2011)

21. Erman, J.E., Vitello, L.B., Mauro, J.M., Kraut, J.: Detection of an oxyferryl porphyrin. pi.-cation-radical intermediate in the reaction between hydrogen peroxide and a mutant yeast cytochrome c peroxidase. Evidence for tryptophan-191 involvement in the radical site of compound I. Biochemistry **28**, 7992–7995 (1989)

22. Zhou, M., Diwu, Z., Panchuk-Voloshina, N., Haugland, R.P.: A stable nonfluorescent derivative of resorufin for the fluorometric determination of trace hydrogen peroxide: applications in detecting the activity of phagocyte NADPH oxidase and other oxidases. Anal. Biochem. **253**, 162–168 (1997)

23. Marcano, D.C., Kosynkin, D.V., Berlin, J.M., Sinitskii, A., Sun, Z., Slesarev, A., et al.: Improved synthesis of graphene oxide. ACS Nano **4**, 4806–4814 (2010)

24. Hassan, R.Y., Bilitewski, U.: Direct electrochemical determination of Candida albicans activity. Biosens. Bioelectron. **49**, 192–198 (2013)

25. Bindu, P., Thomas, S.: Estimation of lattice strain in ZnO nanoparticles: X-ray peak profile analysis. J. Theor. Appl. Phys. **8**, 123–134 (2014)

26. Janotti, A., Van de Walle, C.G.: Fundamentals of zinc oxide as a semiconductor. Rep. Prog. Phys. **72**, 126501 (2009)

27. Butwong, N., Zhou, L., Moore, E., Srijaranai, S., Luong, J.H., Glennon, J.D.: A highly sensitive hydrogen peroxide biosensor based on hemoglobin immobilized on cadmium sulfide quantum dots/chitosan composite modified glassy carbon electrode. Electroanalysis **26**, 2465–2473 (2014)

20

Theoretical study of the adsorption of NOx on TiO2/MoS2 nanocomposites: a comparison between undoped and N-doped nanocomposites

Amirali Abbasi[1,2,3] · Jaber Jahanbin Sardroodi[1,2,3]

Abstract First-principle calculations within density functional theory were performed to investigate the interactions of NO and NO_2 molecules with TiO_2/MoS_2 nanocomposites. Given the need to further comprehend the behavior of the NO_x molecules positioned between the TiO_2 nanoparticle and MoS_2 monolayer, we have geometrically optimized the complex systems consisting of the NO_x molecule oriented at appropriate positions between the nanoparticle and MoS_2 monolayer. The structural properties, such as bond lengths, bond angles, adsorption energies and Mulliken population analysis, and the electronic properties, including the density of states and molecular orbitals, were also analyzed in detail. The results indicate that the interactions between NO_x molecules and N-doped TiO_2 in TiO_2-N/MoS_2 nanocomposites are stronger than those between gas molecules and undoped TiO_2 in TiO_2/MoS_2 nanocomposites, which reveal that the N-doping helps to strengthen the interaction of toxic gas molecules with hybrid TiO_2/MoS_2 nanocomposites. The N-doped TiO_2/MoS_2 nanocomposites have higher sensing capabilities than the undoped ones, and the interaction of NO_x molecules with N-doped nanocomposites is more favorable in energy than the interaction with undoped nanocomposites. Therefore, the obtained results also present a theoretical basis for the potential application of TiO_2/MoS_2 nanocomposite as an extremely sensitive gas sensor for NO and NO_2 molecules.

Graphical Abstract

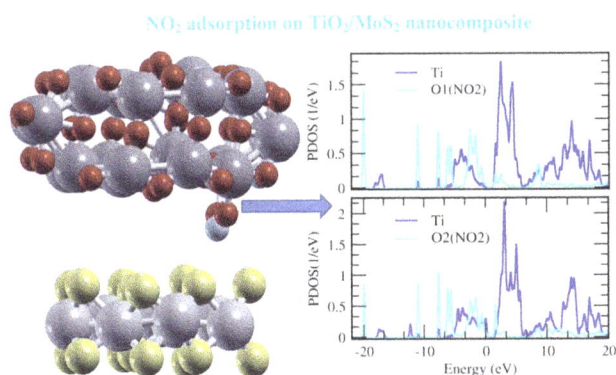

NO2 adsorption on TiO2/MoS2 nanocomposite

Keywords Density functional theory · NO_x · TiO_2/MoS_2 nanocomposite · Density of States · Adsorption

✉ Amirali Abbasi
a_abbasi@azaruniv.edu

1 Molecular Simulation Laboratory (MSL), Azarbaijan Shahid Madani University, Tabriz, Iran

2 Department of Chemistry, Faculty of Basic Sciences, Azarbaijan Shahid Madani University, Tabriz, Iran

3 Computational Nanomaterials Research Group (CNRG), Azarbaijan Shahid Madani University, Tabriz, Iran

Introduction

Titanium dioxide (TiO_2, Titania) has aroused great attentions as an important semiconductor material due to its effectiveness and outstanding properties, such as non-toxicity, low cost, high catalytic efficiency, photoactivity [1], and stability. TiO_2 has been widely utilized in many fields, such as photo-catalysis, gas sensing, organic dye-sensitized solar cells, water splitting, and pollutant degradation [2–5]. Three important polymorphs were found for TiO$_2$, namely, anatase, rutile, and brookite [6], in which anatase and rutile forms are the most widely studied ones in different fields of science and technology. The photocatalytic applications of TiO$_2$ were restricted due to its wide bandgap (3.2 eV), which allows the absorption of the solar spectrum at the ultraviolet region by a lower percentage (3–5 % of the incoming solar light). The doping of TiO$_2$ anatase with some nonmetal elements, such as nitrogen, is a convenient solution, which would enhance the photo-efficiency of TiO$_2$ to the visible region and improve its photocatalytic activity [7, 8]. Two-dimensional (2D) semiconductor

Fig. 1 Representation of NO and NO$_2$ molecules in a large cubic supercell

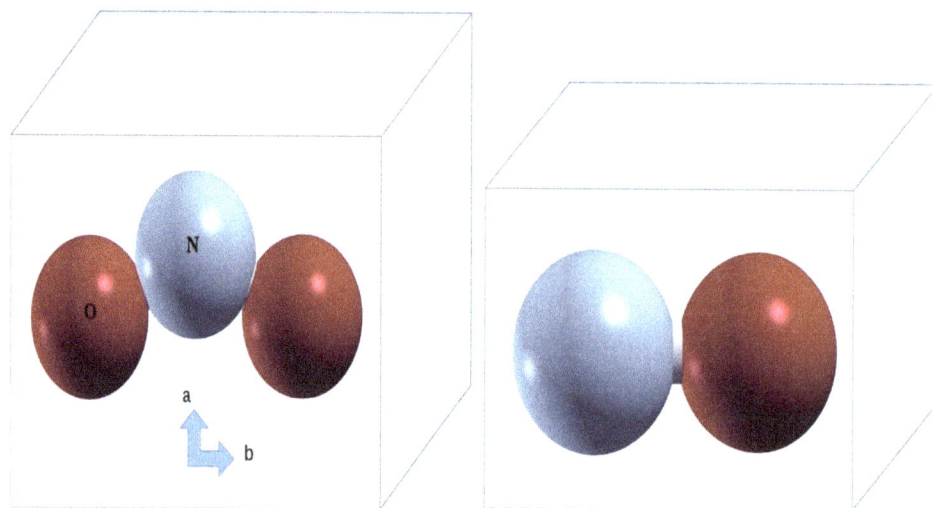

Fig. 2 Optimized structure of the chosen MoS$_2$ monolayer with area values, **a** front view and **b** lateral view. Mo atoms are sketched by *gray balls* and S atoms by *yellow balls*

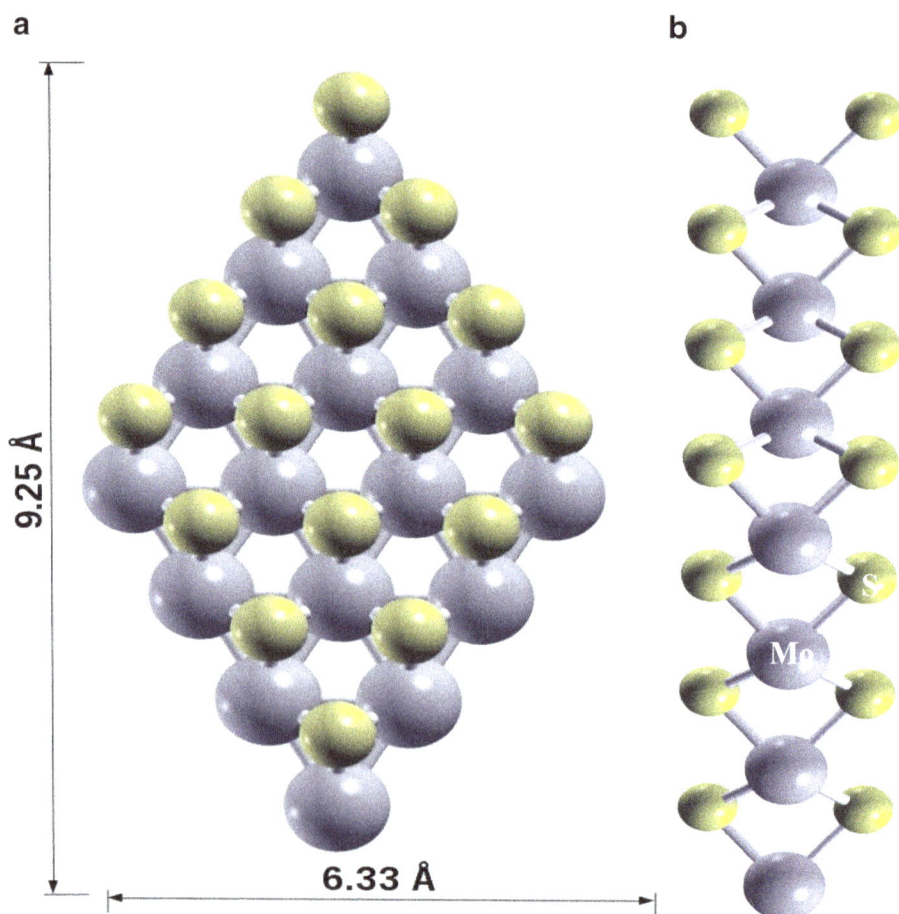

materials, such as MoS_2 [9], and other dichalcogenides consisting of transition metals, such as $MoSe_2$, WS_2, and so on, indicate the final scale for chalcogenide dimension around the vertical direction. MoS_2, a layered structure consisting of Mo and S atoms arranged in hexagonal structure of atomic sheets of molybdenum and sulphur atoms, attracts numerous attentions due to the excellent electrical, mechanical, and optical properties, such as satisfied bandgap, thermal stability, carrier mobility, and so on [10–12]. MoS_2 has been broadly used in abundant applications, such as photocatalysts, nanotribology, lithium battery, dry lubrication, hydrodesulfurization catalyst, and photovoltaic cell because of its unique electronic, photosensitive, and catalytic properties [13–16]. Nanoelectronic devices fabricated on 2D materials, such as MoS_2, suggest also many efficiencies for these layered materials, which cause the further miniaturization of the integrated circuits beyond Moore's law. Recently, numerous electronic devices were made using the few-layer MoS_2 as an important component, such as field-effect transistors [17], sensors [18], etc. However, several computational studies of N-doped TiO_2 anatase nanoparticles and few-layer MoS_2 structures have been separately published, describing some of the main electronic and physical properties of these materials. Particularly, the gas-sensing capabilities of MoS_2-based field-effect transistors and sensing films for NO and NH_3 were experimentally revealed with an enhanced sensitivity in some other works [19, 20]. TiO_2/MoS_2 nanocomposites have been successfully synthesized for different purposes by some experimental methods [21–23]. There are no explanative computational studies on the adsorption behaviors of TiO_2/MoS_2 nanocomposites. NO_x molecules have been characterized as toxic gases which are mainly emitted from power plants and vehicle engines. For the general public, the most outstanding provenance of NO_2 is internal combustion engines that burn fossil fuels to work properly. In indoor places, NO_2 emission mainly stems from cigarette smoke, kerosene heaters, and stoves. Therefore, optimal removal of these harmful molecules is an important subject to human health and environmental protection [24]. In this study, the interaction of NO_x molecules with TiO_2/MoS_2 nanocomposites has been investigated by density functional theory (DFT) computations. We present here the results of calculations of complex systems consisting of NO_x molecule positioned between the TiO_2 anatase nanoparticle and MoS_2 monolayer. The electronic structure of the adsorption systems has also been analyzed, including the projected density of states (PDOSs) and molecular orbitals (MOs). The main aim of this study is to supply an overall understanding on the adsorption behaviors of nano-TiO_2/MoS_2 composites as highly sensitive NO_x sensors.

Computational details and structural models

Methodology

DFT calculations [25, 26] were performed as implemented in the Open-Source Package for Material eXplorer (OPENMX3.8) [27], being a well-organized software package for nano-scale materials simulations based on DFT, PAO basis functions, and VPS pseudopotentials [28, 29]. Pseudo-atomic orbitals were utilized as basis sets in the geometry optimizations. The considered cut-off energy is set to the value of 150 Rydberg in our calculations [29], The PAOs are generated via the basis sets (3-s, 3-p, and 1-d) for Ti atom, (3-s, 3-p, and 2-d) for Mo atom, (2-s and 2-p) for O and N atoms, (3-s and 3-p) for S atom with the chosen cut-off radii of 7 for Ti, 9 for Mo, 5 for O and N, and 8 for S (all in Bohrs). The generalized gradient approximation (GGA) of Perdew–Burke–Ernzerhof (PBE) was used to describe the exchange–correlation energy functional [30]. The convergence criterion of self-consistent field calculations was set at 1.0×10^{-6} Hartree, whereas that of energy calculation was chosen to be 1.0×10^{-4} Hartree/bohr. For the geometry optimization, 'Opt' is used as the geometry optimizer, which is a robust and efficient scheme. The crystalline and molecular structure visualization program, XCrysDen [31], was employed for displaying molecular orbital isosurfaces. The box considered in these computations contains 96 atoms (24 Ti, 48 O, 8 Mo, and 16 S atoms) of undoped or N-doped TiO_2 nanoparticle with MoS_2 monolayer. The Gaussian broadening method for evaluating electronic DOS is used. For NO_x adsorption on the TiO_2/MoS_2 nanocomposite, the adsorption energy is computed via the following formula:

$$\Delta E_{ad} = E_{(composite+adsorbate)} - E_{composite} - E_{adsorbate} \qquad (1)$$

where $E_{(composite + adsorbate)}$ is the total energy of the adsorption system, $E_{composite}$ is the energy of the TiO_2/MoS_2 nanocomposite, and $E_{adsorbate}$ represents the energy of non-adsorbed NO_x molecules. Based on this relation, the most stable configurations would have negative adsorption energies. A higher adsorption energy corresponds to a stronger adsorption between host and adsorbed molecule.

Model building

NO and NO2 molecule model

The chemical formulae of nitric oxide and nitrogen dioxide molecules are NO and NO_2. NO has linear structure, while

NO$_2$ represents a bent geometrical structure. The structures of NO and NO$_2$ molecules were represented in Fig. 1. Distances and angles of the considered molecules were computed in a large cubic supercell. The calculated N–O bond length of free NO molecule is 1.16 Å, while for the bent structure of NO$_2$ molecule, the bond length and bond angles were calculated to be 1.20 Å and 134°, respectively. All these computed values are in comprehensive agreement with the computational results and the experimentally reported data [32].

MoS$_2$ model

Molybdenum disulfide (MoS$_2$) is a layered structure containing molybdenum transition metal, which belongs to the family of two-dimensional dichalcogenides. A hexagonally arrangement of atomic sheets of MoS$_2$ containing Mo and S atoms set as an S–Mo–S sandwich forms MoS$_2$

monolayer. The monolayer of MoS$_2$ model studied here contains 24 atoms in total (8 Mo and 16 S atoms). MoS$_2$ structure is relaxed for calculating the optimized structural parameters. The calculated S–Mo bond length, Mo–Mo distance, and S–S distance in monolayer are 2.43, 3.20, and 3.15 Å, respectively. These computed bond lengths are somewhat consistent with the values of bulk material [33], in reasonable agreement with the reported data [34, 35]. However, there are negligible discrepancies between the results of MoS$_2$ and its bulk material, which can be ignored. The area of the MoS$_2$ slab is 9.25 Å × 6.33 Å. The optimized structure of MoS$_2$ model was displayed in Fig. 2.

TiO$_2$ anatase model

A 3 × 2 × 1 supercell of TiO$_2$ anatase along x, y, and z directions was utilized for constructing the considered TiO$_2$

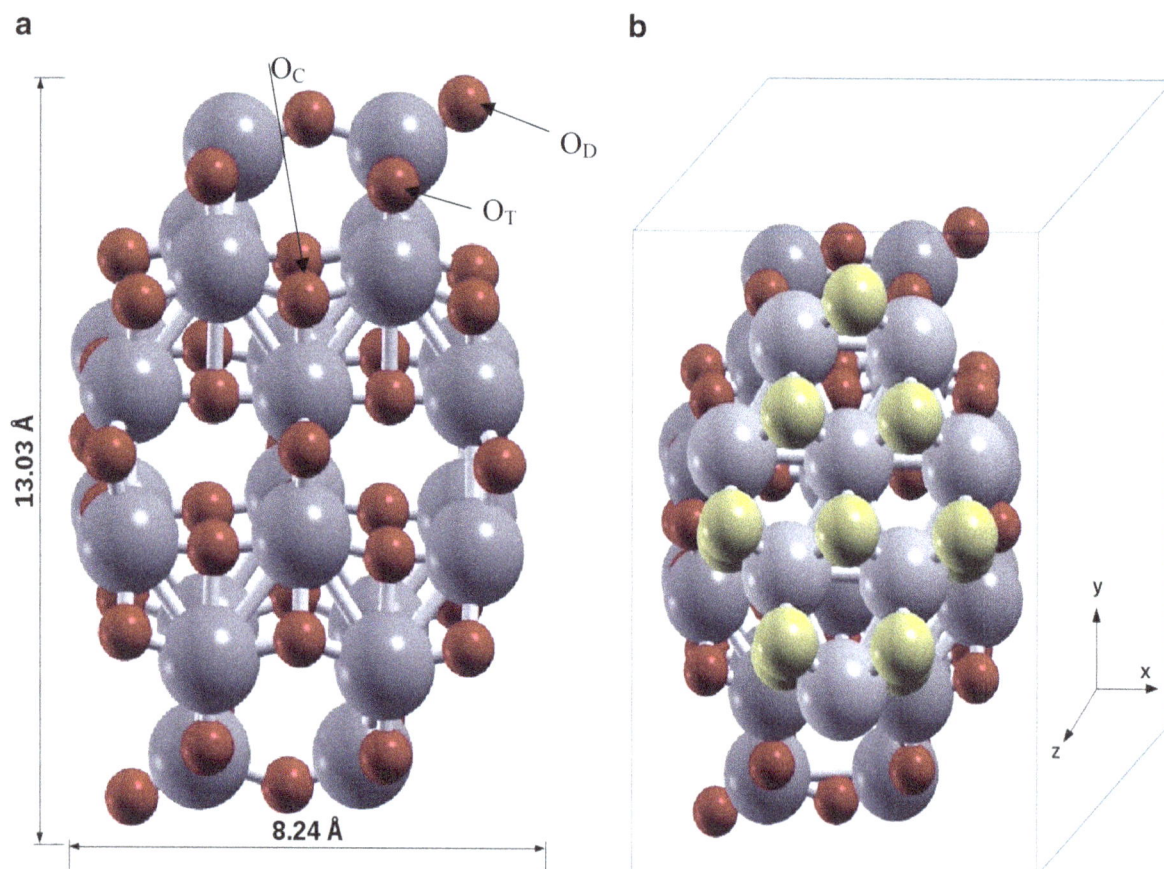

Fig. 3 Optimized structures of **a** undoped 72 atom TiO$_2$ anatase nanoparticle constructed using the 3 × 2 × 1 unit cells (O$_C$: central oxygen; O$_T$: twofold coordinated oxygen; O$_D$: dangling oxygen) and

b TiO$_2$/MoS$_2$ nanocomposite constructed from TiO$_2$ nanoparticle and MoS$_2$ monolayer. Ti atoms are sketched by *dark gray balls*, O atoms by *red balls* and N atoms by *blue balls*

anatase nanoparticles containing 72 atoms. The unit cell is available at "American Mineralogists Database" webpage [36] and was reported by Wyckoff [37]. Two appropriate oxygen atoms of TiO_2 nanoparticle were replaced by nitrogen atoms to model the N-doped particles. Doping of TiO_2 nanoparticle with nitrogen atom is done according to two doping positions. These two doping positions refer to the middle oxygen and twofold coordinated oxygen atoms substitutions illustrated by O_C and O_T in Fig. 3a, respectively. The area of the anatase nanoparticle is 13.03 Å × 8.24 Å. The crystalline structure of TiO_2 contains two kinds of titanium atoms, namely, fivefold coordinated titanium (5f-Ti) and sixfold coordinated one (6f-Ti), and two kinds of oxygen atoms, namely, threefold coordinated oxygen (3f-O) and twofold coordinated one (2f-O) atoms (see Fig. 3a). [38] It was found that the twofold coordinated oxygen and fivefold coordinated titanium atoms are more reactive than the threefold coordinated oxygen and sixfold coordinated titanium atoms due to the undercoordination in twofold coordinated oxygen and fivefold coordinated titanium atoms. The thickness of the vacuum

spacing is 11.5 Å, which is helpful to avoid the additional interactions between the neighbor particles. The optimized structure of TiO_2/MoS_2 nanocomposite was displayed in Fig. 3b. Figure 4 also displays the optimized geometries of the N-doped TiO_2/MoS_2 nanocomposites. The results of geometrical optimizations represent that the O_T-substituted TiO_2/MoS_2 nanocomposite is more favorable in energy than the O_C-substituted one.

Results and discussion

Bond lengths, bond angles, and adsorption energies

NO interacts with TiO_2/MoS_2 nanocomposites

For NO molecule, three adsorption configurations are studied here, including the adsorption configurations of types A, B, and C, as shown in Fig. 5. The calculated adsorption energy values for NO_x molecule adsorbed on the considered nanocomposites have been listed in Table 1.

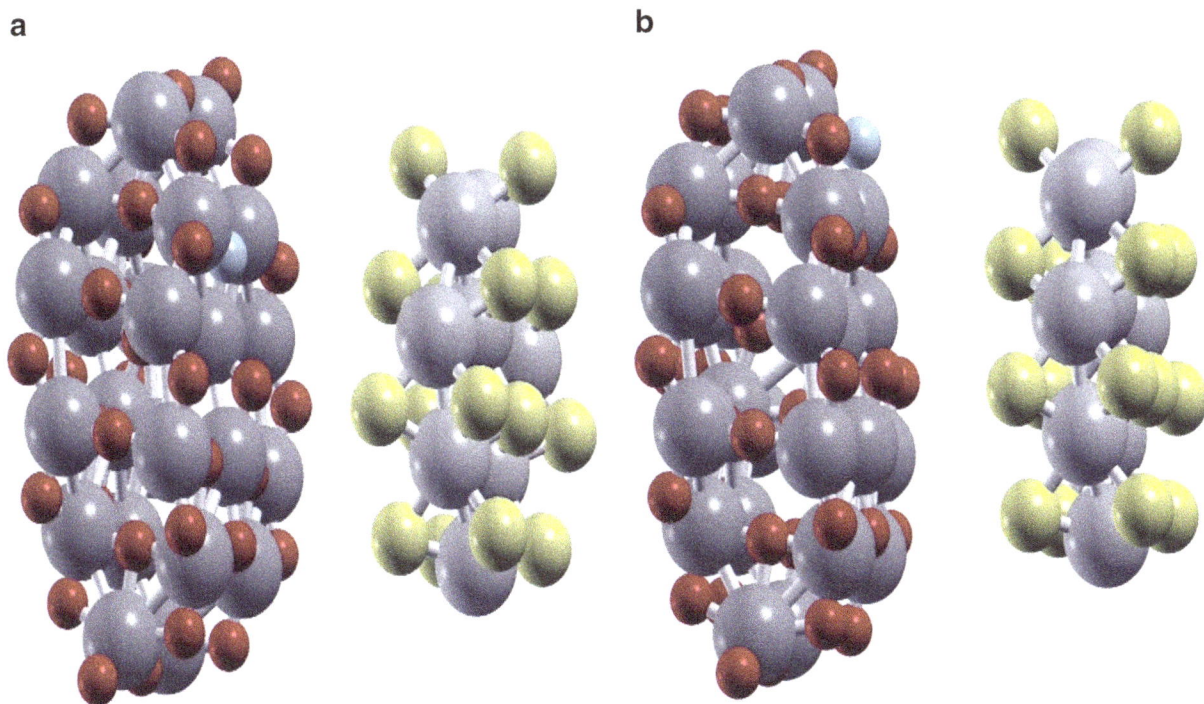

a

b

Fig. 4 Optimized geometries of two types of N-doped TiO_2/MoS_2 nanocomposites. **a** O_C-substituted nanocomposite, which refers to substitution of a threefold coordinated oxygen atom and **b** O_T-substituted nanocomposite, representing the substitution of a twofold coordinated oxygen atom

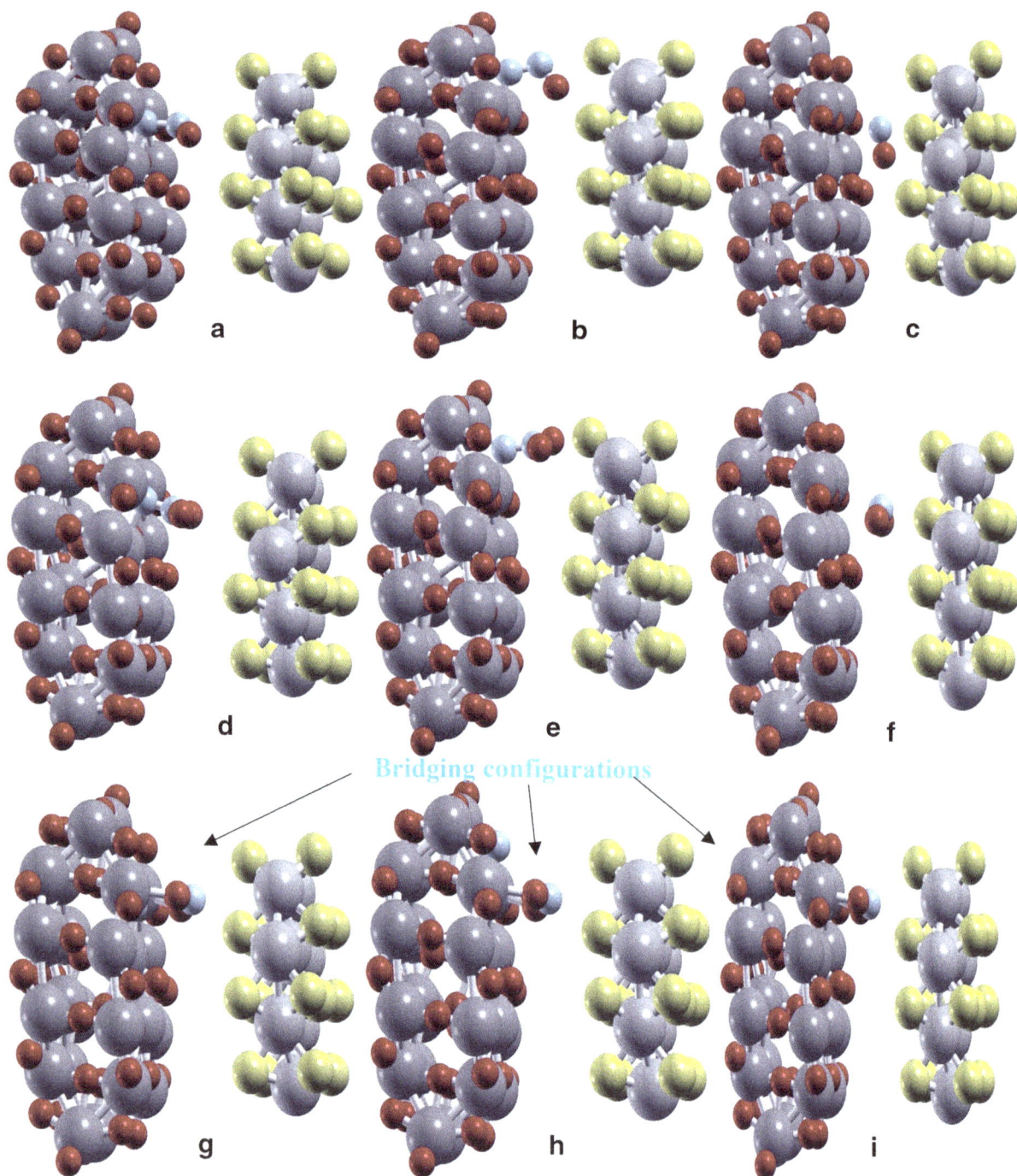

Fig. 5 Optimized geometry configurations of the interaction of NO and NO$_2$ molecules with TiO$_2$/MoS$_2$ nanocomposites. The NO molecule is preferentially adsorbed on the doped nitrogen site of TiO$_2$ nanoparticle, whereas NO$_2$ is adsorbed on both the doped nitrogen atom and fivefold coordinated titanium atoms. Configurations A–C represent the interaction of nanocomposites with NO molecules and the configurations D–I show the interaction between nanocomposites and NO$_2$ molecules

Table 1 Bond lengths (in Å), Mulliken charges, and adsorption energies (in eV) for NO_x molecule adsorbed on TiO_2/MoS_2 nanocomposites

Bond Composite	$N–O_1$ bond length	$N–O_2$ bond length	New N–N bond length	Average Ti–N length	New N–O bond length	Mulliken charge	Adsorption energy
NO adsorption							
Non-adsorbed	1.16	–	–	1.84 1.94	–	−0.122	−2.18
a	1.23		1.63	2.12	–		
b	1.30	–	1.43	1.99	–	−0.066	−3.01
c	1.19	–	–	–	3.16	−0.269	−0.50
NO_2 adsorption							
Non-adsorbed	1.20	1.20	–	1.84 1.94			
d	1.31	1.30	1.45	2.18		−0.064	−1.43
e	1.30	1.30	1.48	2.01		−0.107	−2.12
f	1.28	1.28	–	–	2.50	−0.014	−0.22

Table 2 Bond lengths (in Å), angles (in degrees), Mulliken charges, and adsorption energies (in eV) for complexes providing two contacting point between NO_2 molecule TiO_2/MoS_2 nanocomposites

Composite	New Ti–O1 bond length	New Ti–O2 bond length	N–O1 bond length	N–O2 bond length	O–N–O bond angle	Mulliken charge	Adsorption energy
g	2.41	2.53	1.28	1.29	122.3	−0.101	−1.42
h	2.03	2.28	1.30	1.36	117.4	0.180	−2.22
i	2.14	2.36	1.30	1.34	120.6	0.145	−0.92

Fig. 6 Total density of states for N-doped TiO_2 and two types of N-doped TiO_2/MoS_2 nanocomposites. **a** O_C-substituted nanocomposite and **b** O_T-substituted one

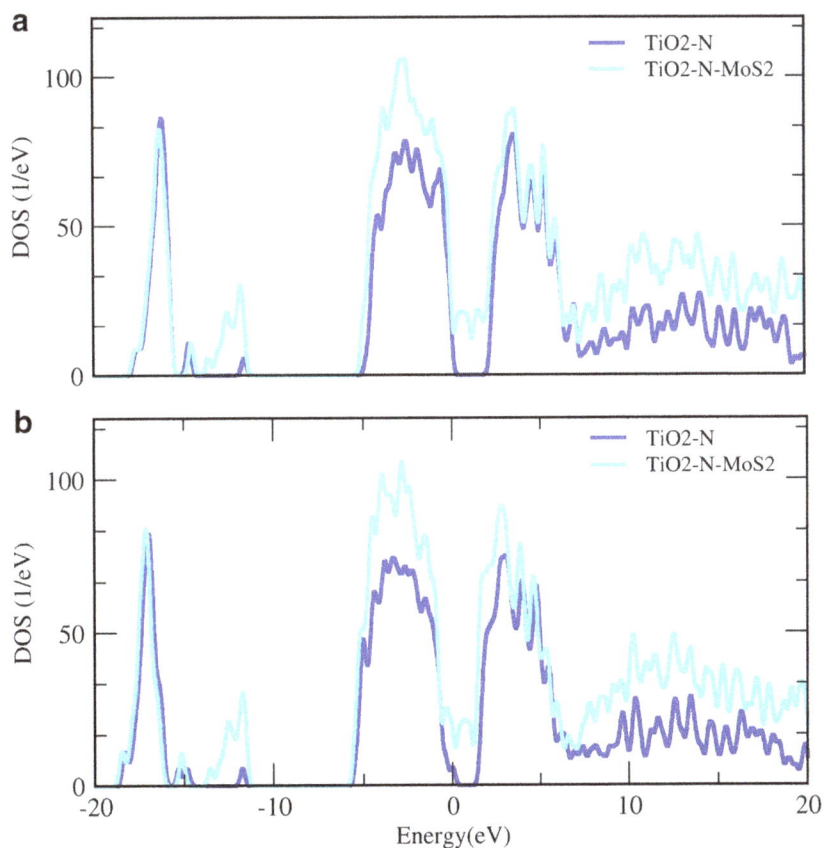

This figure presents the orientations of NO molecule towards the N-doped TiO_2/MoS_2 nanocomposites. For instance, complex A was made from the TiO_2/MoS_2 nanocomposite with O_C-substituted TiO_2 nanoparticle and NO molecule with nearly downward oxygen. NO can stably adsorb on N-doped TiO_2/MoS_2 nanocomposite, compared to the adsorption on the undoped one. The nitrogen atom of NO molecule is preferentially attracted to the doped nitrogen atom of nanocomposite, resulting in the formation of chemical bond between these two nitrogen atoms. The adsorption energy of the NO molecule on the

N-doped TiO_2/MoS_2 nanocomposite (composite A or B) is much higher than that of undoped TiO_2/MoS_2 nanocomposite, which reveals that NO molecule has a stronger interaction with N-doped nanocomposite than with undoped one. These results also indicate that the interaction of NO molecules with N-doped TiO_2/MoS_2 nanocomposites is more energetically favorable than the interaction of NO with undoped nanocomposites. This implies that the N-doped nanocomposite adsorbs NO molecule more effectively compared to the undoped one. Besides this, the configuration B is the most

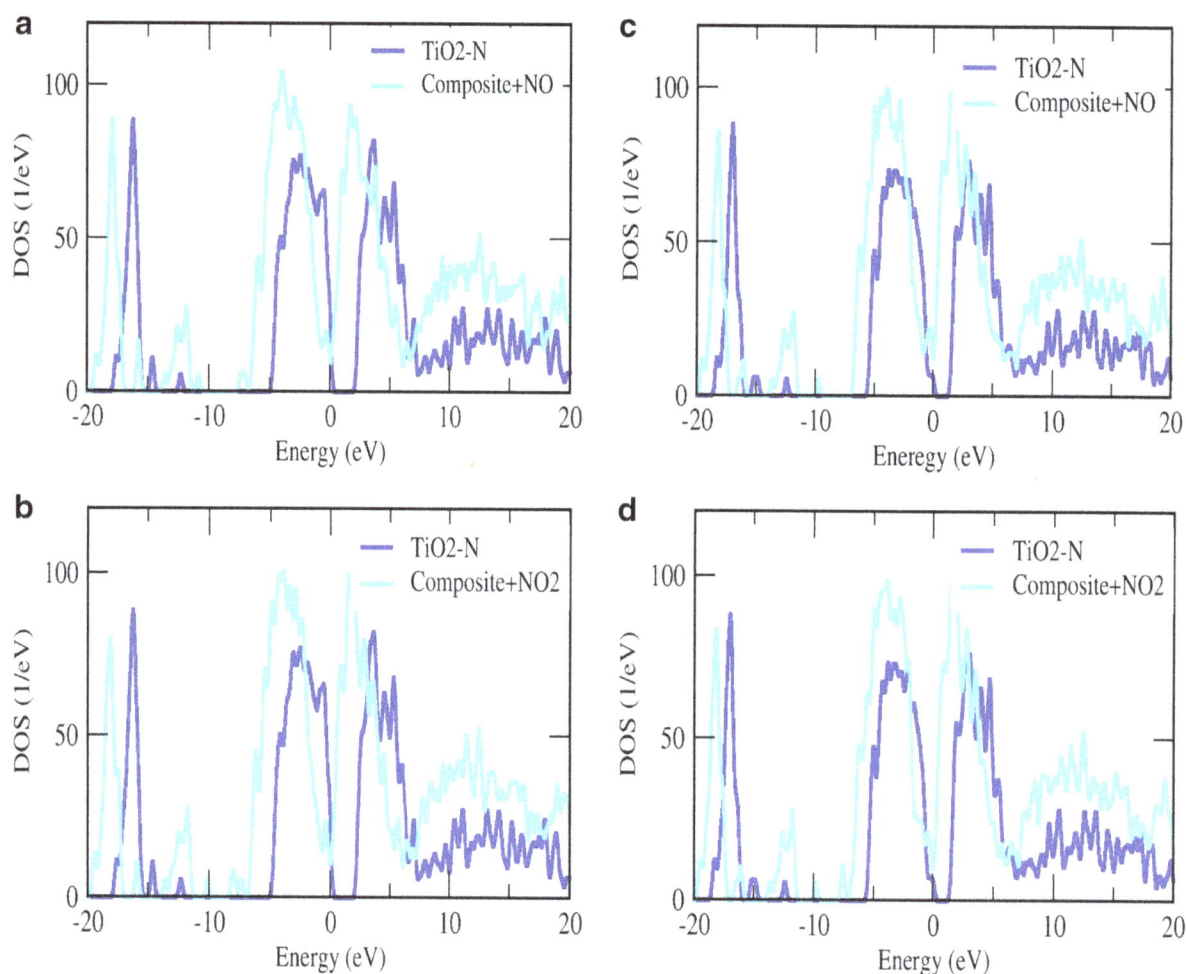

Fig. 7 DOS for the different adsorption configurations of the NO_x molecule on the considered TiO_2/MoS_2 nanocomposites, **a** A complex (NO molecule adsorbed on the O_C-substituted nanocomposite); **b** D complex (NO$_2$ molecule adsorbed on the O_C-substituted

nanocomposite); **c** B complex (NO molecule adsorbed on the O_T-substituted nanocomposite); **d** E complex (NO$_2$ molecule adsorbed on the O_T-substituted nanocomposite)

stable configuration compared to the configuration A due to its more negative adsorption energy. Configuration B contains O_T-substituted nanocomposite with adsorbed NO molecule. The adsorption energy of this configuration is more negative than that of other configurations, which suggests that the adsorption of NO molecule on the O_T-substituted nanocomposite is more energy favorable than the adsorption on the O_C-substituted one (see Table 1). Since a greater value of adsorption energy gives rise to a strong interaction between adsorbate and the adsorbent, it can be seen that there is a stronger interaction between NO and N-doped nanocomposite compared to NO and undoped nanocomposite, implying the dominant effect of N-doping. It means that the nitrogen doping strengthens the interaction of NO with TiO₂/MoS₂ nanocomposites. The greater

the adsorption energy, the higher tendency for adsorption, and, therefore, more efficient adsorption. Table 1 summarizes the bond length values before and after the adsorption of NO molecule on the nanocomposites. The bond lengths given in this table are included N–O bonds of NO_x molecule, average Ti–N distance, and new N–N and N–O distances between the nanocomposite and adsorbed NOx molecule. The values reported in this table show that the Ti–N bonds and N–O bond of the adsorbed NO molecule are elongated, because the electronic density transfers from the Ti–N bonds of N-doped TiO₂ and N–O bond of the adsorbed NO molecule to the newly formed N–N and N–O distances between the nanocomposite and molecule. This transfer of electronic density indicates that the N–O bond of NO molecule is weakened after the adsorption.

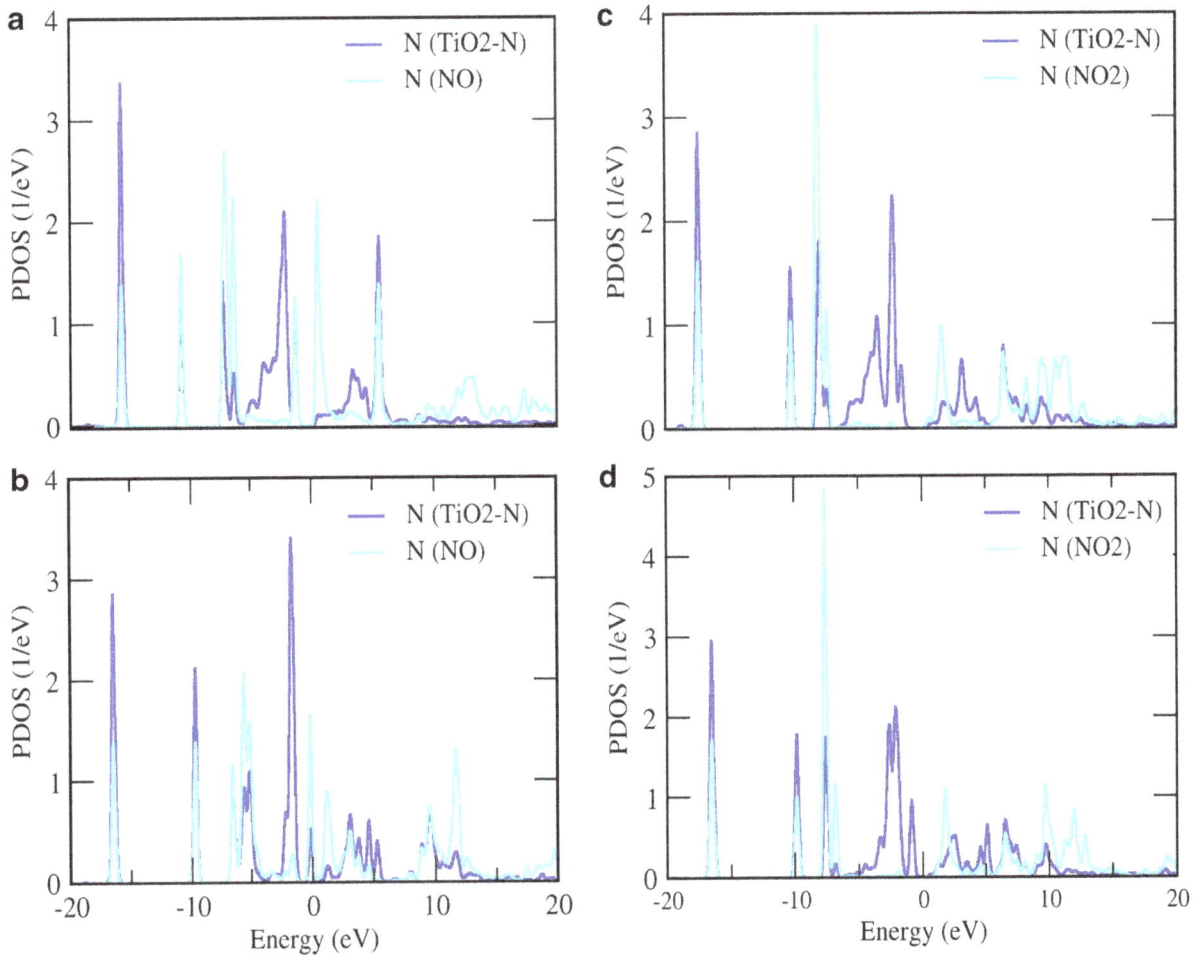

Fig. 8 PDOS for the interaction of NOₓ molecule with TiO₂/MoS₂ nanocomposites, **a** A complex; **b** B complex; **c** D complex; **d** E complex

NO₂ interacts with TiO₂/MoS₂ nanocomposites

The interaction of NO₂ molecule with the substituted nitrogen atom of TiO₂/MoS₂ nanocomposites has also been displayed in Fig. 5 as specified by types D-F adsorption geometries. For the doped nitrogen site, the adsorption process is expected to be more energy favorable than that on the dangling oxygen atom site. The reason can be simply sought using the data collected in Table 1. Similarly, the NO₂ molecule preferentially interacts with the doped nitrogen site on the nanoparticle surface, in comparison with the other surface oxygen atoms. Table 1 also lists the lengths for Ti–N bonds and N–O bonds of the adsorbed NO₂ molecule and the newly formed N–N and N–O distances. The results of this table indicate that the Ti–N bonds of the nanocomposite and the N–O bonds of the NO₂ molecule are stretched after NO₂ adsorption. Because the electronic density transfers from the Ti–N bonds and N–O bonds to the newly formed distances between the nanocomposite and

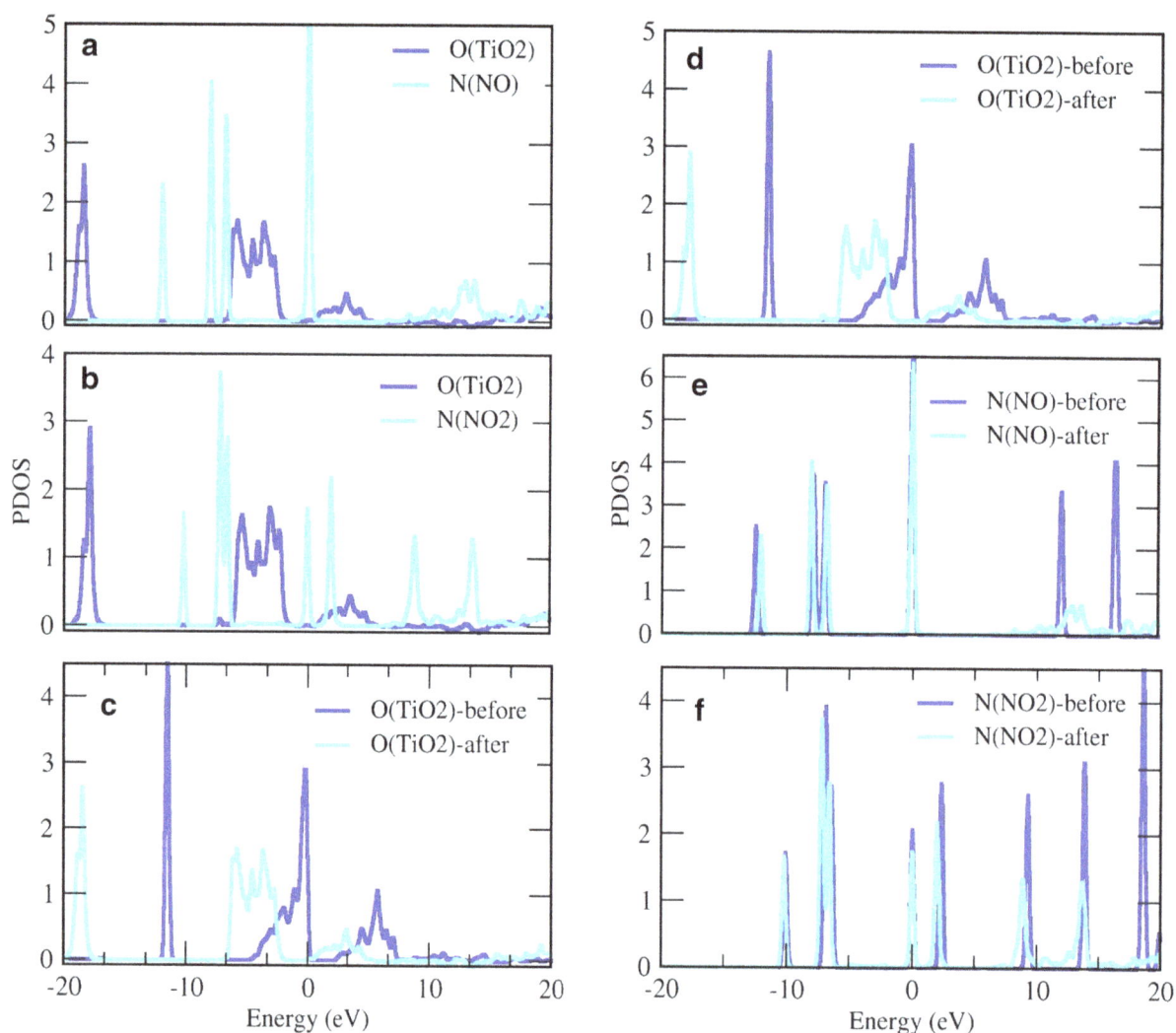

Fig. 9 PDOS for the interaction of NOₓ molecule with TiO₂/MoS₂ nanocomposites, **a** C complex (NO molecule adsorbed on the pristine nanocomposite); **b** F complex (NO₂ molecule adsorbed on the pristine nanocomposite); **c** C complex; **d** F complex; **e** C complex; **f** F complex

adsorbed NO_2 molecule. In configuration D, the nitrogen atom in the NO_2 molecule interacts with the doped nitrogen site on the TiO_2 nanoparticle to form a strong chemical bond and, therefore, strong interaction (1.45 Å N–N bond length). Among three models for NO adsorption, the adsorption configuration in which the nitrogen atom of NO interacts with the doped nitrogen site of TiO_2 at O_T position is the most energy favorable one. In the case of NO_2 adsorption, the adsorption energy of configuration E is much higher (more negative) than that of configuration D and undoped system adsorption (configuration F). It should be noted that the adsorption on the O_T-substituted nanocomposite leads to the

stable configurations (stronger interactions), compared to the adsorptions on the O_C-substituted one. As can be seen from Tables 1 and 2, the adsorption energies on the O_T-substituted nanocomposites (complexes E and H) are more negative than the adsorption energies on the O_C-substituted ones, implying that NO_2 adsorption on the O_T-substituted nanocomposites is energetically more favorable than the adsorption on the O_C-substituted ones. As a result, the adsorption of NO_2 molecule on the N-doped TiO_2/MoS_2 nanocomposite is more energy favorable than the adsorption of NO_2 on the undoped nanocomposite, indicating that the N-doped nanocomposite can adsorb NO_2 molecule more efficiently. Thus, the N-doping

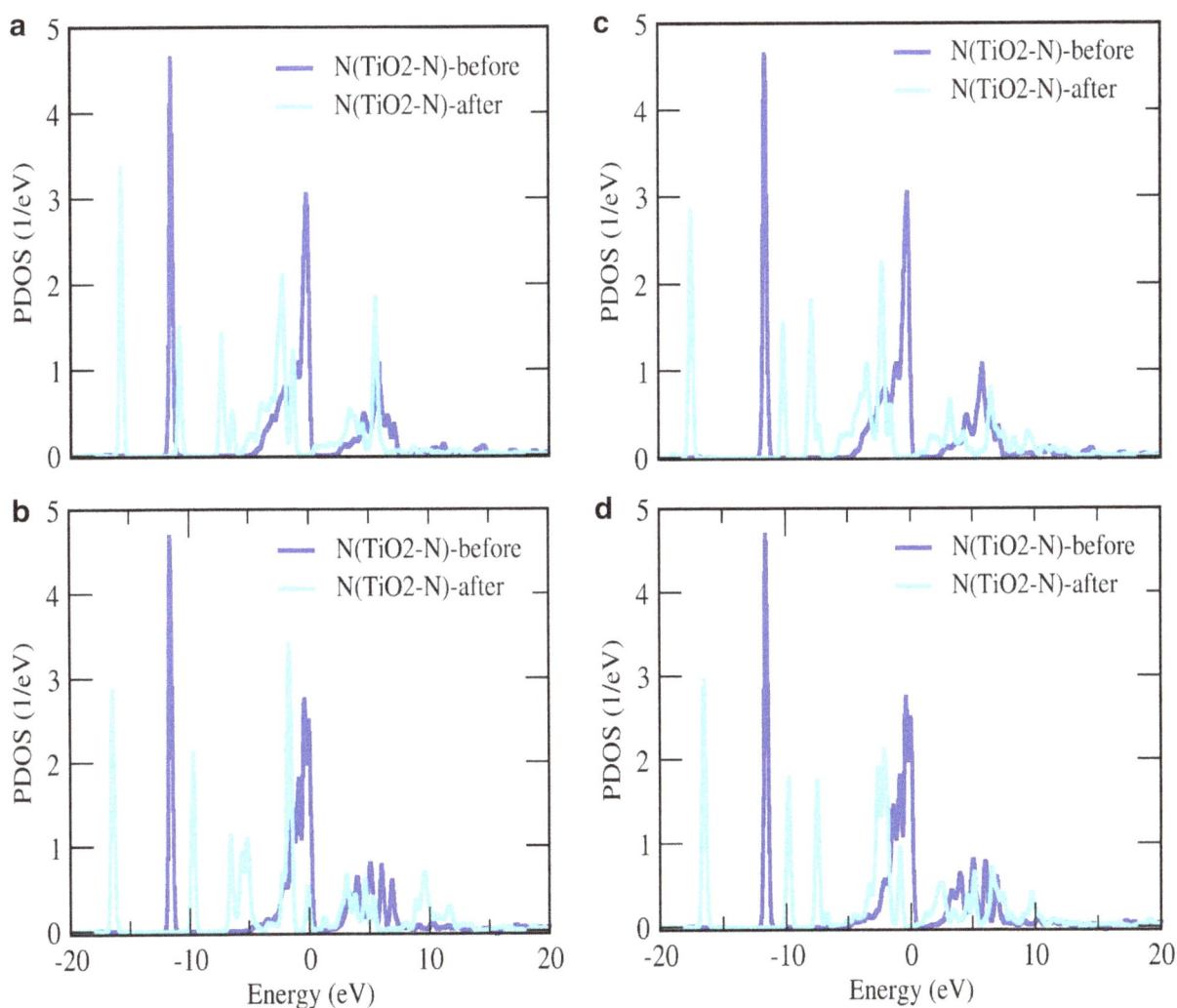

Fig. 10 PDOS of the nitrogen atom of nanocomposite before and after the adsorption process. **a** A complex; **b** B complex; **c** D complex; **d** E complex

strengthens the interaction of NO_2 molecule with TiO_2/MoS_2 nanocomposites. The O–N–O bond angles of the NO_2 molecule have been decreased after the adsorption process because of the formation of new chemical bond between nitrogen atom of NO_2 with nitrogen atom of TiO_2 nanoparticle. This chemical bond formation leads to an increase in the p characteristics of bonding molecular orbitals. Thus, the *sp* hybridization of nitrogen in the NO_2 molecule converts to near-*sp³* hybridization. For the case of TiO_2 adsorption on the fivefold coordinated titanium site presenting two contacting point between the nanoparticle and NO_2 molecule, the bond length and bond angle values have been reported in Table 2. We can see

two newly formed Ti–O bonds between the titanium atoms of the nanoparticle with oxygen atoms of NO_2 molecule. This adsorption configuration is referred to as bridge configuration, and their complexes were illustrated in Fig. 5 (G–I complexes). The N–O bonds of NO_2 molecule have been lengthened after the adsorption process, suggesting the weakening of the N–O bonds. In addition, the adsorption process results in a decrease in the O–N–O bond angle values of NO_2. The adsorption energy analysis reveals that the complex H is the most energy favorable complex in comparison with complex G and both are more stable than the pristine system adsorption. These results show that the NO_x adsorption on

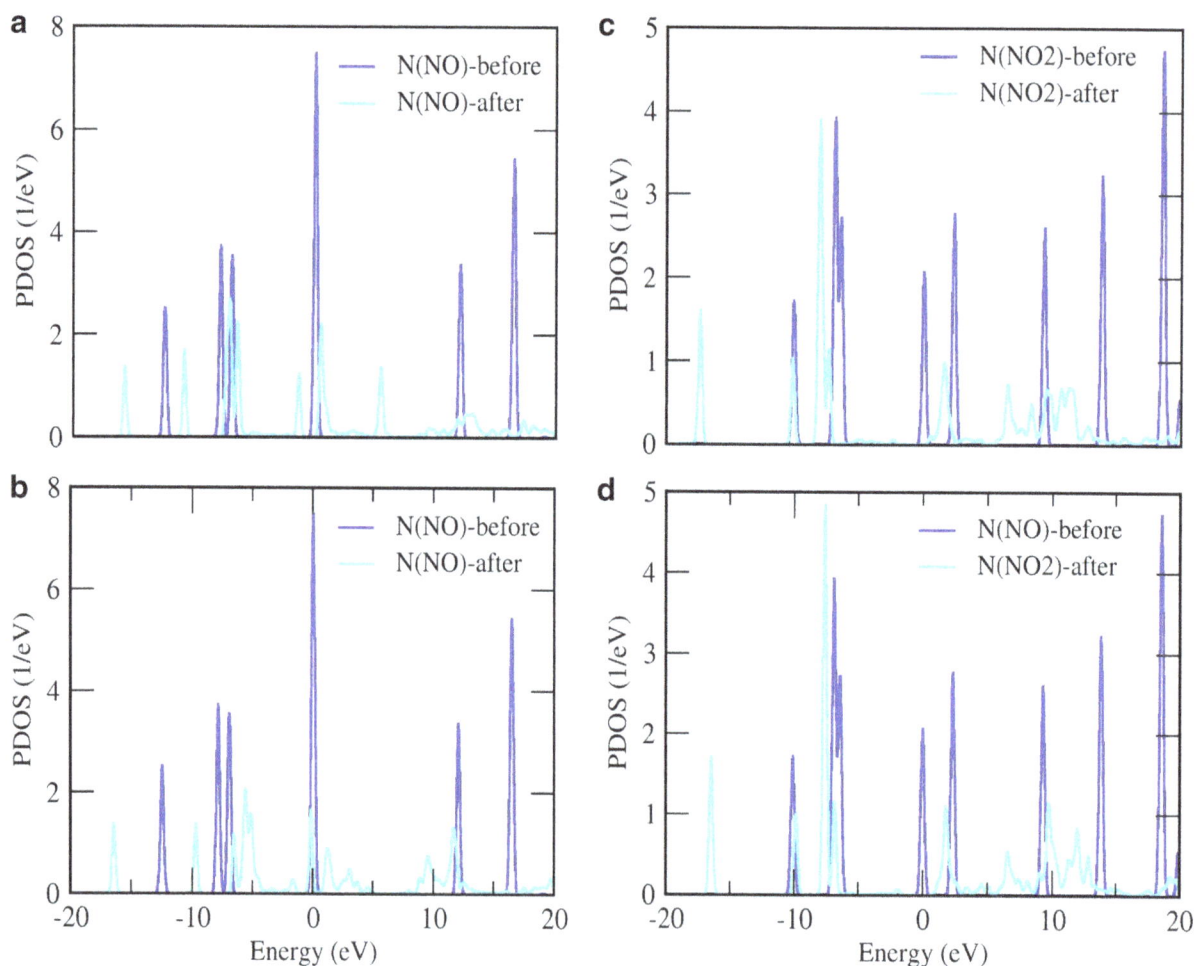

Fig. 11 PDOS of the nitrogen atom of NOx molecule before and after the adsorption process, **a** A complex; **b** B complex; **c** D complex; **d** E complex

the N-doped nanocomposite is more favorable in energy than the NO_x adsorption on the pristine nanocomposite. By considering this, we found that the nitrogen doping strengthens the interaction of NO_x molecule with TiO_2/MoS_2 nanocomposites. The obtained improvements on the structural and electronic properties of TiO_2/MoS_2 nanocomposites here represent that the N-doped TiO_2-based nanocomposite can be efficiently utilized in the removal and sensing of toxic NO_x molecule.

Electronic structures

Figure 6 presents the total density of states (TDOS) for N-doped TiO_2 anatase nanoparticles and corresponding TiO_2/MoS_2 nanocomposites. This figure reveals a creation of small peak in the density of states (DOSs) of N-doped nanocomposite at the energy ranges near to -12 eV. TDOSs of adsorption configurations were also displayed in Fig. 7. A closer inspection of these figures indicates the increase of the discrepancies between DOS of N-doped TiO_2 and nanocomposite by adding the MoS_2 monolayer and adsorption of NO_x. These differences included considerable shifts in the energies of the peaks and appearance of some peaks in the DOS of the studied systems. As distinct from these figures, the DOSs of the considered nanocomposites were mainly shifted to the lower energy values after the adsorption process. Therefore, the resultant variations in the energy of the

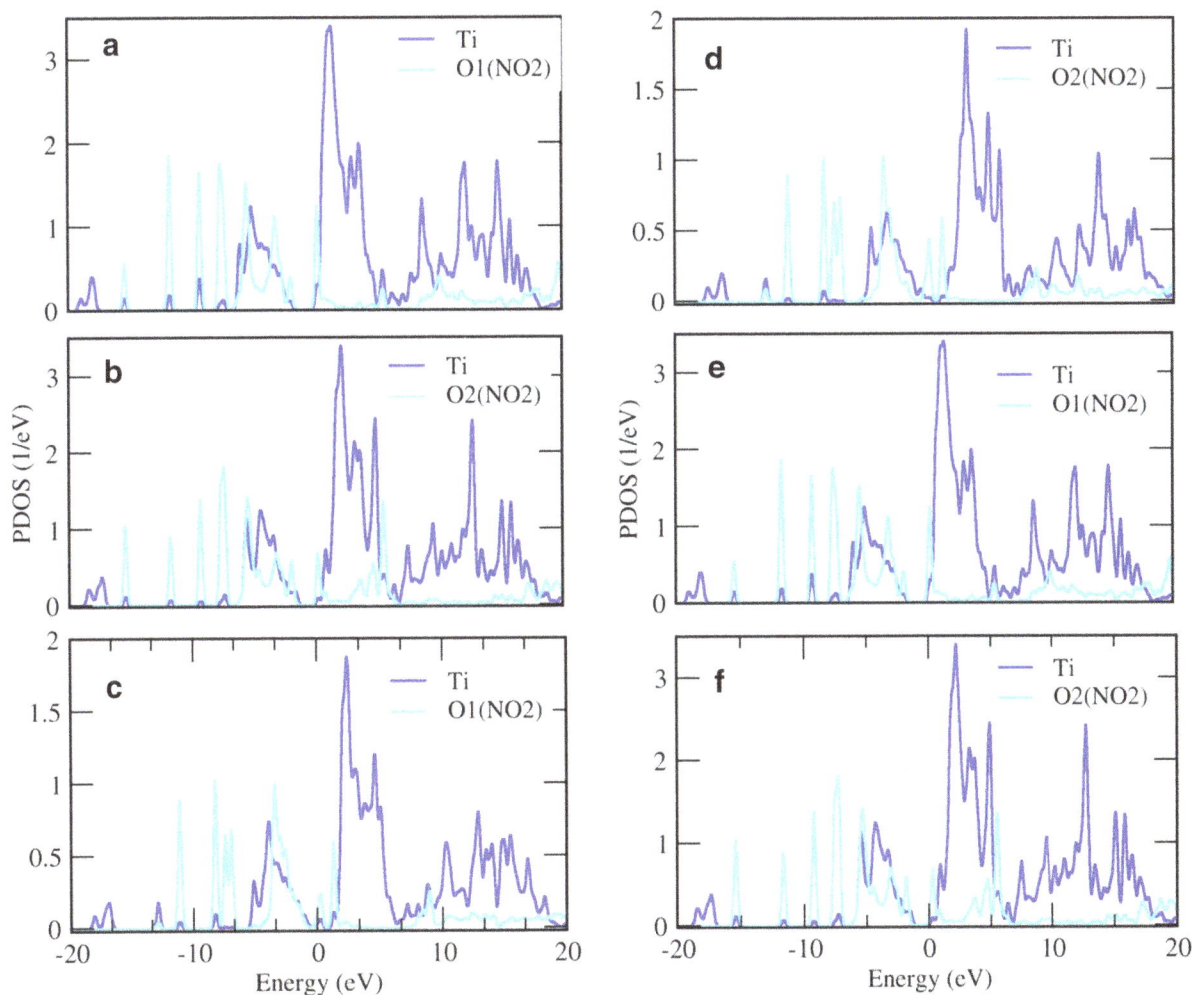

Fig. 12 PDOS of the titanium and oxygen atoms in complexes providing two contacting point between the nanocomposite and adsorbed molecule. **a** G complex (NO_2 adsorbed on the O_C-substituted nanocomposite in a bridge geometry); **b** G complex; **c** H complex (NO_2 adsorbed on the O_T-substituted nanocomposite in a bridge geometry); **d** H complex; **e** I complex (NO_2 adsorbed on the pristine nanocomposite in a bridge geometry); **f** I complex

states can have positive effects on the electronic transport properties of the nanocomposites and in turn can provide a helpful procedure for designing and engineering NO_x sensors based on N-doped TiO_2 and two-dimensional transition metal dichalcogenides (i.e., MoS_2 monolayer). The projected partial density of states (PDOSs) for the interaction of NO_x molecule with TiO_2/MoS_2 nanocomposites have been displayed in Fig. 8a–d. Panels (a, b) present the PDOS of the nitrogen atom of NO molecule and the doped nitrogen atom of N-doped nanocomposite. The large overlap between the PDOS of the mentioned atoms exhibits that the nitrogen atom of NO molecule interacts with the doped nitrogen atom of nanocomposite, suggesting the formation of new N–N bond. The PDOSs for NO_2 adsorption on the doped nitrogen site have also been shown as panels (c, d), which indicate a high overlap between the PDOS of nitrogen atom of NO_2 molecule and the nitrogen atom of

nanocomposite and consequently forming a chemical bond. For NO and NO_2 adsorption on the middle oxygen (O_C site), the calculated PDOSs have been displayed in Fig. 9 (panels a, b), representing a low PDOS overlap between the nitrogen atom of NO and NO_2 molecules and the O_C atom of nanocomposite. This means a weak interaction between NO_x and nanocomposite. The other panels of Fig. 9 represent the PDOS of oxygen atom of nanocomposite before and after the adsorption on the undoped nanocomposite, as well as the PDOS of nitrogen atoms of NO and NO_2 molecules. As can be seen, the main difference is the creation a small peak in the PDOS curves and also shifting the position of the peaks to the lower lying energies. Figure 10a–d shows the PDOS of the nitrogen atom of nanocomposite before and after the adsorption on the N-doped nanocomposite, which also suggests a shifting of the PDOS of nitrogen atom to the lower energy values. To further discover the

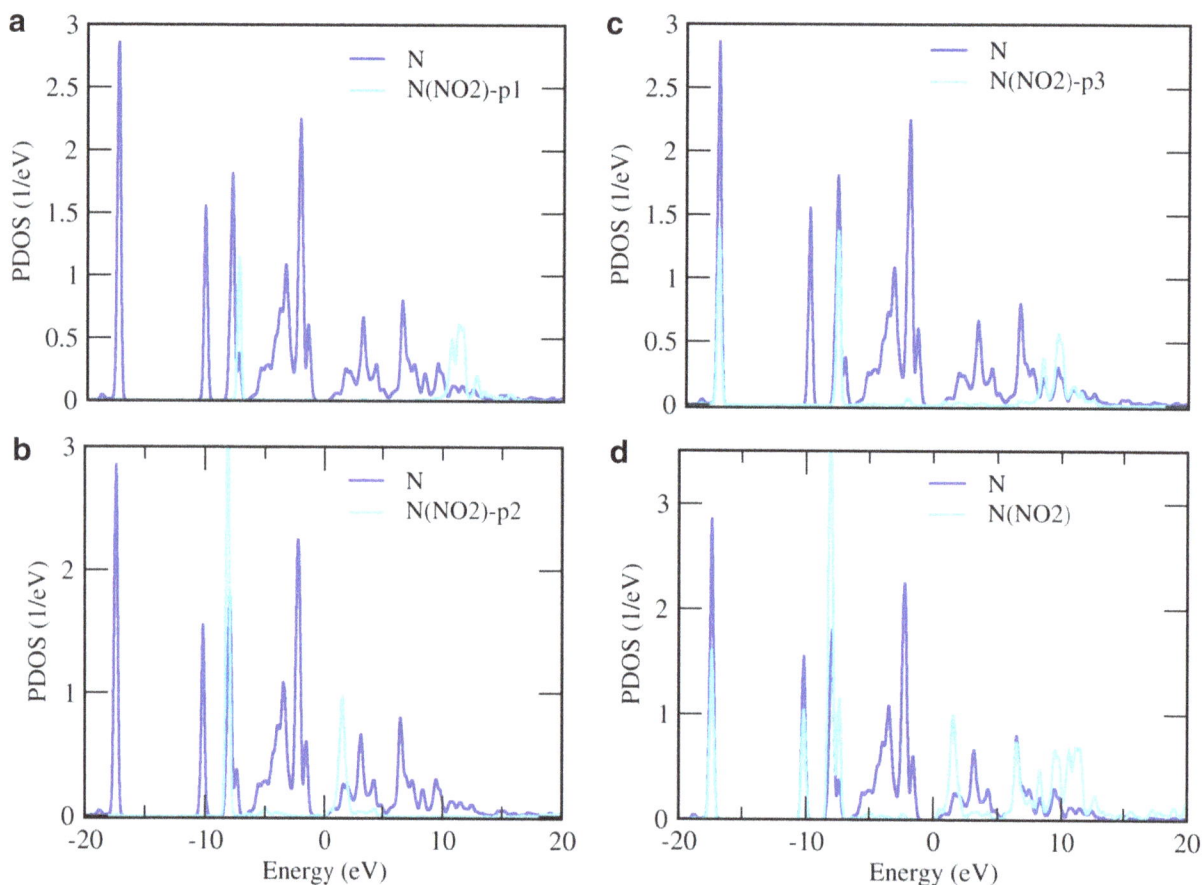

Fig. 13 PDOS of the nitrogen atoms and their related p orbitals after the adsorption process for D complex

electronic variations at the adsorption site, the PDOSs of nitrogen atom of NO and NO_2 molecule before and after the adsorption were also presented in Fig. 11a–d. Similarly, it can be seen from these PDOS plots that the biggest difference is the creation of some small peaks and also the state changing to the smaller energy values. The PDOSs of the complexes providing two contacting point (G, H and I) have also been displayed in Fig. 12, which indicate a higher overlap between the PDOS of Titanium atoms with two left and right oxygen atoms in all six panels. This means a formation of two chemical bonds between the titanium and oxygen atoms. Figures 13 and 14 represent the PDOSs of nitrogen atoms and their related p orbitals for complex D. As can be seen from these figures, p^3 orbital of the nitrogen atom of nanocomposite and NO_2 represents a considerable overlap with the other atom participating in chemical bond formation. This is an indication of the higher

contribution of p^3 orbital in chemical bond in comparison with the other orbitals. The PDOSs of nitrogen atoms and their p orbitals for complex E have also been displayed in Fig. 15, indicating a high overlap between the PDOS of nitrogen atom with p^1 atomic orbital compared to the other orbitals. The HOMO and LUMO molecular orbitals were also displayed in Figs. 16 and 17, respectively. A closer inspection reveals that the HOMOs are strongly located on the nanoparticle, whereas the electronic densities in the LUMOs are mainly dominant on the NO_x molecules. As can be seen from Fig. 17, the electronic density in the LUMOs seems to be distributed over the NO_x molecules and on the middle of newly formed bonds. The accumulation of the electronic density at the middle of the newly formed bonds confirms the formation of new bonds and consequently the transfer of electronic density from the Ti–N bonds and N–O bonds to the newly formed bonds.

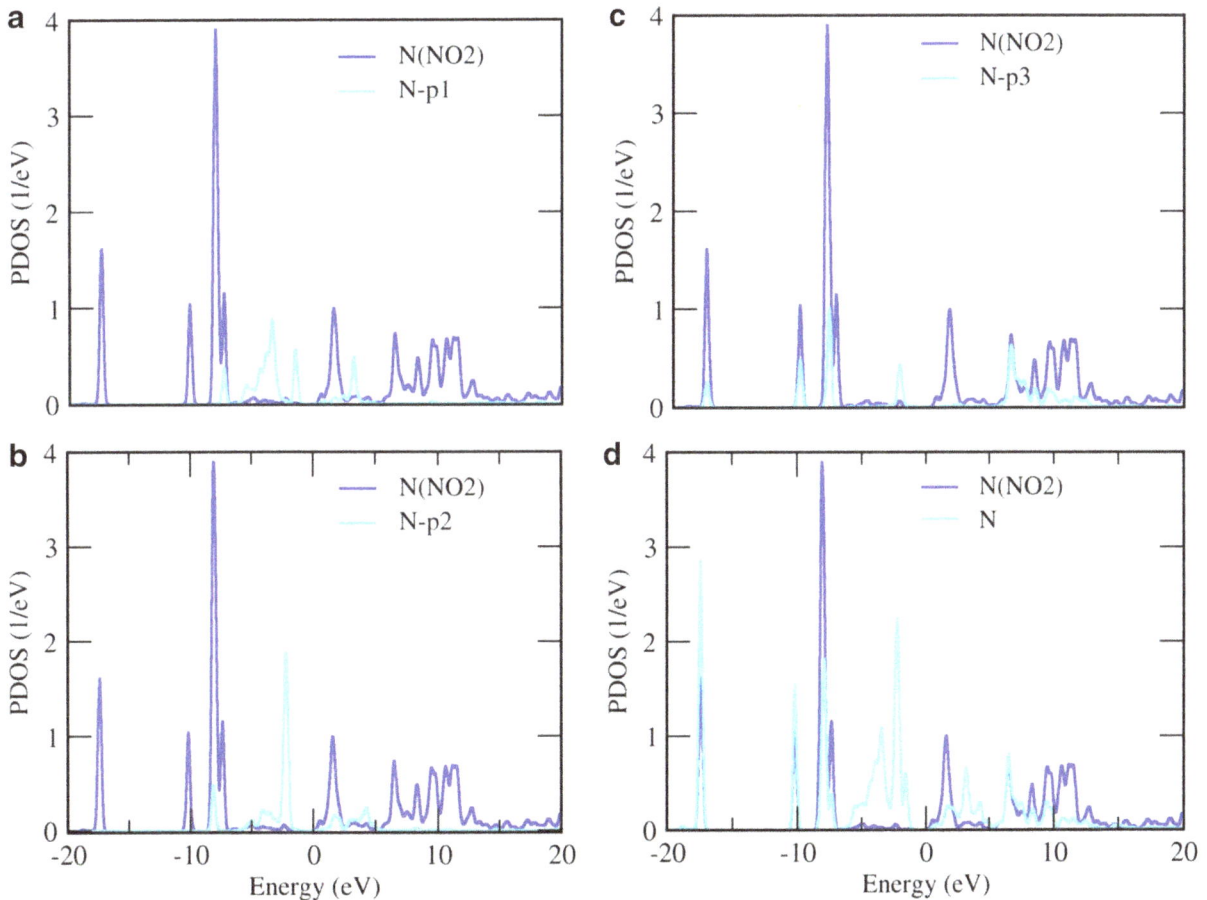

Fig. 14 PDOS of the nitrogen atoms and their related p orbitals after the adsorption process for D complex

However, the resultant improvements on the electronic properties of TiO$_2$/MoS$_2$ nanocomposites obtained by N-doping here demonstrate that the N-doped TiO$_2$/MoS$_2$ nanocomposites have stronger sensing capabilities than the pristine ones. To fully analyze the NO$_x$ adsorption on the considered TiO$_2$/MoS$_2$ nanocomposites, the Mulliken population analysis has been conducted to analyze the charge distribution of the atoms and bonds in a complex system. The calculated Mulliken charge values for studied complexes were collected in Tables 1 and 2. For complex A, NO$_x$ adsorption induces a considerable charge transfer of about -0.122 e from NO$_x$ molecule to the nanoparticle, suggesting that NO$_x$ acts as an acceptor. In other words, the NO$_x$ molecule receives electros from nanocomposite. This leads to the changes in the

conductivity of the system, which would be an efficient property to aid in the design and fabrication of novel sensor devices for nitrogen oxides recognition.

Conclusions

DFT calculations were conducted to investigate the interaction of NO$_x$ molecules with undoped and N-doped TiO$_2$/MoS$_2$ nanocomposites to effectively understand the sensing properties of these nanocomposites in adsorption processes. The bond angles of the adsorbed NO$_2$ molecule are decreased compared to those in the isolated gas phase NO$_2$, which lead to an increase in the p characteristics of bonding molecular orbitals of nitrogen in the NO$_2$ molecule. The

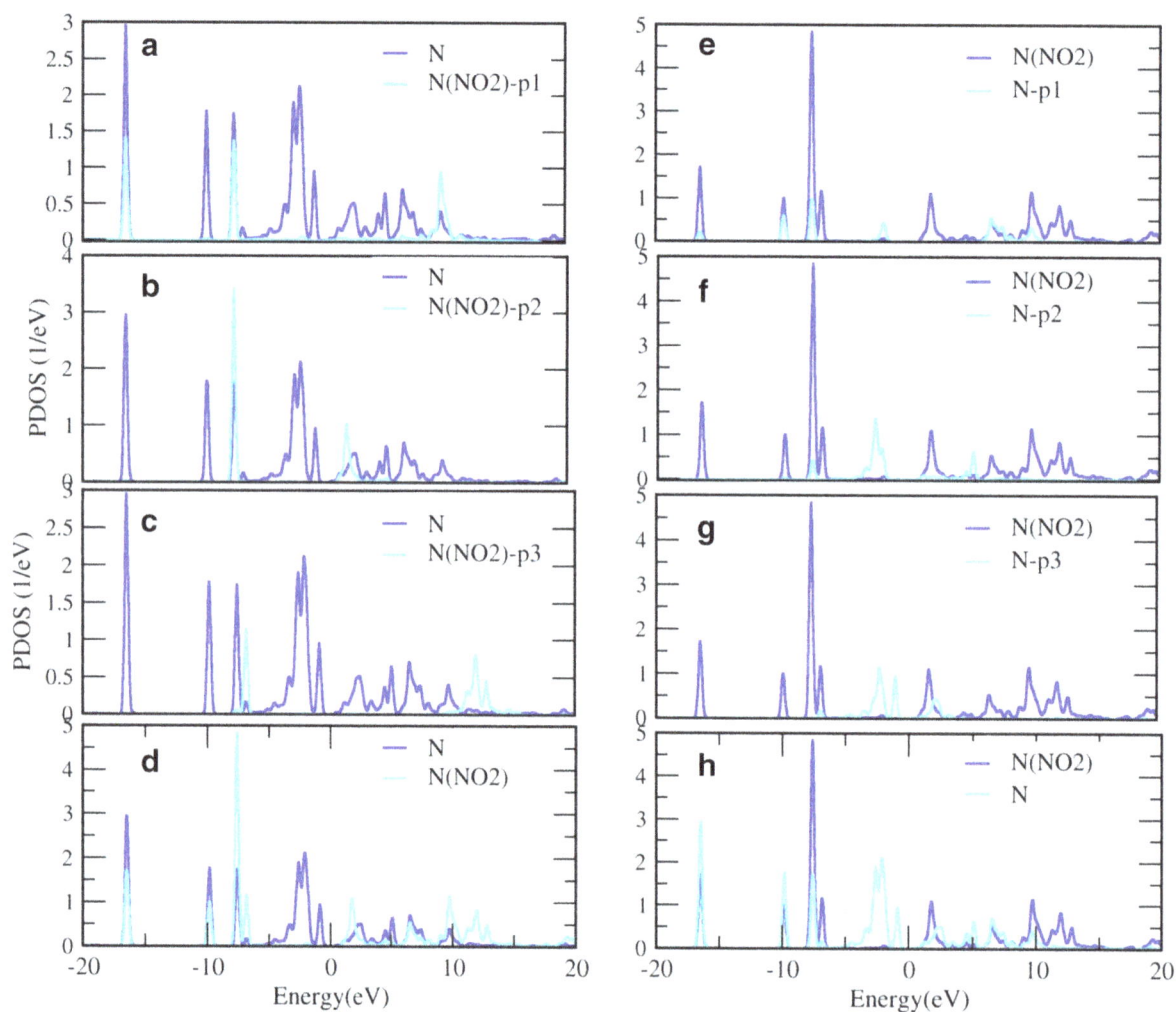

Fig. 15 PDOS of the nitrogen atoms and their related p orbitals after the adsorption process for E complex

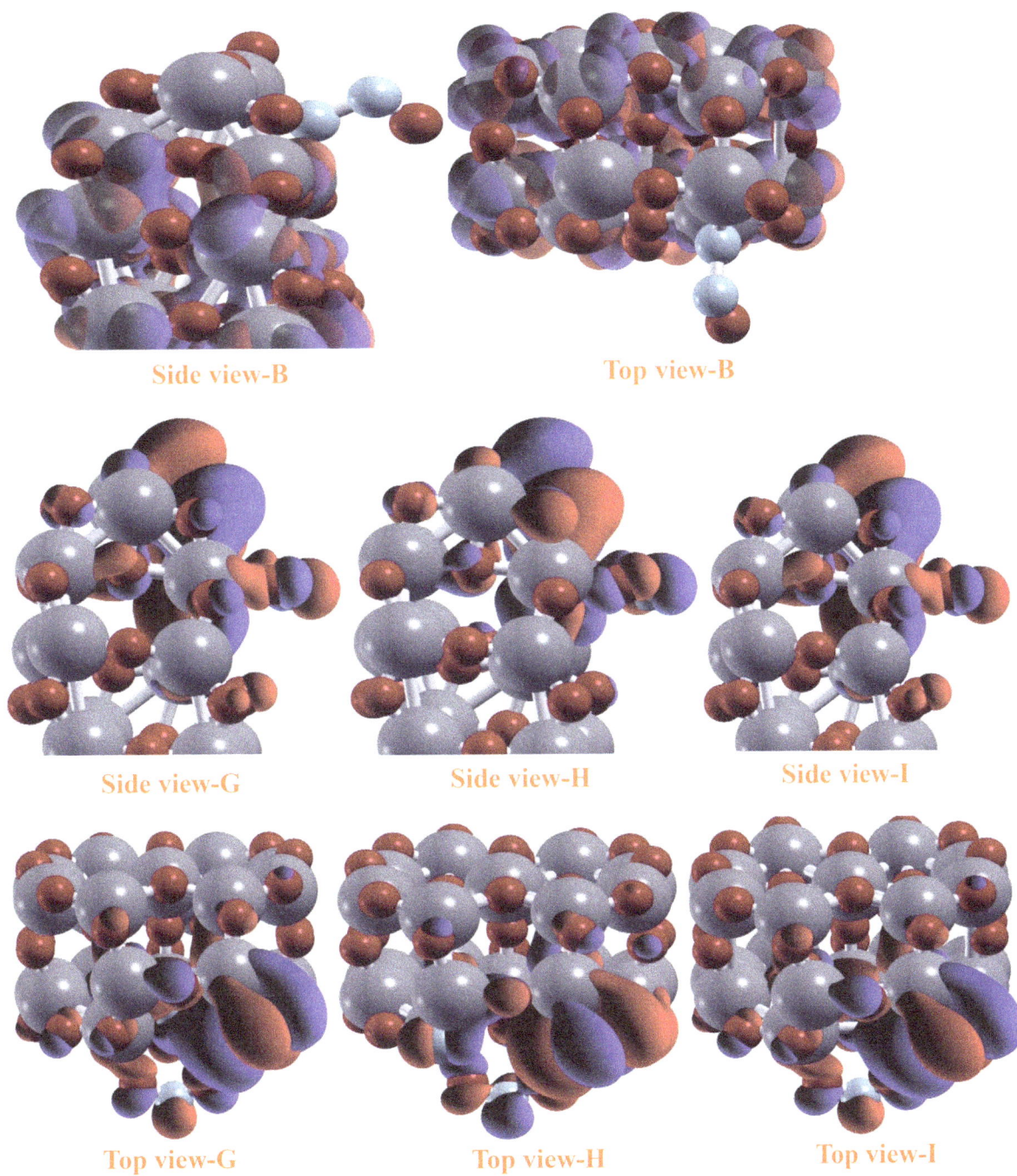

Side view-B

Top view-B

Side view-G

Side view-H

Side view-I

Top view-G

Top view-H

Top view-I

Fig. 16 Isosurfaces of HOMO molecular orbitals for the adsorption of NO and NO$_2$ molecules on the TiO$_2$/MoS$_2$ nanocomposites, where |0.05| was used as an isovalue of the molecular orbital

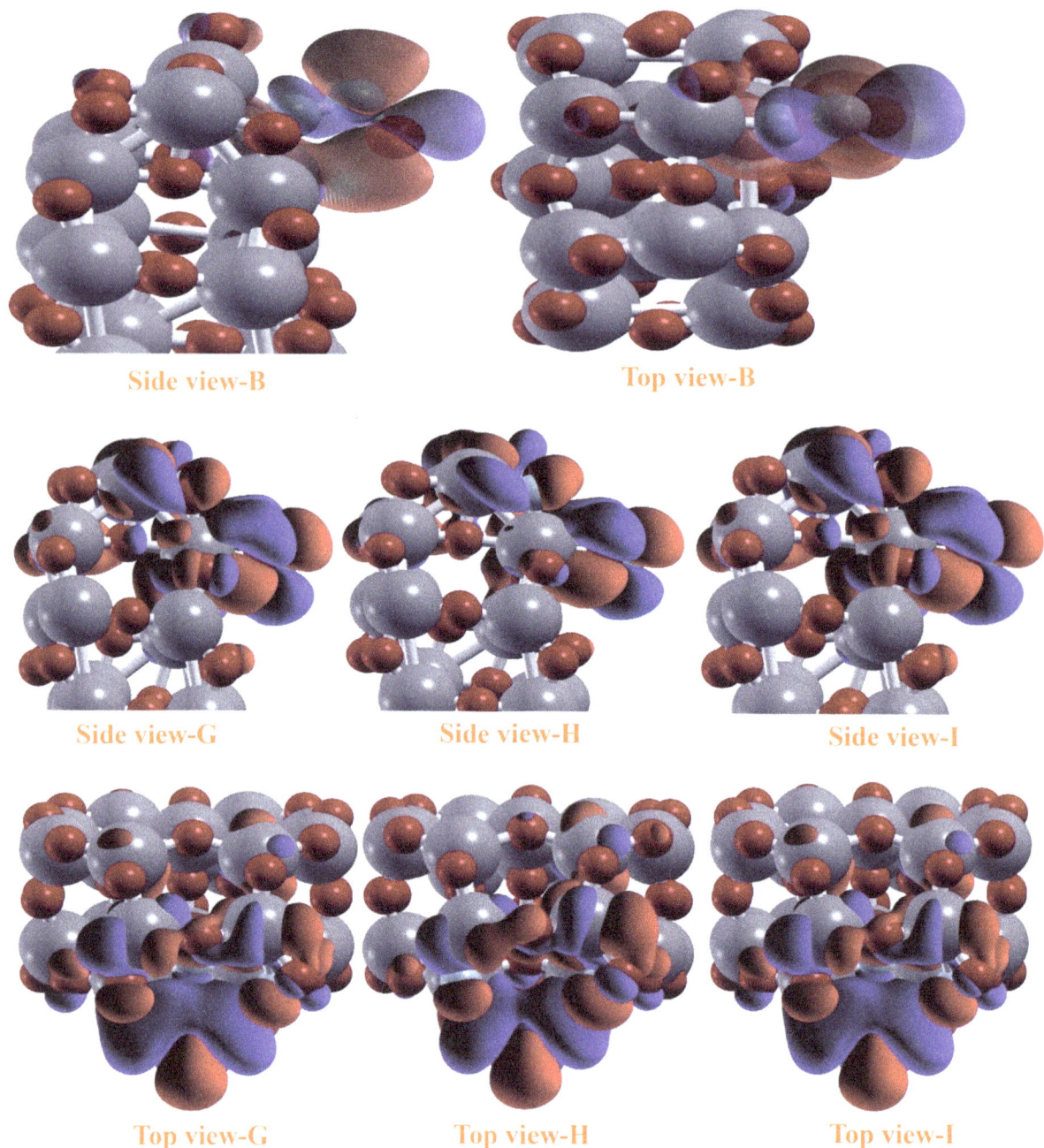

Fig. 17 Isosurfaces of LUMO molecular orbitals for the adsorption of NO and NO_2 molecules on the TiO_2/MoS_2 nanocomposites, where |0.05| was used as an isovalue of the molecular orbital

results also suggest that the N-doped nanocomposites have a higher efficiency to interact with harmful NO_x molecules in the environment. In other words, the doping of nitrogen atom provides an increased affinity for the TiO_2/MoS_2 nanocomposites to interact with NO_x molecules. The analysis of adsorption energies reveals that the adsorption of NO_x molecules on the N-doped TiO_2/MoS_2 nanocomposites is more favorable in energy than the adsorption of NO_x on the undoped ones. The variation in the electronic

structure and molecular orbitals induced by N-doping is found to be responsible for the conductivity of the nanocomposite system. Our calculated results, therefore, suggest a theoretical basis for the prospective application of TiO_2/MoS_2 hybrid nanostructures as gas sensors for important air pollutants, such as NO and NO_2, in the environment.

Acknowledgments This work was supported by Azarbaijan Shahid Madani University [217/D/14271].

References

1. Fujishima, A., Zhang, X., Tryk, D.A.: TiO_2 photocatalysis and related surface phenomena. J. Surf. Sci. Rep. **63**, 515–582 (2008)

2. Fernandez-Garcia, M., Martinez-Arias, A., Hanson, J.C., Rodriguez, J.A.: Nanostructured oxides in chemistry: characterization and properties. J. Chem. Rev. **104**, 4063–4104 (2004)

3. Tang, S., Cao, Z.: Adsorption of nitrogen oxides on graphene and graphene oxides: insights from density functional calculations. J. Chem. Phys. **134**, 044710 (1-14) (2011)

4. Tang., S, Zhu, J.: Structural and electronic properties of Pd decorated graphene oxides and their effects on the adsorption of nitrogen oxides: insights from density functional calculations. J. RSC Adv. **4**, 23084–23096 (2014)

5. Topalian, Z., Niklasson, G.A., Granqvist, C.G., Österlund, L.: Spectroscopic study of the photofixation of SO_2 on anatase TiO_2 thin films and their oleophobic properties. ACS Appl. Mater. Interf. **4**, 672–679 (2012)

6. Li, W.K., Gong, X.Q., Lu, G., Selloni, A.: Different reactivities of TiO_2 polymorphs: comparative DFT calculations of water and formic acid adsorption at anatase and brookite TiO_2 surfaces. J. Phys. Chem. C **112**(17), 6594–6596 (2008)

7. Nambu, A., Graciani, J., Rodriguez, J.A., Wu, Q., Fujita, E., Sanz, J.F.: N doping of TiO_2 (110) Photoemission and density-functional studies. J. Chem. Phys. **125**, 094706(1–8) (2006)

8. Rumaiz, A.K., Woicik, J.C., Cockayne, E., Lin, H.Y., Jaffari, G.H., Shah, S.I.: Oxygen vacancies in N doped anatase TiO_2: experiment and first-principles calculations. J. Appl. Phys. Lett. **95**(26), 262111 (1–3) (2009)

9. Liu, D., Chen, X., Li, D., Wang, F., Luo, X., Yang, B.: Simulation of MoS_2 crystal structure and the experimental study of thermal decomposition. J. Mol. Struct. **980**, 66–71 (2010)

10. Wang, H., Yu, L., Lee, Y.H., Shi, Y., Hsu, A., Chin, M.L., Li, L.J., Dubey, M., Kong, J., Palacios, T.: Integrated circuits based on bilayer MoS_2 transistors. Nano Lett. **12**, 4674–4680 (2012)

11. Kou, L., Tang, C., Zhang, Y., Heine, T., Chen, C., Frauenheim, T.: Tuning magnetism and electronic phase transitions by strain and electric field in Zigzag MoS2 nanoribbons. J. Phys. Chem. Letts. **3**, 2934–2941 (2012)

12. Bertolazzi, S., Brivio, J., Kis, A.: Stretching and breaking of ultrathin MoS_2. ACS Nano **5**(12), 9703–9709 (2011)

13. Dolui, K., Pemmaraju, C.D., Sanvito, S.: Electric field effects on armchair MoS_2 nanoribbons. ACS Nano **6**(6), 4823–4834 (2012)

14. Yang, S., Li, D., Zhang, T., Tao, Z., Chen, J.: First-principles study of zigzag MoS_2 nanoribbon as a promising cathode material for rechargeable Mg batteries. J. Phys. Chem. C **116**, 1307–1312 (2012)

15. Frame, F.A., Osterloh, F.E.: CdSe-MoS2: a quantum size-confined p CdSe-MoS2: a quantum size-confined photocatalyst for hydrogen evolution from water under visible light. J. Phys. Chem. C **114**, 10628–10633 (2010)

16. Li, T.S., Galli, G.L.: Electronic properties of MoS2 nanoparticles. J. Phys. Chem. C **111**, 16192–16196 (2007)

17. Radisavljevic, B., Radenovic, A., Brivio, J., Giacometti, V., Kis, A.: Single-layer MoS2 transistors. Nat. Nanotechnol. **6**, 147–150 (2011)

18. Lembke, D., Kis, A.: Breakdown of high-performance monolayer MoS_2 transistors. ACS Nano **6**, 10070–10075 (2012)

19. Li, H., Yin, Z., He, Q., Li, H, Huang, X., Lu, G., Fam, D.W.H., Tok, A.I.Y., Zhang, H.: Fabrication of single-and multilayer MoS2 film-based field-effect transistors for sensing NO at room temperature. Small **8**(1), 63–67 (2012)

20. He, Q., Zeng, Z., Yin, Z., Li, H., Wu, S., Huang, X., Zhang, H.: Fabrication of flexible MoS2 thin-film transistor arrays for practical gas-sensing applications. Small **8**(19), 2994–2999 (2012)

21. Liu, H., Neal, A.T., Ye, P.D.: Channel length scaling of MoS_2 MOSFETs. ACS Nano **6**, 8563–8569 (2012)

22. Zhou, W., Yin, Z., Du, Y., Huang, X., Zeng, Z., Fan, Z., Liu, H., Wang, J., Zhang, H.: Synthesis of few-layer MoS2 nanosheet-coated TiO2 nanobelt heterostructures for enhanced photocatalytic activities. Small **9**(1), 140–147 (2013)

23. Hu, K.H., Hu, X.G., Xu, Y.F., Sun, J.D.: Synthesis of nano-MoS2/TiO2 composite and its catalytic degradation effect on methyl orange. J. Mater. Sci. **45**(10), 2640–2648 (2010)

24. Shokuhi-Rad, A., Esfahanian, M., Maleki, S., Gharati, G.: Application of carbon nanostructures toward SO_2 and SO_3 adsorption: a comparison between pristine graphene and N-doped graphene by DFT calculations. J. Sulfur Chem. **37**(2), 176–188 (2016)

25. Hohenberg, P., Kohn, W.: Inhomogeneous electron gas. J. Phys. Rev. **136**, B864–B871 (1964)

26. Kohn, W., Sham, L.: Self-consistent equations including exchange and correlation effects. J. Phys. Rev. **140**, A1133–A1138 (1965)

27. Ozaki, T., Kino, H., Yu, J., Han, M.J., Kobayashi, N., Ohfuti, M., Ishii, F., et al.: User's manual of OPENMX version 3.8. http://www.openmxsquare.org

28. Ozaki, T., Kino, H.: Numerical atomic basis orbitals from H to Kr. J. Phys. Rev. B. **69**(1–19), 195113 (2004)

29. Ozaki, T., Kino, H.: Variationally optimized basis orbitals for biological molecules. J. Phys. Rev. B. **121**(22), 10879–10888 (2005)

30. Perdew, J.P., Burke, K., Ernzerhof, M.: Generalized gradient approximation made simple. J. Phys. Rev. Lett. **78**, 1396 (1997)

31. Koklj, A.: Computer graphics and graphical user interfaces as tools in simulations of matter at the atomic scale. J Comput Mater Sci. **28**, 155–168 (2003)

32. Schneider, W.F.: Qualitative differences in the adsorption chemistry of acidic (CO_2, SO_x) and amphiphilic (NO_x) species on the alkaline earth oxides. J. Phys. Chem. B. **108**, 273–282 (2004)

33. Liu, Q., Li, L., Li, Y., Gao, Z., Chen, Z., Lu, J.: Tuning electronic structure of bilayer MoS_2 by vertical electric field: a first-principles investigation. J. Phys. Chem. C **116**, 21556–21562 (2012)

34. Li, Y., Zhou, Z., Zhang, S., Chen, Z.: MoS_2 nanoribbons: high stability and unusual electronic and magnetic properties. J. Am. Chem. Sci. **130**(49), 16739–16744 (2012)

35. Pan, H., Zhang, Y.W.: Tuning the electronic and magnetic properties of MoS2 nanoribbons by strain engineering. J. Phys. Chem. C **116**, 11752–11757 (2012)

36. Web page at: http://rruff.geo.arizona.edu/AMS/amcsd.php

37. Wyckoff, R.W.G.: Crystal structures. (2nd Eds.) Interscience Publishers, USA (1963)

38. Wu, C., Chen, M., Skelton, A.A., Cummings, P.T., Zheng, T.: Adsorption of arginine–glycine–aspartate tripeptide onto negatively charged rutile (110) mediated by cations: the effect of surface hydroxylation. ACS Appl. Mater. Interf. **5**, 2567–2579 (2013)

21

Synthesis of conductive polymer-coated mesoporous MCM-41 for textile dye removal from aqueous media

Ali Torabinejad[1] · Navid Nasirizadeh[1] · Mohammad Esmail Yazdanshenas[1] · Habib-Allah Tayebi[2]

Abstract In this paper, we aimed to evaluate Acid Blue 62 removal from aqueous media, using mesoporous silicate MCM-41, loaded with polypyrrole (PPy) and polyaniline (PAni) composites. PPy/MCM-41 nanocomposite showed higher performance than PAni/MCM-41 due to its smaller molecule size. For characterizing the synthesized composites, different methods were applied. The Langmuir model showed the greatest agreement with the experimental findings (q_m, 55.55 mg g^{-1}). The kinetic study also confirmed the compatibility between the pseudo-second-order model and adsorption. Moreover, we measured Gibbs free energy changes (ΔG^o) and enthalpy changes (ΔH^o). Considering the negative ΔG^o and positive ΔH^o, AB62 adsorption on PPy/MCM-41 nanocomposite can be considered a spontaneous, endothermic reaction.

Keywords Adsorption · Acid Blue 62 · Polypyrrole · Polyaniline · MCM-41

Introduction

The presence of dyes in industrial wastewaters is a major issue in different countries. Different industries, including textile printing, plastic, pharmaceutical, and food industries, use dyes in their procedures. About 20% of dye production in the world is lost during the dyeing process [1]; therefore, the presence of dyes in wastewater is inevitable.

Dye removal from wastewater is significant, as the quality of water greatly depends on color; even a very low dosage of dye (<1 ppm) can be easily seen in water and is found to be unfavorable. Besides, most dyes result in the occurrence of diseases, such as skin inflammation, skin irritation, carcinomas, and mutations in humans [1]. Removal of colored contaminants from wastewater is carried out using various techniques, including biological and physicochemical technologies.

Oxidation [2], coagulation–flocculation [3], separation of membranes [4], and adsorption [5, 6] are among the major processes involved in treatment. Among the mentioned processes, adsorption is the most frequently used technique, showing feasibility, high yield rate, and less expenditure [5]. Species in the adsorption process are moved to the solid phase and can minimize the effluent volume [1].

Researchers have recently introduced favorable adsorbents [5–7]. Highly porous materials such as mobile crystalline material-41 (MCM-41) [8, 9], modified hexagonal mesoporous silica (HMS) [5, 10], and nanoporous silica (SBA-15) [11] seem to be appropriate for removing different dyes. Different types of surfactants are used for the preparation of these materials. Surfactants act as templates throughout sol–gel and hydrothermal processes. The main characteristics of mesoporous materials include great specific surface area, uniform and limited pore size distribution, and great thermal stability [12, 13].

Recently, MCM-41 has been used for the adsorption of dyes from wastewater. Considering the presence of SiO and SiOH (known to adsorb cationic dyes and inhibit anionic dye adsorption), the structure of MCM-41 is negatively charged [14]. The importance of interaction

✉ Navid Nasirizadeh
nasirizadeh@iauyazd.ac.ir

[1] Department of Textile and Polymer Engineering, Yazd Branch, Islamic Azad University, Yazd, Iran

[2] Department of Textile Engineering, Islamic Azad University of Qaemshahr, Qaemshahr, Iran

between MCM-41 and safranin (a large positively charged dye) was examined in a previous study [15]. To enhance the MCM-41 capacity in adsorbing specific substances, it is necessary to make surface modifications. Considering the adsorbent–adsorbate interactions, surface modifications also improve the selectivity of MCM-41 [16].

Amines, as well-known functional groups, have been used in many studies for surface modification of different types of adsorbents. To eliminate mercury ions from water, MCM-41 was functionalized with diethylenetriamine (DETA) [9]. In a previous study, aminopropyltrimethoxysilane (APTS) was applied for MCM-41 ($NH_3 +$ -MCM-41) modification to remove four types of anionic dyes from the aqueous media [16].

Transition metals including nickel are extensively applied for the modification of MCM-41 structure [17]. In a previous study, to remove methyl blue from aqueous solutions, highly ordered nickel-supplemented MCM-41 adsorbents were synthesized (with varying nickel contents) [14]. Overall, according to our literature research, few studies have been conducted on the use of polypyrrole (PPy)/MCM-41 and polyaniline (PAni)/MCM-41 in anionic dye adsorption.

With this background in mind, PPy/MCM-41 and PAni/MCM-41 nanocomposites were synthesized in the current study. Nanocomposites were described using different methods, including transmission electron microscopy (TEM), scanning electron microscopy (SEM), Fourier transform infrared spectroscopy (FTIR), X-ray diffraction (XRD) analysis, and BET method. Through performing batch experiments, the efficiency of PPy/MCM-41 and PAni/MCM-41 in AB62 elimination from aqueous media was also examined. Moreover, the significance of solution pH, contact time, temperature, and adsorbent dose was evaluated. Finally, thermodynamic and kinetic evaluations were performed to measure the parameters.

Materials and methods

Substances

The reagents used for the preparation of samples and experimental tests included cetyl-trimethyl-ammonium bromide (CTMABr), ammonium hydroxide, tetraethyl ortho silicate (TEOS, $SiC_8H_{20}O_4$), aniline, pyrrole, potassium iodate (KIO_3), ferric chloride, sulfuric acid, acetone, hydrochloric acid (HCl), NaOH, Na_2HPO_4, NaH_2PO_4, and deionized water (Merck Co., Germany). Moreover, Dystar Co. (Germany) provided anionic dye (Acid Blue 62 or AB62, $\lambda_{max} = 595$ nm). Figure 1 illustrates the chemical composition of AB62.

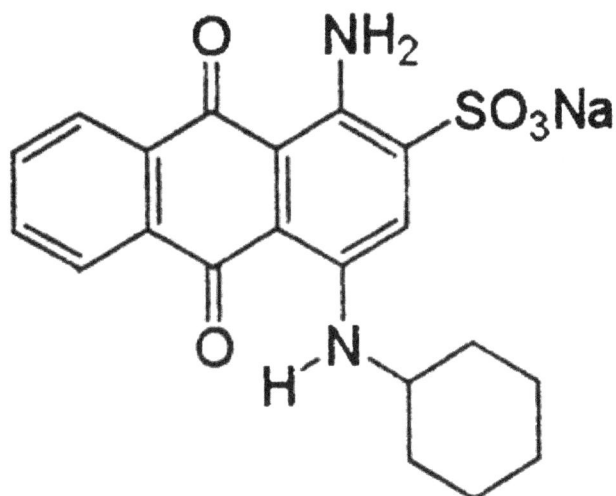

Fig. 1 Molecular structure of dye Acid Blue 62

MCM-41 synthesis

MCM-41 synthesis was performed in line with the technique proposed by Kamarudin et al. [18]. In brief, 2.4 g of cetyltrimethylammonium bromide (CTMABr) as the template was added to deionized water (120 g) and stirred to form a uniform solution. Afterward, ammonium hydroxide (8 mL) was added and stirred over 5 min. Following that, 10 mL of tetraethyl orthosilicate (TEOS), as the silicon source, was added and stirred for 24 h.

The solution was moved to a steel autoclave and stored at 145 °C for 2 days. The pH was adjusted typically during 48–72 h to attain stability. The obtained product was filtered, washed, and stored at 100 °C for 1 day. In the final step, calcination was performed at 600 °C for 5 h.

PAni/MCM-41 nanocomposite preparation

For the preparation of PAni/MCM-41, potassium iodate (KIO_3, 1 g) and 1 M sulfuric acid (100 mL) were mixed and stirred during 10 min. Afterward, 0.2 g of cetyl trimethylammonium bromide surfactant, along with 1 g of MCM-41, was added. Following a 20-min interval, distilled aniline monomer (1 mL) was mixed in the solution. The color of the solution changed from light to dark violet, indicating the occurrence of polymeric reaction and formation of PAni.

The reaction was performed at room temperature during 5 h. Finally, the composite was filtered and impurities were eliminated. PAni/MCM-41 composite was rinsed several times using distilled water and acetone and placed in an oven during 1 day at a temperature of 70 °C. Finally, PAni/MCM-41 composite was stored in a desiccator for subsequent experiments [6, 7].

PPy/MCM-41 preparation

For preparing PPy/MCM-41 nanocomposite, 5.4 g of ferric chloride was added to distilled water and stirred for 20 min. Afterward, MCM-41 (1 g) and pyrrole (1 mL) were mixed in the filtered solution. For fulfilling the polymerization reaction, the final mixture was stirred for 5 h. The synthetic composite was filtered and dried for 24 h [11].

Characterization

XRD patterns were obtained by a refractometer (35 kV, 28.5 mA, and 298 K; Philips Instruments, Australia). The BET surface area was estimated with respect to nitrogen adsorption at −196 °C. The BJH method (Quantachrome NovaWin2, USA) was utilized to measure the size and volume distribution of pores, based on the N_2 adsorption–desorption isotherm curve. The existence of functional groups on MCM-41 surface was determined with FTIR (8400S, Shimadzu, Japan) at wavenumbers of $400-4000$ cm^{-1} based on the KBr method. We used UV–Vis spectrophotometry for determining dye concentration before and after adsorption. TEM images of the adsorbents were obtained by a Philips camera (CM120, the Netherlands) operating at 150 kV. Also, SEM analysis (HITACHI S-4160) was used for identifying the sample morphology.

Batch experiments

Batch experiments were performed with 100 mL of various concentrations of dye solution (20, 40, 60, 80, and 100 mg L^{-1}). The experiments were performed in an incubator at 200 rpm. The pH of dye solutions was adjusted to 2–10 with HCl and NaOH solutions (0.1 M). Also, the impact of adsorbent dosage, primary dye dosage, duration of contact, and temperature was examined.

To obtain the equilibrium concentrations, a UV–Vis spectrophotometer was used (6310, JENWAY, UK). After contact durations of 15, 30, 45, 60, 90, 120, and 180 min, sampling was performed; the samples were centrifuged at a speed of 6000 rpm during 30 min. Equation (1) was used to measure the adsorption potential of the adsorbent:

$$q_t = (C_o - C_t)V/M. \tag{1}$$

In this equation, q_t (mg g^{-1}) represents the adsorption efficiency at time t until equilibrium, C_o and C_t (mg L^{-1}) denote dye dosage at baseline and time t, respectively, V denotes dye volume (L), and M represents the concentration of the adsorbent (g). In addition, Eq. (2) was used to calculate the dye removal efficiency:

$$\text{Removal efficiency} = (C_o - C_t)/C_o \times 100. \tag{2}$$

Results and discussion

Characterization

Figure 2 demonstrates the N_2 adsorption/desorption isotherms related to MCM-41, PAni/MCM-41, and PPy/MCM-41 nanocomposites. Isotherm type IV was recognizable for MCM-41 with hysteresis. An obvious rise in the adsorbed N_2 was reported at P/P_0 of 0.6–0.8, which characterizes mesoporous materials [19–21]. It was concluded from the BJH method that MCM-41 has a limited pore size distribution (average, 6.7 nm). Table 1 demonstrates parameters including the average pore diameter, as well as pore volume in MCM-41 and synthesized nanocomposites.

After pyrrole and aniline polymerization, the inflection point of the nanocomposite isotherm changed to a lower P/P_0; based on this finding, PPy and PAni were in the MCM-41 channels [21, 22]. Moreover, the reduction in BET surface area, as well as pore volume and size, in PPy/MCM-41 and PAni/MCM-41 nanocomposites clearly shows that the polymers penetrated into the MCM-41 channels [19, 21, 22].

According to the findings, the MCM-41 BET surface area reduced from 1003.61 to 434.89 and 535.34 m^2 g^{-1} after loading with PPy and PAni, respectively. PPy with smaller molecules could rapidly fill the MCM-41 pores in comparison with PAni; therefore, it caused pore obstruction and decreased the BET surface area.

Figure 3 indicates the XRD analysis of synthesized MCM-41, PAni/MCM-41, and PPy/MCM-41 in the range of $0 < 2\theta < 10$. There was a sharp peak at $2 < 2\theta < 3$, as well as two weaker peaks at $3 < 2\theta < 5$, which are

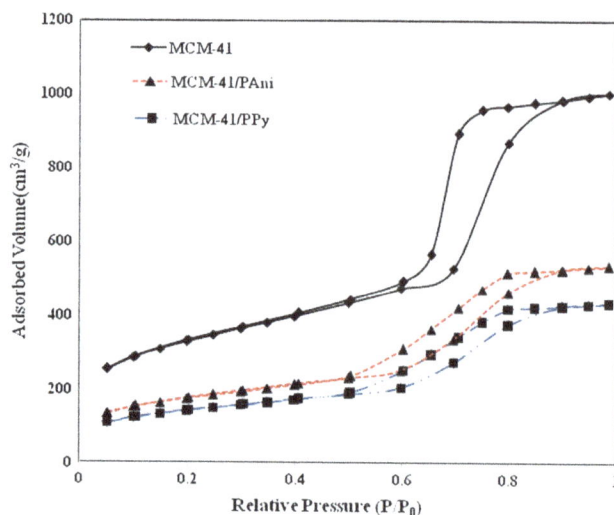

Fig. 2 N_2 adsorption–desorption isotherms of MCM-41, PAni/MCM-41, and PPy/MCM-41

Table 1 Characterization results of MCM-41, PAni/MCM-41, and PPy/MCM-41

Sample	Surface area (BET) ($m^2 g^{-1}$)	Pore diameter (BJH) (nm)	Pore volume (BJH) ($cm^3 g^{-1}$)
MCM-41	1003.61	6.7	0.78
MCM-41/PAni	535.34	4.8	0.47
MCM-41/PPy	434.89	3.6	0.32

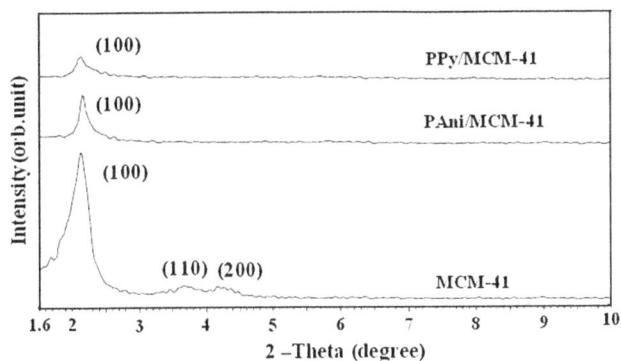

Fig. 3 XRD patterns of MCM-41, PAni/MCM-41, and PPy/MCM-41

attributed to (100), (110), and (200) planes of the MCM-41 structure [23]. According to Fig. 3, following PPy and PAni loading, the XRD pattern of MCM-41 was comparable with pure MCM-41. Nevertheless, the diffraction intensity of PPy/MCM-41 was below that of MCM-41 and PAni/MCM-41. MCM-41 pore filling with PPy might be the probable reason for the reduction in peak intensity.

Coating of MCM-41 outer surface by amine groups did not majorly affect MCM-41 diffraction intensity [24]. According to this finding, it can be concluded that quick pore filling in the presence of PPy is attributed to the smaller molecular size of PPy, compared to PAni. On the other hand, as for the small size of PPy particles, they can penetrate into the MCM-41 structure, generating more positions to absorb a constant volume. Since a larger number of molecules can penetrate into the pore throughout MCM-41 loading, the structure rapidly collapses and leads to the reduction of diffraction intensity [19, 22, 25].

In Fig. 4, SEM results are presented. The MCM-41 and composites are formed with small particle sizes (almost uniform). The increase in MCM-41 particle size after loading indicates the contribution of PPy and PAni to the formation of composites. In addition, TEM findings are presented in Fig. 5. It can be observed that both MCM-41 and composites had hexagonal well-ordered mesoporous structures, where organic compounds could easily access the active sites. Therefore, MCM-41 pore structure after PPy and PAni loading remained undamaged, and PPy and PAni were evenly scattered on the MCM-41 surface [14]. Also, the light and black points on the surface of the composite are representative of MCM-41 nanopowder and PAni-PPy, respectively.

To explain the surface properties of MCM-41 for identification of functional groups, the FTIR technique was applied. Figure 6 demonstrates the MCM-41 and composite spectra (range 400–4000 cm^{-1}). The band at 3400 cm^{-1} was attributed to the surface hydroxyl groups (SiOH) [14]. The observed peaks at 1030–1080 cm^{-1} indicated asymmetric stretching vibrations pertaining to the Si–O–Si bridges. In addition, the absorption bands observed at 780–800 cm^{-1} were a result of symmetric stretching vibrations of Si–O–Si. Also, Si–O bending vibrations were observed at 450–460 cm^{-1} [14].

The C=C bonds at 1584 and 1492 cm^{-1} were related to quinonoid and benzenoid rings, respectively, in the PAni/MCM-41 composite; the peaks at 1300 cm^{-1} were related to C–N [26]. The peak at 810 cm^{-1} indicated out-of-plane C–H deformation in the Π-disubstituted benzene ring. The strong broad band at 3427 cm^{-1} for PPy/MCM-41 was attributed to PPy stretching vibrations. Also, C–H vibrations accounted for the bands at 2918 cm^{-1}, whereas bands at 2356 cm^{-1} were related to stretching vibrations of C–N.

C=C ring stretching of pyrrole accounted for the absorption band at 1635 cm^{-1}. The band at 1305 cm^{-1} was attributed to C–H vibrations. Also, the peak at 1091 cm^{-1} was related to C–O symmetric stretching and in-plane O–H deformation. C–H deformation vibrations in the CH=CH group could account for the peak at 910 cm^{-1} [21, 25, 27].

Comparison of MCM-41 and PPy/MCM-41 FTIR patterns indicated that some bands had disappeared and, therefore, PPy was incorporated in MCM-41 particles. In addition, the distinguished absorption peak of PPy/MCM-41 near 3427 cm^{-1} decreased following AB62 adsorption. This finding showed that PPy/MCM-41 pore structure changed with dye adsorption probably because of inherent disorder, and not the MCM-41 structure collapse.

Adsorption assessments

pH

Overall, pH is the most effective parameter during adsorption, which suppresses the quantity of dye ions on adsorbent active sites [28]. The removal efficiency of AB62 was examined in different pH ranges (2–10), whereas other variables including the adsorbent quantity, dye dosage, and temperature remained constant. The

Fig. 4 SEM images of MCM-41, PAni/MCM-41, and PPy/MCM-41

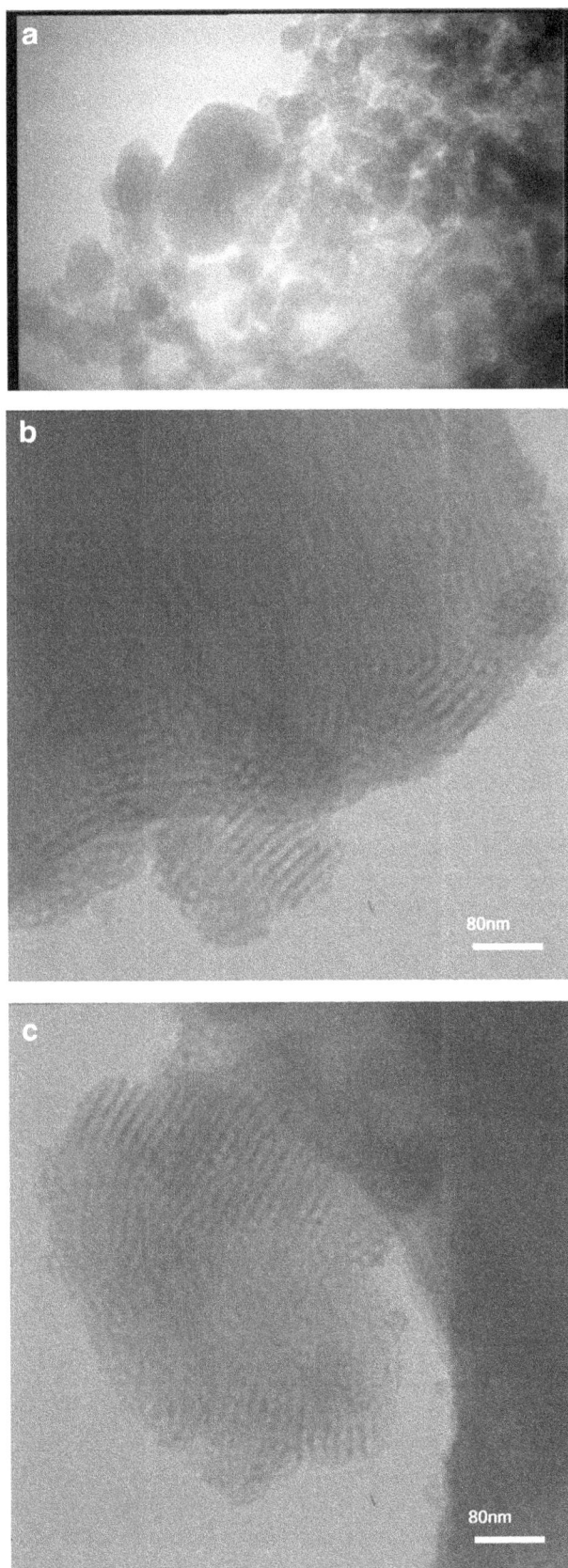

Fig. 5 TEM images of MCM-41, PAni/MCM-41 and PPy/MCM-41

importance of pH in AB62 adsorption on the origin and modified MCM-41 is presented in Fig. 7.

The primary doses of dye and composites were 40 mg L^{-1} and 0.02 g, respectively. As can be seen, a rise in pH from 2 to 10 led to a decline in dye adsorption. The composite surface could attract positive charges at a lower pH [29], and the powerful electrostatic bond between the positive and negative charges of composite and dye molecules, respectively, could enhance dye adsorption. In addition, lower AB62 adsorption, observed at alkaline pH, might be associated with electrostatic repulsion among anionic dye molecules and OH ions [5].

In comparison with PPy, PAni has a larger molecular size and a lower dye adsorption considering the steric hindrance. In fact, steric hindrance enhances, as the molecular size of the polymer increases [30]. In addition, unmodified MCM-41 had a lower dye removal efficiency than MCM-41 modified with PPy; also, PAni/MCM-41 showed lower efficiency in acidic conditions.

Adsorbent concentration

The significance of adsorbent concentration was examined, using 0.02–0.2 g of both composites and 40 mg L^{-1} of dye solution in 100 mL solution (pH 2). As presented in Fig. 8, by increasing the concentration of PPy/MCM-41 and PAni/MCM-41 to 0.1 and 0.15 g, respectively, we could enhance the removal efficiency (up to 81 and 38%, respectively). However, further increase of composites had no effects on removal efficiency due to aggregation of composites [31, 32]. Therefore, 0.1 g of PPy/MCM-41 and 0.15 g of PAni/MCM-41 were considered as the optimal adsorbent dosages for AB62 removal.

Temperature

For evaluating the effect of temperature on PPy/MCM-41 and PAni/MCM-41 removal, assessments were performed at various temperatures with primary dye doses of 40–100 mg L^{-1}, 0.1 g of PPy/MCM-41, and 0.15 g of PAni/MCM-41 during 90 min of equilibrium (pH 2). As Fig. 9 presents, removal reduced by increasing the dye dosage and enhanced by increasing the temperature to 323 K; this finding might be attributed to the increased activity on the surface [5, 6]. Therefore, AB62 adsorption on nanocomposite adsorbents is both a spontaneous and an endothermic reaction.

Contact time

Dye adsorption by PPy/MCM-41 and PAni/MCM-41 was evaluated at 20 °C; the findings are presented in Fig. 10. At the beginning of the adsorption process, rapid dye removal

Fig. 6 FTIR spectra of MCM-41, PAni/MCM-41, and PPy/MCM-41 before and after AB62 adsorption

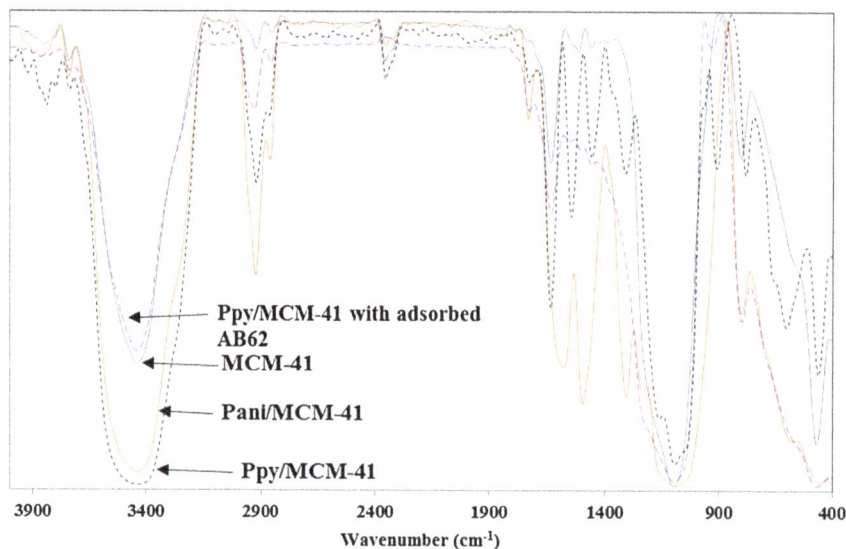

Ppy/MCM-41 with adsorbed AB62
MCM-41
Pani/MCM-41
Ppy/MCM-41

Wavenumber (cm⁻¹)

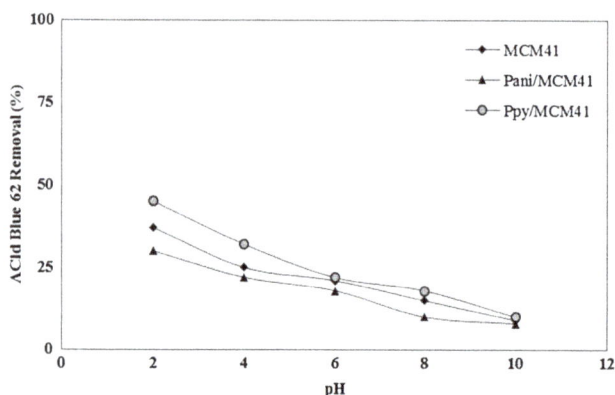

Fig. 7 Effect of pH on the percentage of dye removal by MCM-41, PAni/MCM-41, and PPy/MCM-41 (adsorbent dosage, 0.02 g, 100 mL of dye solutions, initial dye concentration, 40 mg L⁻¹, contact time, 90 min, and T 20 °C)

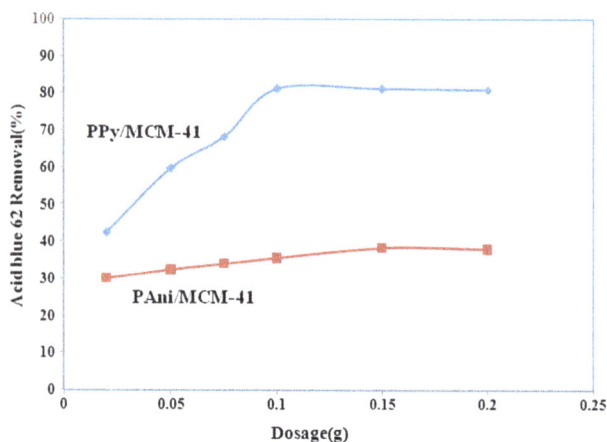

Fig. 8 Effect of adsorbent dosage on the percentage of dye removal (initial dye concentration 40 ppm; pH 2; contact time 90 min; T 20 °C)

occurred, given more access to the active sites. Afterward, adsorption gradually slowed down until equilibrium at 90 min. In the equilibrium state, the adsorption sites of the composites were filled [33]. Since PPy/MCM-41 nanocomposite showed higher removal efficiency than PAni/MCM-41 in all the experiments, 0.15 g of PPy/MCM-41 was considered as the optimal dosage and the subsequent experiments were carried out with 0.15 g of PPy/MCM-41.

Adsorption thermodynamics

We used thermodynamic analysis to understand the characteristics and mechanisms of adsorption. ΔG°, ΔH°, and ΔS° values were measured using the following formulae:

$$k_{c} = C_{Ae}/C_{e}, \tag{3}$$

$$\Delta G^{\circ} = -RT \ln kc, \tag{4}$$

$$\Delta G^{\circ} = \Delta H^{\circ} - T\Delta S^{\circ}, \tag{5}$$

$$\log k_{c} = \Delta S^{\circ}/2.303R - \Delta H^{\circ}/2.303RT, \tag{6}$$

where ΔH (kJ mol⁻¹) is measured with respect to the slope of $\log k_{c}$ vs. $1/T$ plot and ΔS (J mol⁻¹ K⁻¹) is measured relative to the plot's intercept. Also, ΔG° was measured using Eq. (6). The results of thermodynamic analysis are shown in Table 2.

According to Table 2, increasing the temperature could enhance adsorption. Therefore, dye adsorption on PPy/MCM-41 was endothermic, as confirmed by the positive value of ΔH°. On the other hand, the heat of physical adsorption was less than 21 kJ mol⁻¹. Based on the ΔH° values, AB62 adsorption on PPy/MCM-41 was in the form of physical adsorption [5, 34, 35]. The negative ΔG° values confirmed the feasibility of dye adsorption. Also, positive

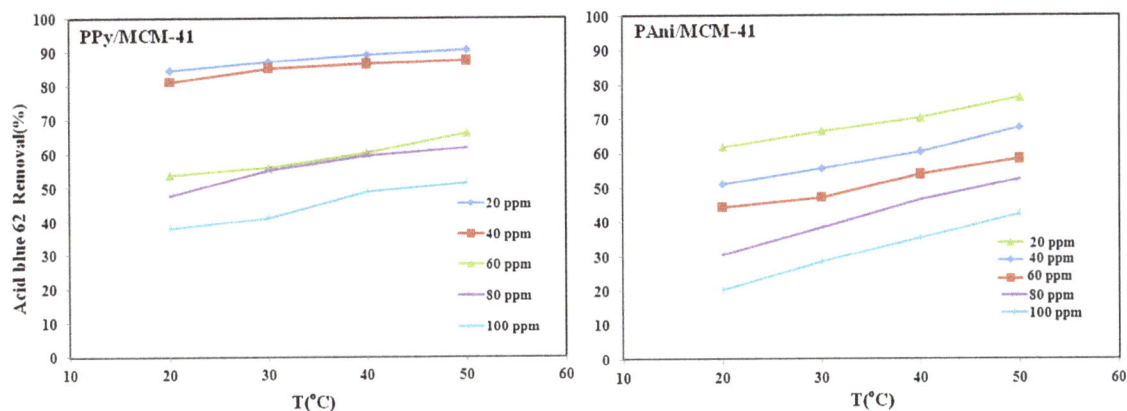

Fig. 9 Effect of temperature on dye removal (initial dye concentration, 40 up to 100 ppm; nanocomposite dosage; 0.1 g of PPy/MCM-41 and 0.15 g of PPy/MCM-41; pH of 2; and contact time 90 min)

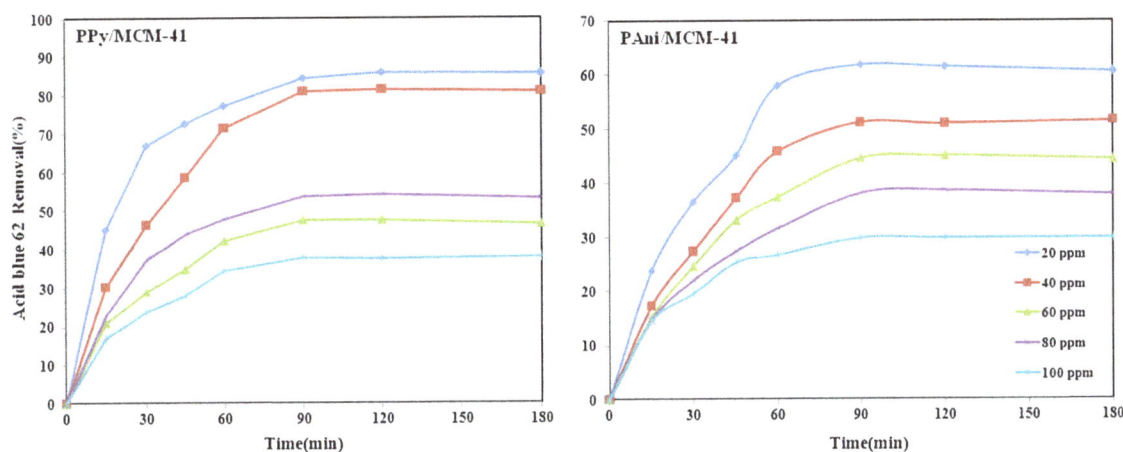

Fig. 10 Effect of contact time on the percentage of dye removal (initial dye concentration, 40 up to 100 ppm; 0.15 g of PAni/MCM-41 and 0.1 g of PPy/MCM-41 in 100 mL of dye solutions; pH 2; and T 20 °C)

Table 2 Thermodynamic parameters for the adsorption of Acid Blue 62 onto the PPy/MCM-41

Initial dye concentration (mg L^{-1})	ΔH° (kJ mol^{-1})	ΔS° (J mol^{-1} K)	20 °C	ΔG° (kJ mol^{-1}) 30 °C	40 °C	50 °C
20	14.80	64.75	−4.14	−4.82	−5.51	−6.07
40	13.44	54.01	−3.55	−4.41	−4.83	−5.19
60	12.12	46.77	−0.36	−0.60	−1.08	−1.78
80	11.88	40.16	−0.22	−0.52	−1.00	−1.27
100	10.12	38.21	−0.20	−0.41	−0.88	−0.91

ΔS° values indicated the reversible adsorption of dye on PPy/MCM-41 [5, 11].

Adsorption isotherm equations

Adsorption isotherm equations describe the adsorbent–adsorbate interactions. Isotherm models are applied for designing the adsorption process. To evaluate dye adsorption isotherms, equilibrium data were prepared with 0.1 g of PPy/MCM-14 nanocomposite at 50 °C (pH 2). In this

study, five important isotherms were applied to investigate the dye adsorption.

Langmuir isotherm equation

This equation, which is based on monolayer adsorption, was applied to determine the adsorption efficiency. Adsorption was observed on the same active sites on the surface, without any interactions among the adsorbate molecules [11].

The nonlinear Langmuir isotherm equation is as follows:

$$q_e = q_m k_L C_e/(1 + k_L C_e), \qquad (7)$$

where q_e denotes the adsorption efficiency in equilibrium (mg g^{-1}), C_e the equilibrium dosage (mg L^{-1}), q_m the highest adsorption potential (mg g^{-1}), and k_L the Langmuir constant. The binding affinity of dye molecules is indicated by high k_L values.

Freundlich isotherm equation

This equation is derived from empirical information and is generally applied for the expression of heterogeneous surfaces [36]:

$$q_e = K_f C_e^{1/n}, \qquad (8)$$

where K_f represents the adsorption capacity (L g^{-1}), n (a constant) the intensity of adsorption, and $1/n$ the heterogeneity of the composite surface; n values between 2 and 10 exhibit favorable adsorption.

Redlich–Peterson model

In this three-parameter equation, Langmuir and Freundlich isotherms are combined [37]:

$$q_e = k_R C_e/\left(1 + \alpha_R C_e^\beta\right), \qquad (9)$$

where k_R (L g^{-1}) and α_R (L mg^{-1}) denote the Redlich–Peterson constants and β represents the equation exponent. The Freundlich isotherm is the supreme isotherm when β is

near 0, while the Langmuir model is predominant when β is close to the unit.

Dubinin–Radushkevich equation

It is written as [38]:

$$q_e = q_m e^{-\beta \varepsilon 2}, \qquad (10)$$

where q_m denotes the monolayer efficiency (mg g^{-1}), β the adsorption energy constant, and ε the Polanyi adsorption potential:

$$\varepsilon = RT \ln(1 + 1/C_e). \qquad (11)$$

In this equation, R represents the gas constant (8.314 J^{-1} mol K^{-1}), T the absolute temperature, β the mean free energy, and ε the dye molecules moved to the solid surface. The relation of ε with β is described as follows:

$$\varepsilon = 1/[(2\beta)0.5]. \qquad (12)$$

Temkin isotherm

For investigating multilayer adsorption, the Temkin isotherm was applied [39, 40]:

$$q_e = B \ln(A C_e). \qquad (13)$$

Parameters of the isotherms were measured, based on the linear regression of nonlinear forms in the models, as shown in Table 3. Comparison was performed between the

Table 3 The values of parameters for isotherm models

Isotherm	Nonlinear equation	Linear form	Parameter	R^2
Langmuir 1	$q_e = q_m K_L C_e/(1 + K_L C_e)$	$C_e/q_e = 1/k_1 q_m + (1/q_m) C_e$ C_e/q_e vs. C_e	$q_m = 55.55$ mg L^{-1} $K_1 = 0.227$ L mg^{-1}	0.988
Langmuir 2		$1/q_e = 1/k_1 q_m C_e + 1/q_m$ $1/q_e$ vs. $1/C_e$	$q_m = 55.55$ mg L^{-1} $K_1 = 0.818$ L mg^{-1}	0.972
Langmuir 3		$q_e = q_m - q_e/k_L C_e$ q_e vs. q_e/C_e	$q_m = 52.47$ mg L^{-1} $K_1 = 0.878$ L mg^{-1}	0.878
Langmuir 4		$q_e/C_e = q_m/k_1 - q_e/k_1$ q_e/C_e vs. q_e	$q_m = 54.47$ mg L^{-1} $K_1 = 3.73$ L mg^{-1}	0.878
Freundlich	$q_e = K_f C_e^{1/n}$	$\ln q_e = \ln K_f + 1/n \ln C_e$ Ln q_e vs. ln C_e	$n = 3.41$ $K_f = 17.38$ (mg g^{-1}) (L mg^{-1})$^{1/n}$	0.887
Redlich–Peterson	$q_e = k_R C_e/\left(1 + a_R C_e^\beta\right)$	Ln($K_R * (C_e/q_e) - 1$) vs. $\beta \ln C_e + \ln \alpha_R$	$K_R = 17.85$ L g^{-1} $\alpha_R = 0.38$ (L mg^{-1})$^\beta$ $\beta = 0.35$	0.911
Temkin	$q_e = B \ln(A C_e)$	q_e vs. ln C_e	$B = 19.63$ J mole^{-1} $A = 11.78$ L g^{-1}	0.901
Dubinin-Radushkevich	$q_e = q_m e^{-\beta \varepsilon 2}$	Ln q_e vs. ε^2	$q_m = 46.3$ mg g^{-1} $\beta = 0.13$ $E = 1/\sqrt{2\beta} = 1.9$ kJ mole^{-1}	0.881

Table 4 Pseudo-first order, pseudo-second-order, and intraparticle diffusion model parameters for Acid Blue 62 adsorption on PPy/MCM-41 adsorbent

Initial DY86 Concentration (mg L^{-1})	Pseudo-first order			Pseudo-second-order R^2				Intraparticle diffusion		
	k_{ad} (1 min^{-1})	$q_{e,cal.}$ (mg g^{-1})	R^2	k_h (g mg^{-1} min)	$q_{e,cal.}$ (mg g^{-1})	$q_{e,ex.}$ (mg g^{-1})	R^2	K_p (mg g^{-1} min$^{0.5}$)	C (mg g^{-1})	R^2
20	0.036	22.20	0.97	0.0028	20.40	16.91	0.99	1.33	5.00	0.90
40	0.036	38.70	0.96	0.0003	52.63	32.45	0.99	3.24	1.87	0.98
60	0.032	45.38	0.99	0.0007	43.47	32.66	0.99	3.73	2.12	0.94
80	0.034	56.68	0.94	0.0004	55.55	38.17	0.98	3.93	2.80	0.98
100	0.034	54.24	0.93	0.0005	52.63	40.14	0.98	4.14	2.95	0.97

experimental information and model findings with respect to R^2 values:

Langmuir(type 1) > Redlich–Peterson >

Temkin > Dubinin–Radushkevich > Freundlich.

Based on the obtained results, Langmuir equation showed the greatest correlation coefficient indicative of monolayer adsorption. Regarding the Freundlich isotherm, n was measured to be 3.41, indicating suitable adsorption. The Redlich–Peterson isotherm correlation coefficient (0.911) was below the value obtained by the Langmuir model; as a result, the Langmuir model was introduced as the supreme isotherm.

To continue the experiments, the Dubinin–Radushkevich model was used. The correlation coefficient of this model (0.881) was lower than that of Langmuir. We also applied Temkin isotherm to evaluate multilayer adsorption. The correlation coefficient was measured to be 0.901, which is below that reported in the Langmuir model. Therefore, it can be concluded that monolayer adsorption happened, and the Langmuir model is the predominant one.

Adsorption kinetics

Using different models, the mechanism of adsorption was evaluated. The pseudo-first and pseudo-second-order linear models are commonly used to fit the kinetic information via Eqs. (14) and (15), respectively:

$$\ln(q_e - q) = \ln(q_e) - k_1 t, \tag{14}$$

$$t/q = 1/k_h q_e^2 + t/q_e, \tag{15}$$

where q_t denotes the adsorbate amount in each adsorbent unit at time t (mg g^{-1}), k_1 the pseudo-first-order constant, and k_h the second-order constant. The rate constants, as well as the adsorption efficiencies in the pseudo-first-order model, were measured at various doses with respect to the $\ln(q_e - q_t)$ vs. t plot. In addition, in the other model, the rate constants and adsorption capacities were measured at various doses, based on the t/q_t versus t plot. The findings are demonstrated in Table 4.

According to Table 4, the evaluated models showed great correlation (R^2), although q_e values measured by the pseudo-second-order model (q_e, cal.) showed greater agreement with the experimental values (q_e, exp.). Therefore, the results demonstrated that AB62 adsorption on PPy/MCM-41 is best characterized, using this model. In fact, it could describe the process of adsorption and indicated that chemical adsorption controls AB62 adsorption on the composite surface.

The adsorption rate can be managed using intraparticle diffusion, along with other kinetic mechanisms including boundary layer or film diffusion effects. Weber and Morris first presented the model of intraparticle diffusion [41]:

$$q_t = k_p t^{0.5} + C. \tag{16}$$

In this equation, C denotes the intercept, k represents the intraparticle rate constant, and k_p is calculated with respect to the slope of q_t (mg g^{-1}) versus $t^{0.5}$ plot. The k_p values are presented in Table 4. Based on the findings, given the great driving force, the increase in diffusion rate was associated with an increase in the primary dye dosage.

Intraparticle diffusion characterizes the rate-control stage if the linear regression of q_t versus $t^{0.5}$ plot crosses

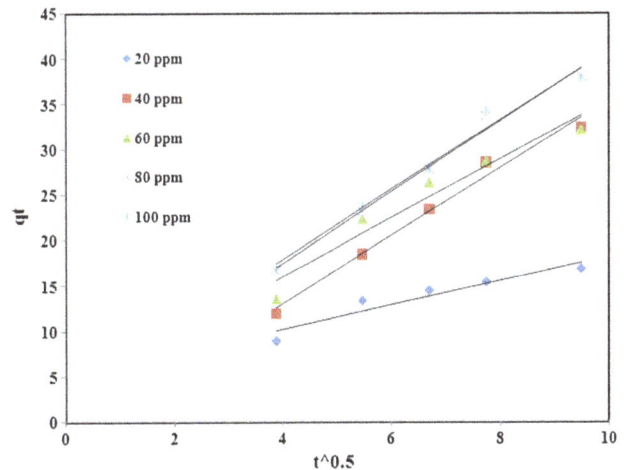

Fig. 11 Intraparticle diffusion model for the adsorption of Acid Blue 62 onto PPy/MCM-41

Table 5 A comparison
between the results of the
present work and some reported
results in literature

Adsorbent	Dye	q_e	Refs.
Activated carbon	Acid Red 97	52.08	[43]
Activated red mud	Acid Blue 113	83.33	[44]
Spinel ferrite nanoparticles	Acid Red 88	111.1	[45]
copper oxide nanoparticle loaded on activated carbon	Acid Blue 129	65.36	[46]
Ppy/MCM-41	Acid Blue 62	55.55	This work

Fig. 12 Adsorption and
desorption behaviors of AB62
dye on PPy/MCM-41

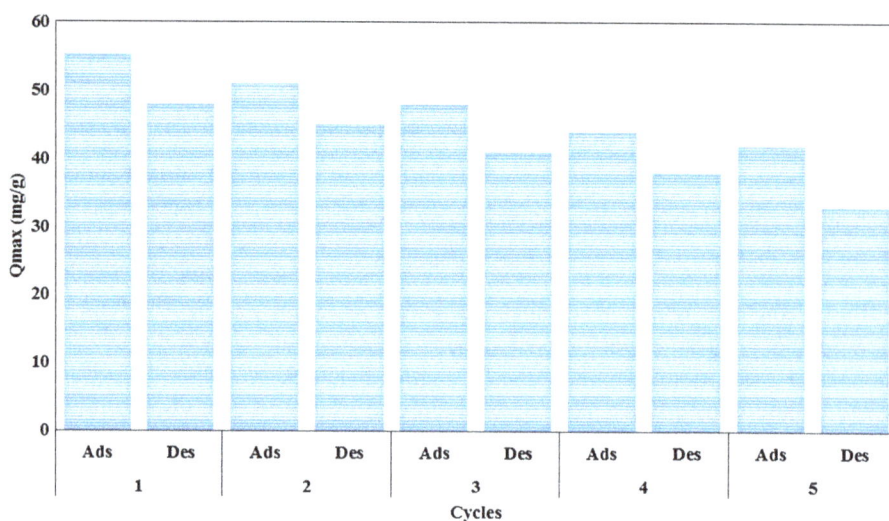

the origin. Nevertheless, it is obvious from Fig. 11 that the linear graphs did not cross the origin. As a result, other mechanisms could have affected the adsorption rate (it is not only dependent on intraparticle diffusion) [11, 42]. Table 5 presents the comparison of AB62 adsorption on PPy/MCM-41 with other adsorbents used in previous studies [43–46]. As can be seen, the presented adsorbent generally has a good adsorption capacity for AB62. The prepared adsorbent showed a higher adsorption capacity than other composites.

Adsorption mechanism

With respect to AB62 adsorption on different adsorbents (i.e., MCM-41/PPy and MCM-41/PAni), various mechanisms can be convoluted including ionic bonding between cationic functional groups of the adsorbents and anionic functional group(s) in dissolved dye molecules. During adsorption, different phenomena may occur. First, in the aqueous medium, AB62 is dissolved and sulfonate groups of dye transform into anionic dye ions (17). In the second step, the adsorbent amino groups (PPy and PAni) are protonated in the acidic medium (18). According to Eq. (19), adsorption continues, given the electrostatic bond between the counter ions:

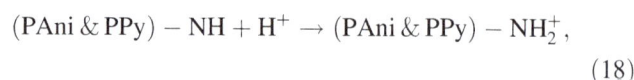

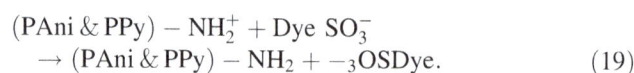

$$Dye\text{-}SO_3Na \rightarrow DyeSO_3^- + Na^+, \tag{17}$$

$$(PAni\,\&\,PPy) - NH + H^+ \rightarrow (PAni\,\&\,PPy) - NH_2^+, \tag{18}$$

$$(PAni\,\&\,PPy) - NH_2^+ + Dye\,SO_3^- \\ \rightarrow (PAni\,\&\,PPy) - NH_2 + {}_{-3}OSDye. \tag{19}$$

Desorption studies

By performing desorption assessments, we can analyze the adsorption mechanism and the possibility of adsorbent reuse. The interaction of the adsorbent surface with dye molecules (either a strong or a weak bond) determines the reversibility of adsorption. Alkaline solutions (pH 10) were applied to desorb AB62 dye from PPy/MCM-41. The results of desorption study for five cycles are presented in Fig. 12. As can be seen, PPy/MCM-41 showed desorption ability.

Based on our findings, weak forces mainly accounted for the bonding between the adsorbent surface and AB62 molecules and could lead to high desorption capacity and produce adsorbents; the activation energy and thermodynamic parameters also support this finding. The adsorbent showed acceptable adsorption efficiency for AB62 and might be an acceptable adsorbent for treatment purposes.

Conclusion

PPy/MCM-41 and PAni/MCM-41 nanocomposites were successfully synthesized through chemical polymerization and were described using different techniques. The adsorption of AB62 on both synthesized nanocomposites was evaluated in aqueous solutions. Batch experiments demonstrated that the dye removal efficiency of PPy/MCM-41 nanocomposite was more favorable than PAni/MCM-41. The smaller size of PPy molecules in comparison with PAni makes it more capable of removing dyes. Moreover, the significance of pH, adsorbent dosage, duration of contact, and temperature in dye removal was studied. The removal capacity showed adverse dependence on pH, which enhanced by increasing the duration of contact, adsorbent dosage, and temperature. The highest removal efficiency was 90.57% at a pH of 2 and contact time of 90 min, using 20 mg L^{-1} of AB62 and 0.1 g of PPy/MCM-41 nanocomposite. Thermodynamic studies showed that adsorption is both an endothermic and a spontaneous reaction. Moreover, the findings reported in equilibrium were compatible with the Langmuir isotherm, and adsorption capacity on PPy/MCM-41 was 55.55 mg g^{-1} at a primary dye dosage of 20 mg L^{-1}. Based on the intraparticle diffusion model, adsorption is influenced by a boundary layer, as well as intraparticle diffusion. Also, the pseudo-second-order model showed the greatest agreement with the experimental findings.

References

1. Yang, X., Guan, Q., Li, W.: Effect of template in MCM-41 on the adsorption of aniline from aqueous solution. J. Environ. Manag. **92**, 2939–2943 (2011)
2. Castillo, M., Perez, N.A., Ramos, R.O., et al.: Removal of diethyl phthalate from water solution by adsorption, photo–oxidation, ozonation and advanced oxidation process (UV/H_2O_2, O_3/H_2O_2 and O_3/activated carbon). Sci. Total Environ. **442**, 26–35 (2013)
3. Saini, R., Kumar, P.: Simultaneous removal of methyl parathion and chlorpyrifos pesticides from model wastewater using coagulation/flocculation. Central composite design. J. Environ. Chem. Eng. **4**, 673–680 (2016)
4. Shirzad Kebria, M.R., Jahanshahi, M., Rahimpour, A.: SiO_2 modified polyethyleneimine-based nanofiltration membranes for dye removal from aqueous and organic solutions. Desalination **367**, 255–264 (2015)
5. Binaeian, E., Seghatoleslami, N., Chaichi, M.J.: Synthesis of oak gall tannin-immobilized hexagonal mesoporous silicate (OGT-HMS) as a new super adsorbent for the removal of anionic dye from aqueous solution. Desalin. Water Treat. **57**, 8420–8436 (2016)
6. Tayebi, H.A., Dalirandeh, Z., Shokuhi Rad, A., Mirabi, A., Binaeian, E.: Synthesis of polyaniline/Fe_3O_4 magnetic nanoparticles for removal of reactive red 198 from textile waste water: kinetic, isotherm, and thermodynamic studies. Desalin. Water Treat. **57**, 22551–22563 (2016)
7. Shabandokht, M., Binaeian, E., Tayebi, H.A.: Adsorption of food dye acid red 18 onto polyaniline modified rice husk composite: isotherm and kinetic analysis. Desalin. Water Treat. **57**, 1–13 (2016)
8. Sun, X.Y., Li, P.Z., Ai, B., Wang, Y.B.: Surface modification of MCM-41 and its application in DNA adsorption. Chin. Chem. Lett. **27**, 139–144 (2016)
9. Idris, S.A., Harvey, S.R., Gibson, L.T.: Selective extraction of mercury (II) from water samples using mercapto functionalised-MCM-41 and regeneration of the sorbent using microwave digestion. J. Hazard. Mater. **193**, 171–176 (2011)
10. Peng, X., Huang, D., Odoom-Wubah, T., Fu, D., Huang, J., Qin, Q.: Adsorption of anionic and cationic dyes on ferromagnetic ordered mesoporous carbon from aqueous solution, equilibrium, thermodynamic and kinetics. J. Colloid Interf. Sci. **430**, 272–282 (2014)
11. Shafiabadi, M., Dashti, A., Tayebi, H.A.: Removal of Hg(II) from aqueous solution using polypyrrole/SBA-15 nanocomposite. Synth. Met. **212**, 154–160 (2016)
12. Prida, K., Mishra, K.G., Dash, S.K.: Adsorption of toxic metal ion Cr(VI) from aqueous state by TiO_2-MCM-41: equilibrium and kinetic studies. J. Hazard. Mater. **241**, 395–403 (2012)
13. Wongsakulphasatch, S., Kiatkittipong, W., Saiswat, J., Oonkhanond, B., Striolo, A., Assabumrungrat, S.: The adsorption aspect of Cu^{2+} and Zn^{2+} on MCM-41 and SDS-modified MCM-41. Inorg. Chem. Commun. **46**, 301–304 (2014)
14. Shu, Y., Shao, Y., Wei, X., Wang, X., Sun, Q., Zhang, Q., Li, L.: Synthesis and characterization of Ni-MCM-41 for methylene blue adsorption. Micropor. Mesopor. Mater. **214**, 88–94 (2015)
15. Kaur, S., Rani, S., Mahajan, R.K., Asif, M., Gupta, V.K.: Synthesis and adsorption properties of mesoporous material for the removal of dye safranin: kinetics, equilibrium, and thermodynamics. J. Ind. Eng. Chem. **22**, 19–27 (2015)
16. Anbia, M., Salehi, S.: Removal of acid dyes from aqueous media by adsorption onto amino-functionalized nano-mesoporous silica SBA-3. Dyes Pigments **94**, 1–9 (2012)
17. Tanaka, M., Itadani, A., Kuroda, Y., Iwamoto, M.: Effect of pore size and nickel content of Ni-MCM-41 on catalytic activity for ethylene dimerization and local structures of nickel ions. J. Phys. Chem. C **116**, 5664–5672 (2012)
18. Kamarudin, K.S.N., Alias, N.: Adsorption performance of MCM-41 impregnated with amine for CO_2 removal. Fuel Process. Technol. **106**, 332–337 (2013)
19. Fang, F.F., Choi, H.J., Ahn, W.S.: Electrorheology of a mesoporous silica having conducting polypyrrole inside expanded pores. Micropor. Mesopor. Mater. **130**, 338–343 (2010)
20. Esfandiyari, T., Nasirizadeh, N., Ehrampoosh, M., Tabatabaee, M.: Characterization and absorption studies of cationic dye on multi walled carbon nanotube–carbon ceramic composite. J. Ind. Eng. Chem. **46**, 35–43 (2017)
21. Radi, M.A., Nasirizadeh, N., Rohani-Moghadam, M., Dehghani, M.: The comparison of sonochemistry, electrochemistry and sonoelectrochemistry techniques on decolorization of CI Reactive Blue 49. Ultrason. Sonochem. **27**, 609–615 (2015)
22. Liu, Z., Teng, Y., Kai, Z., Yan, C., Pan, W.: CO_2 adsorption properties and thermal stability of different amine-impregnated MCM-41 materials. J. Fuel. Chem. Technol. **41**, 469–476 (2013)
23. Zareyee, D., Tayebi, H., Javadi, S.H.: Preparation of polyaniline/activated carbon composite for removal of reactive red 198 from aqueous solution. Iranian J. Org. Chem. **4**(1), 799–802 (2012)
24. Gil, M., Tiscornia, I., de la Iglesia, O., Mallada, R., Santamaria, J.: Monoamine-grafted MCM-48: an efficient material for CO_2 removal at low partial pressures. Chem. Eng. J. **175**, 291–297 (2011)
25. Tao, Q., Xu, Z., Wang, J., Liu, F., Wan, H., Zheng, S.: Adsorption of humic acid to aminopropyl functionalized SBA-15. Micropor. Mesopor. Mater. **131**, 177–185 (2010)
26. Javadian, H., Sorkhrodi, F.Z., Koutenaei, B.B., et al.: Experimental investigation on enhancing aqueous cadmium removal via

nanostructure composite of modified hexagonal type mesoporous silica with polyaniline/polypyrrole nanoparticles. J. Ind. Eng. Chem. **20**, 3678–3688 (2014)

27. Idris, S.A., Davidson, C.M., Manamon, C., et al.: Large pore diameter MCM-41 and its application for lead removal from aqueous media. J. Hazard. Mater. **185**, 898–904 (2011)

28. Ballav, N., Maity, A., Mishra, S.B.: High efficient removal of chromium (VI) using glycine doped polypyrrole adsorbent from aqueous solution. J. Chem. Eng. **198–199**, 536–546 (2012)

29. Shahbazi, A., Younesi, H., Badiei, A.: Functionalized SBA-15 mesoporous silica by melamine-based dendrimer amines for adsorptive characteristics of Pb(II), Cu(II) and Cd(II) heavy metal ions in batch and fixed bed column. J. Chem. Eng. **168**, 505–518 (2011)

30. Belmabkhout, Y., Serna Guerrero, R., Sayari, A.: Adsorption of CO_2-containing gas mixtures over amine-bearing pore-expanded MCM-41 silica: application for gas purification. Ind. Eng. Chem. Res. **49**(1), 359–365 (2010)

31. Arshadi, M.: Manganese chloride nanoparticles: a practical adsorbent for the sequestration of Hg(II) ions from aqueous solution. J. Chem. Eng. **259**, 170–182 (2015)

32. Kakavandi, B., Rezaei Kalantary, R., Farzadkia, M., Mahvi, A.H., Esrafili, A., Azari, A., Yari, A.R., Javid, A.B.: Enhanced chromium(VI) removal using activated carbon modified by zero valent iron and silver bimetallic nanoparticles. J. Environ. Health Sci. Eng. **12**, 115–124 (2014)

33. Zhang, M., Zhu, W., Li, H., et al.: One-pot synthesis, characterization and desulfurization of functional mesoporous W-MCM-41 from POM-based ionic liquids. J. Chem. Eng. **243**, 386–393 (2014)

34. Rahman, M.M., Akter, N., Karim, M.R., Ahmad, N., Rahman, M.M., Siddiquey, I.A., Bahadur, N.M., Hasnat, M.A.: Optimization, kinetic and thermodynamic studies for removal of Brilliant Red (X-3B) using Tannin gel. J. Environ. Chem. Eng. **2**(1), 76–83 (2014)

35. Kumar, P.S., Ramalingam, S., Senthamarai, C., et al.: Adsorption of dye from aqueous solution by cashew nut shell: studies on equilibrium isotherm, kinetics and thermodynamics of interactions. Desalination **261**, 52–60 (2010)

36. Galve, A., Sieffert, D., Staudt, C., et al.: Combination of ordered mesoporous silica MCM-41 and layered titanosilicate JDF-L1 fillers for 6FDA-basedcopolyimide mixed matrix membranes. J. Membr. Sci. **431**, 163–170 (2013)

37. Maneesuwan, H., Longloilert, R., Chaisuwan, Th, Wongkasemjit, S.: Synthesis and characterization of Fe–Ce-MCM-48 from silatrane precursor via sol–gel process. Mater. Lett. **94**, 65–68 (2013)

38. Braga, R.M., Barros, J.M.F., Melo, D.M.A., et al.: Kinetic study of template removal of MCM-41 derived from rice husk ash. J. Therm. Anal. Calorim. **111**, 1013–1018 (2013)

39. Pouretedal, H.R., Ahmadi, M.: Preparation, characterization and determination of photocatalytic activity of MCM-41/ZnO and MCM-48/ZnO nanocomposites. Iranian J. Catal. **3**(3), 149–155 (2013)

40. Benhamou, A., Basly, J.P., Baudu, M., Derriche, Z., Hamacha, R.: Amino-functionalized MCM-41 and MCM-48 for the removal of chromate and arsenate. J. Colloid Interf. Sci. **404**, 135–139 (2013)

41. Raji, F., Pakize, M.: Study of Hg(II) species removal from aqueous solution using hybrydeZnCl2-MCM-41 adsorbent. Appl. Surf. Sci. **282**, 415–424 (2013)

42. Ghaedi, M., Sadeghian, B., Kokhdan, S.N., et al.: Study of removal of Direct Yellow 12 by cadmium oxide nanowires loaded on activated carbon. Mater. Sci. Eng. **33**, 2258–2265 (2013)

43. Gomez, V., Larrechi, M.S., Callao, M.P.: Kinetic and adsorption study of acid dye removal using activated carbon. Chemosphere **69**(7), 1151–1158 (2007)

44. Shirzad Siboni, M.: Removal of acid blue 113 and reactive black 5 dye from aqueous solutions by activated red mud. J. Ind. Eng. Chem. **20**, 1432–1437 (2014)

45. Konicki, W.: Equilibrium and kinetic studies on acid dye Acid Red 88 adsorption by magnetic $ZnFe_2O_4$ spinel ferrite nanoparticles. J. Colloid Interf. Sci. **398**, 152–160 (2013)

46. Nekouei, F.: Kinetic, thermodynamic and isotherm studies for Acid Blue 129 removal from liquids using copper oxide nanoparticle-modified activated carbon as a novel adsorbent. J. Mol. Liq. **201**, 124–133 (2015)

Permissions

The contributors of this book come from diverse backgrounds, making this book a truly international effort. This book will bring forth new frontiers with its revolutionizing research information and detailed analysis of the nascent developments around the world.

We would like to thank all the contributing authors for lending their expertise to make the book truly unique. They have played a crucial role in the development of this book. Without their invaluable contributions this book wouldn't have been possible. They have made vital efforts to compile up to date information on the varied aspects of this subject to make this book a valuable addition to the collection of many professionals and students.

This book was conceptualized with the vision of imparting up-to-date information and advanced data in this field. To ensure the same, a matchless editorial board was set up. Every individual on the board went through rigorous rounds of assessment to prove their worth. After which they invested a large part of their time researching and compiling the most relevant data for our readers.

The editorial board has been involved in producing this book since its inception. They have spent rigorous hours researching and exploring the diverse topics which have resulted in the successful publishing of this book. They have passed on their knowledge of decades through this book. To expedite this challenging task, the publisher supported the team at every step. A small team of assistant editors was also appointed to further simplify the editing procedure and attain best results for the readers.

Apart from the editorial board, the designing team has also invested a significant amount of their time in understanding the subject and creating the most relevant covers. They scrutinized every image to scout for the most suitable representation of the subject and create an appropriate cover for the book.

The publishing team has been an ardent support to the editorial, designing and production team. Their endless efforts to recruit the best for this project, has resulted in the accomplishment of this book. They are a veteran in the field of academics and their pool of knowledge is as vast as their experience in printing. Their expertise and guidance has proved useful at every step. Their uncompromising quality standards have made this book an exceptional effort. Their encouragement from time to time has been an inspiration for everyone.

The publisher and the editorial board hope that this book will prove to be a valuable piece of knowledge for researchers, students, practitioners and scholars across the globe.

List of Contributors

Abdol Mohammad Attaran and Somaye Abdol-Manafi
Department of Chemistry, Payame Noor University, Delijan, Iran

Mehran Javanbakht
Department of Chemistry, Amirkabir University of Technology, Tehran, Iran

Morteza Enhessari
Department of Chemistry, Naragh Branch, Islamic Azad University, Naragh, Iran

Mohamed Khairy and Farouk A. Rashwan
Chemistry Department, Faculty of Science, Sohag University, Sohag 82524, Egypt

Hanan F. Abdel-Hafez
Plant Protection Research Institute, ARC, Dokki, Giza, Egypt

Haytham A. Ayoub
Chemistry Department, Faculty of Science, Sohag University, Sohag 82524, Egypt
Plant Protection Research Institute, ARC, Dokki, Giza, Egypt

Mahdi Rezaei-Sameti and Neda Javadi Jukar
Department of Applied Chemistry, Faculty of Science, Malayer University, Malayer 65174, Iran

Zulhelmi Ismail
Faculty of Manufacturing Engineering, Universiti Malaysia Pahang, 26600 Pekan, Pahang, Malaysia

Abu Hannifa Abdullah, Anis Sakinah Zainal Abidin and Kamal Yusoh
Faculty of Chemical Engineering and Natural Resources, Universiti Malaysia Pahang, 26300 Kuantan, Pahang, Malaysia

Mahdi Rezaei-Sameti and Negin Hemmati
Department of Applied Chemistry, Faculty of Science, Malayer University, Malayer 65174, Iran

Farzad Molani
Department of Chemistry, Sanandaj Branch, Islamic Azad University, P. O. Box 618, Sanandaj, Iran

Mahboobeh Balar and Zahra Azizi
Department of Chemistry, Karaj Branch, Islamic Azad University, P.O. Box 31485-313, Karaj, Iran

Mohammad Ghashghaee
Faculty of Petrochemicals, Iran Polymer and Petrochemical Institute, P.O. Box 14975-112, Tehran, Iran

Zahra Ghiamaty
Research Laboratory of Real Samples Analysis, Faculty of Chemistry, Iran University of Science and Technology, Tehran 1684613114, Iran

Ali Ghaffarinejad
Research Laboratory of Real Samples Analysis, Faculty of Chemistry, Iran University of Science and Technology, Tehran 1684613114, Iran
Electroanalytical Chemistry Research Centre, Iran University of Science and Technology, Tehran 1684613114, Iran

Mojtaba Faryadras
Faculty of Chemistry, Iran University of Science and Technology, Tehran 1684613114, Iran

Abbas Abdolmaleki
Department of Chemistry, Malek-Ashtar University of Technology, Tehran, P.O. Box 16765-34543, Iran

Hojjat Kazemi
Research Institute of Petroleum Industry, Tehran 1485733111, Iran

Ali Maleki, Pedram Zand and Zahra Mohseni
Catalysts and Organic Synthesis Research Laboratory, Department of Chemistry, Iran University of Science and Technology, Tehran 16846-13114, Iran

Shalaleh Gilani
Department of Food Science, Islamic Azad University, Sarab Branch, Sarab, Iran

Mohammad Ghorbanpour
Chemical Engineering Department, University of Mohaghegh Ardabili, Ardabil, Iran

Aiyoub Parchehbaf Jadid
Department of Chemistry, Islamic Azad University, Ardabil Branch, Ardabil, Iran

Rahmatollah Rahimi, Shabnam Pordel and Mahboubeh Rabbani
Department of Chemistry, Iran University of Science and Technology, 16846-13114 Tehran, Iran

Ramin Mostafalu, Akbar Heydari, Marzban Arefi and Maryam Kazemi
Chemistry Department, Tarbiat Modarres University, P.O. Box 14155-4838, Tehran, Iran

Abbas Banaei and Fatemeh Ghorbani
Padideh Shimi Jam Co., Karaj, Iran

Ali Shokuhi Rad
Department of Chemical Engineering, Qaemshahr Branch, Islamic Azad University, Qaemshahr, Iran

N. V. Abramov, S. P. Turanska, A. L. Petranovska and P. P. Gorbyk
Chuiko Institute of Surface Chemistry, National Academy of Sciences of Ukraine, 17 General Naumov Str., Kiev 03164, Ukraine

A. P. Kusyak
Chuiko Institute of Surface Chemistry, National Academy of Sciences of Ukraine, 17 General Naumov Str., Kiev 03164, Ukraine
Ivan Franko Zhytomyr State University, 40V. Berdychevska Str., Zhytomyr 10008, Ukraine

Roya Payami
Department of Food Science, Sarab Branch, Islamic Azad University, Sarab, Iran

Mohammad Ghorbanpour
Chemical Engineering Department, University of Mohaghegh Ardabili, Ardabil, Iran

Aiyoub Parchehbaf Jadid
Department of Chemistry, Ardabil Branch, Islamic Azad University, Ardabil, Iran

Kamel A. Abd-Elsalam
Plant Pathology Research Institute, Agricultural Research Center (ARC), Giza, Egypt
Unit of Excellence in Nano-Molecular Plant Pathology Research Center, Plant Pathology Research Institute, Giza, Egypt

Margarita S. Rubina, Alexander Yu. Vasil'kov, Alexander V. Naumkin and Sergey S. Abramchuk
A.N. Nesmeyanov Institute of Organoelement Compounds (INEOS), Russian Academy of Sciences, Moscow, Russia

Mousa A. Alghuthaymi
Biology Department, Science and Humanities College, Shaqra University, Alquwayiyah, Saudi Arabia

Eleonora V. Shtykova
A.V. Shubnikov Institute of Crystallography, Russian Academy of Sciences, Moscow, Russia

T. Navaei Diva, K. Zare and M. Yousefi
Department of Chemistry, Science and Research Branch, Islamic Azad University, Tehran, Iran

F. Taleshi
Department of Physics, Qaemshahr Branch, Islamic Azad University, Qaemshahr, Iran

A. A. Farghali
Materials Science and Nanotechnology Department, Faculty of Postgraduate Studies for Advanced Sciences, Beni-Suef University, Beni Suef, Egypt

H. A. Abdel Tawab and S. A. Abdel Moaty
Materials Science Lab, Chemistry Department, Faculty of Science, Beni-Suef University, Beni Suef, Egypt

Rehab Khaled
Department of Chemistry, Faculty of Science, Beni-Suef University, Beni Suef, Egypt

Ehab Salih, Ibrahim M. El-Sherbiny and Moataz Mekawy
Center for Materials Science, Zewail City of Science and Technology, 6th October City, Giza 12588, Egypt

Rabeay Y. A. Hassan
Microanalysis Lab, Applied Organic Chemistry Department, National Research Centre (NRC), El-Buhouth St., Dokki, Cairo 12622, Egypt

Amirali Abbasi and Jaber Jahanbin Sardroodi
Molecular Simulation Laboratory (MSL), Azarbaijan Shahid Madani University, Tabriz, Iran
Department of Chemistry, Faculty of Basic Sciences, Azarbaijan Shahid Madani University, Tabriz, Iran
Computational Nanomaterials Research Group (CNRG), Azarbaijan Shahid Madani University, Tabriz, Iran

Ali Torabinejad, Navid Nasirizadeh and Mohammad Esmail Yazdanshenas
Department of Textile and Polymer Engineering, Yazd Branch, Islamic Azad University, Yazd, Iran

Habib-Allah Tayebi
Department of Textile Engineering, Islamic Azad University of Qaemshahr, Qaemshahr, Iran

Index

www.ingramcontent.com/pod-product-compliance
Lightning Source LLC
Chambersburg PA
CBHW082034190326
41458CB00010B/3361